# Cengage Student Workbook for Algebra Activities

# Intermediate Algebra

## FIFTH EDITION

## Ron Larson

The Pennsylvania State University, The Behrend College

## Maria Haverhals Andersen

Muskegon Community College

BROOKS/COLE
CENGAGE Learning

Australia • Brazil • Japan • Korea • Mexico • Singapore • Spain • United Kingdom • United States

ISBN-13: 978-0-538-49356-7
ISBN-10: 0-538-49356-9

**Brooks/Cole**
20 Channel Center Street
Boston, MA 02210
USA

Cengage Learning is a leading provider of customized learning solutions with office locations around the globe, including Singapore, the United Kingdom, Australia, Mexico, Brazil, and Japan. Locate your local office at: **www.cengage.com/global**

Cengage Learning products are represented in Canada by Nelson Education, Ltd.

To learn more about Brooks/Cole, visit **www.cengage.com/brookscole**

Purchase any of our products at your local college store or at our preferred online store **www.CengageBrain.com**

Printed in the United States of America
1 2 3 4 5    14 13 12 11 10

ED027

Algebra is weightlifting for the Brain!

# Table of Contents: Algebra Activities

# RNUM: Real Numbers

# Student Activity
## Building Blocks

**Fill In The Blanks:** Fill in the missing numerators with whole numbers to build equivalent fractions to the fraction in the "Goal" box. If there is not a whole number numerator that will work, then cross out the fraction.

**Example:**

| Goal: $\dfrac{1}{2}$ | $\dfrac{\boxed{15}}{30}$ | $\dfrac{\times}{7}$ | $\dfrac{\boxed{5}}{10}$ |
|---|---|---|---|

| Goal: $\dfrac{1}{4}$ | $\dfrac{\quad}{24}$ | $\dfrac{\quad}{36}$ | $\dfrac{\quad}{16}$ |
|---|---|---|---|

| Goal: $\dfrac{3}{16}$ | $\dfrac{\quad}{32}$ | $\dfrac{\quad}{4}$ | $\dfrac{\quad}{80}$ |
|---|---|---|---|

| Goal: $\dfrac{3}{2}$ | $\dfrac{\quad}{50}$ | $\dfrac{\quad}{8}$ | $\dfrac{\quad}{1000}$ |
|---|---|---|---|

| Goal: $\dfrac{5}{8}$ | $\dfrac{\quad}{48}$ | $\dfrac{\quad}{32}$ | $\dfrac{\quad}{60}$ |
|---|---|---|---|

| Goal: $\dfrac{3}{3}$ | $\dfrac{\quad}{9}$ | $\dfrac{\quad}{15}$ | $\dfrac{\quad}{99}$ |
|---|---|---|---|

| Goal: $\dfrac{7}{9}$ | $\dfrac{\quad}{99}$ | $\dfrac{\quad}{81}$ | $\dfrac{\quad}{54}$ |
|---|---|---|---|

| Goal: $\dfrac{2}{3}$ | $\dfrac{\quad}{18}$ | $\dfrac{\quad}{15}$ | $\dfrac{\quad}{1}$ |
|---|---|---|---|

| Goal: $\dfrac{7}{5}$ | $\dfrac{\quad}{25}$ | $\dfrac{\quad}{35}$ | $\dfrac{\quad}{14}$ |
|---|---|---|---|

# Student Activity
## Factor Pairings

**Directions:** In each diagram, there is a number in the top box and exactly enough spaces beneath it to write all the possible factor-pairs involving whole numbers. See if you can find all the missing factor-pairs. The number 30 has been done for you.

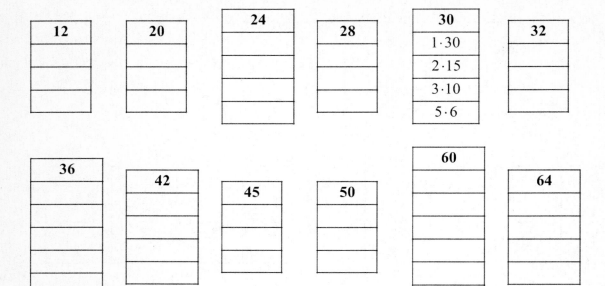

1. What is the largest number that is a factor of 20 and 30? _____

   Simplify: $\dfrac{20}{30}$

2. What is the largest number that is a factor of 28 and 36? _____

   Simplify: $\dfrac{28}{36}$

3. What is the largest number that is a factor of 36 and 60? _____

   Simplify: $\dfrac{36}{60}$

4. What is the largest number that is a factor of 24 and 42? _____

   Simplify: $\dfrac{24}{42}$

# Student Activity

## Match Up on Fractions

**Match-up:** Match each of the expressions in the squares of the table below with its simplified value at the top. If the solution is not found among the choices A through D, then choose E (none of these).

**A** 1    **B** $\dfrac{3}{4}$    **C** $\dfrac{7}{8}$    **D** 0    **E** None of these

| | | | |
|---|---|---|---|
| $\left(\dfrac{3}{2}\right)\left(\dfrac{1}{2}\right)$ | $\dfrac{7}{6} \div \dfrac{4}{3}$ | $\dfrac{7}{8} \div 0$ | $2\dfrac{1}{8} - \dfrac{5}{4}$ |
| $\dfrac{15}{8} - 1$ | $6\left(\dfrac{1}{6}\right)$ | $\dfrac{2}{3} \div \dfrac{8}{9}$ | $0 \div \dfrac{3}{4}$ |
| $\dfrac{1}{2} + \dfrac{3}{8}$ | $\dfrac{19}{12} - \dfrac{5}{6}$ | | |
| $\dfrac{1}{5} + \dfrac{8}{10}$ | $\dfrac{3}{4}(0)$ | | |
| $\dfrac{1}{2} + \dfrac{2}{2}$ | $\dfrac{1}{3} \div \dfrac{1}{3}$ | | |

I thought we shared a common denominator but he was only a fraction of the person I thought he was.

# Student Activity
## Fractions Using a Calculator

When you input fractions into a calculator, you must be careful to tell the calculator which parts are fractions. Each calculator has a set of algorithms that tell it what to do first (later on, we will learn the mathematical order of operations, which is similar). In order to ensure that fractions are treated as fractions, for now, you need to tell your calculator which parts ARE fractions.

**1.** For example, first show that $\dfrac{3}{4} \div \dfrac{2}{5}$ is $\dfrac{15}{8}$ by hand:

**2.** To get the decimal value of $\dfrac{15}{8}$ on the calculator, we type $15/8$ or $15 \div 8$ (depending on the calculator). Practice by finding the decimal values for:

$$\dfrac{15}{8} \qquad\qquad \dfrac{3}{20} \qquad\qquad \dfrac{1}{4} \qquad\qquad \dfrac{7}{8} \qquad\qquad \dfrac{2}{3}$$

**3.** Now try using your calculator to evaluate $\dfrac{3}{4} \div \dfrac{2}{5}$, but do it without using any parentheses. Do you get the decimal value equal to 15/8?

**4.** Find the button(s) on your calculator that allow you to input parentheses and write down how to use them on your calculator.

**5.** Try it on your calculator like this now: $\left(\dfrac{3}{4}\right) \div \left(\dfrac{2}{5}\right)$

On my calculators, I type $(3/4)/(2/5)$ or $(3 \div 4) \div (2 \div 5)$ to enter this expression. But each calculator is a little different. When you have done it correctly, you should get 1.875.

Write down how to do it on your calculator:

**6.** Now try these fraction problems *using parentheses* to tell your calculator which numbers represent fractions:

$$\dfrac{1}{2} \cdot \dfrac{4}{9} \qquad\qquad \dfrac{3}{4} \div \dfrac{2}{15} \qquad\qquad \dfrac{3}{8} + \dfrac{4}{5} \qquad\qquad \dfrac{7}{12} - \dfrac{1}{5}$$

**7.** The operation in mixed numbers is **addition**, so when you input $2\dfrac{3}{5}$ into your calculator, you must treat it like $2 + (3/5)$. What is $2\dfrac{3}{5}$ as a decimal? _____

# Guided Learning Activity
## Charting the Real Numbers

The set of **natural numbers** is $\{1, 2, 3, 4, 5, ...\}$.

The set of **whole numbers** is $\{0, 1, 2, 3, 4, 5, ...\}$.

The set of **integers** is $\{..., -4, -3, -2, -1, 0, 1, 2, 3, 4, ...\}$.

The set of **rational numbers** consists of all numbers that can be expressed as a fraction (or *ratio*) of *integers* (except when zero is in the denominator). Note that all rational numbers can also be written as decimals that either terminate or repeat.

The set of **irrational numbers** consists of all *real* numbers that are *not* rational numbers.

The set of **positive numbers** consists of all the numbers *greater* than zero.

The set of **negative numbers** consists of all the numbers *less* than zero.

**Part I:** Using the definitions above, we will categorize each number below. For each of the numbers in the first column, place an "X" in any set to which that number belongs.

|     |                | Natural | Whole | Integer | Rational | Irrational | Positive | Negative |
|-----|----------------|---------|-------|---------|----------|------------|----------|----------|
| a.  | 5              | X       | X     | X       | X        |            | X        |          |
| b.  | 1              |         |       |         |          |            |          |          |
| c.  | 0              |         |       |         |          |            |          |          |
| d.  | −2             |         |       |         |          |            |          |          |
| e.  | $\sqrt{3}$     |         |       |         |          |            |          |          |
| f.  | $\frac{2}{3}$  |         |       |         |          |            |          |          |
| g.  | $-1.\overline{4}$ |      |       |         |          |            |          |          |
| h.  | 0.75           |         |       |         |          |            |          |          |
| i.  | $\frac{\pi}{3}$ |        |       |         |          |            |          |          |

**Part II:** Now we'll do it backwards. Given the checked properties, find a number (try to use one that is different from one of the numbers in the previous table) that fits the properties. If it is not possible to find a number with all these properties, write "impossible" instead.

|     | Number | Natural | Whole | Integer | Rational | Irrational | Positive | Negative |
|-----|--------|---------|-------|---------|----------|------------|----------|----------|
| a.  |        |         |       | X       | X        |            |          | X        |
| b.  |        |         |       |         | X        |            | X        |          |
| c.  |        |         | X     | X       | X        |            |          |          |
| d.  |        |         |       |         |          | X          |          | X        |
| e.  |        |         |       |         | X        |            |          | X        |
| f.  |        |         |       | X       |          | X          | X        |          |
| g.  |        | X       | X     | X       | X        |            | X        |          |
| h.  |        |         |       |         |          | X          | X        |          |

# Student Activity
## Venn Diagram of the Real Numbers

**Directions:** Place each number below in the *smallest* set in which it belongs. For example, −1 is a real number, a rational number, and an integer, so we place it in the "Integers" box, but not inside the whole numbers or natural numbers.

$$8 \qquad \frac{7}{3} \qquad -1.3 \qquad \pi \qquad 2.175 \qquad 0 \qquad -7 \qquad \sqrt{2} \qquad 1000 \qquad 0.00005$$

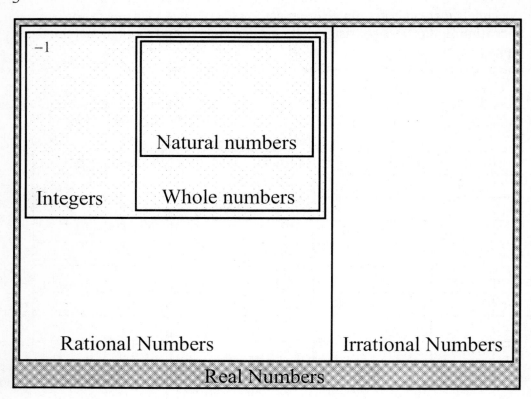

1. Given all possible real numbers, name at least one number that is a whole number, but not a natural number: _____

2. Can a number be both rational and irrational? _____ If yes, name one: _____

3. Can a number be both rational and an integer? _____ If yes, name one: _____

4. Given all possible real numbers, name at least one number that is an integer, but not a whole number: _____

**Side note:** *Just for the record, this diagram in no way conveys the actual size of the sets. In mathematics, the number of elements that belong to a set is called the* **cardinality** *of the set. Technically (and with a lot more mathematics classes behind you) it can be proven that the cardinality of the irrational numbers (uncountable infinity) is actually larger than the cardinality of the rational numbers (countable infinity). Another interesting fact is that the cardinality (size) of the rational numbers, integers, whole numbers, and natural numbers are all equal. This type of mathematics is studied in a course called Real Analysis (that comes after the Calculus sequence).*

# Student Activity
## Linking Rational Numbers with Decimals

Let's investigate why we say that decimals that terminate and repeat are really rational numbers. You will need a calculator and some colored pencils for this activity.

**Rational numbers** consist of all numbers that can be expressed as a fraction (or *ratio*) of *integers* (except when zero is in the denominator).

In the grid below are a bunch of fractions of integers.
1.  Work out the decimal equivalents using your calculator. If the decimals are repeating decimals, use an overbar to indicate the repeating sequence (like in the example that has been done for you).
2.  Shade the grid squares in which fractions were equivalent to repeating decimals in one color and indicate the color here: _____.
3.  Shade the grid squares in which fractions were equivalent to terminating decimals in another color and indicate the color here: _____.
4.  In the last row of the grid, write some of your own fractions built using integers and repeat the steps above.

| | | | |
|---|---|---|---|
| $\frac{2}{3} = 0.\overline{6}$ | $\frac{1}{4} =$ | $\frac{7}{8} =$ | $\frac{5}{9} =$ |
| $\frac{1}{2} =$ | $\frac{4}{9} =$ | $\frac{8}{5} =$ | $\frac{17}{25} =$ |
| $\frac{1}{1000} =$ | $\frac{5}{12} =$ | $\frac{3}{4} =$ | $\frac{7}{27} =$ |
| $\frac{1}{5} =$ | $\frac{12}{5} =$ | $\frac{4}{3} =$ | $\frac{1}{6} =$ |
| | | | |

5.  Are there any fractions in the grid that were not shaded as either terminating or repeating?

6.  If you write one of these fractions as its decimal equivalent, what kind of decimal do you get?

# Guided Learning Activity
## Using Addition Models

**Part I:** The first model for addition of real numbers that we look at is called the "colored counters" method. Traditionally, this is done with black and red counters, but we make a slight modification here to print in black and white.

 Solid counters (black) represent positive integers, +1 for each counter.

 Dashed counters (red), represent negative integers, −1 for each counter.

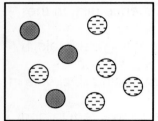 When we look at a collection of counters (inside each rectangle) we can write an addition problem to represent what we see. We do this by counting the number of solid counters (in this case 3) and counting the number of dashed counters (in this case 5). So the addition problem becomes $3+(-5)=$ _____ .

To perform the addition, we use the Additive Inverse Property, specifically, that $1+(-1)=0$. By matching up pairs of positive and negative counters until we run out of matched pairs, we can see the value of the remaining result. In this example, we are left with two dashed counters, representing the number −2. So the collection of counters represents the problem $3+(-5)=-2$.

Now try to write the problems that represent the collections below.

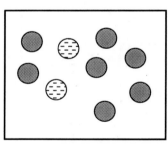

a. _____ + _____ = _____

b. _____ + _____ = _____

c. _____ + _____ = _____

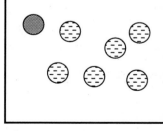

d. _____ + _____ = _____

e. _____ + _____ = _____

f. _____ + _____ = _____

**Part II:** The second model for addition of real numbers that we look at is called the "number line" method. We use directional arcs to represent numbers that are positive and negative. The length that the arc represents corresponds to the magnitude of the number.

When a directional arc indicates a positive direction (to the right), it represents a positive number. In the diagram below, each arc represents the number 2, because each arc represents a length of two and each arrow points to the right.

In the next diagram, each directional arc represents the number −5, since each arc represents a length of five and each arrow points to the left.

When we want to represent an addition problem, **we start at zero**, and travel from each number to the next using a new directional arc. Thus, the following number line diagram represents $2 + (-5) = -3$. The final landing point is the answer to the addition problem.

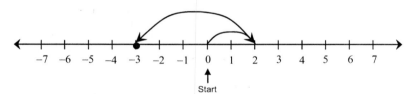

Now try to solve these addition problems on a number line using directional arcs.

**a.** $-3 + 7 =$ _____

**d.** $-6 + 2 =$ _____

**b.** $-2 + (-3) =$ _____

**e.** $-2 + (-2) + (-1) =$ _____

**c.** $5 + (-5) =$ _____

**f.** $4 + (-8) + 3 =$ _____

11

# Student Activity
## Match Up on Addition of Real Numbers

**Match-up:** Match each of the expressions in the squares of the table below with its simplified value at the top. If the solution is not found among the choices A through D, then choose E (none of these).

**A** 3      **B** −6      **C** −1      **D** 1      **E** None of these

| | | | |
|---|---|---|---|
| $-8+2$ | $-9+8$ | $-7+10$ | $-3+(-3)$ |
| $-2+(-1)+6$ | $3+(-3)$ | $-10+(-4)+8$ | $-\dfrac{1}{2}+\left(-\dfrac{1}{2}\right)$ |
| $6+(-7)$ | $5+(-6)$ | $-1+(-1)+(-1)+4$ | $-11+12$ |
| $-4+\left(-\dfrac{1}{2}\right)+\dfrac{1}{2}$ | $8+(-5)+(-9)$ | $10+(-6)+(-1)$ | $-6+12$ |
| $0+(-6)$ | $-1+7$ | $30+(-27)$ | $1+\dfrac{1}{4}+\dfrac{1}{2}+\left(-\dfrac{3}{4}\right)$ |

# Student Activity
## Scrambled Addition Tables

Here is a simple addition tables with natural number inputs.

**Addition:**

| + | 1 | 2 | 3 | 4 |
|---|---|---|---|---|
| 1 | 2 | 3 | 4 | 5 |
| 2 | 3 | 4 | 5 | 6 |
| 3 | 4 | 5 | 6 | 7 |
| 4 | 5 | 6 | 7 | 8 |

**Directions:** The first table that follows is an addition table involving integer inputs. The second table is a *scrambled* addition table with integer inputs (this means that the numbers in the first row and column do not increase nicely like 1, 2, 3, 4). Fill in the missing squares with the appropriate numbers.

| + | –3 | –2 | –1 | 0 | 1 | 2 | 3 |
|---|---|---|---|---|---|---|---|
| –3 | | | | | | | |
| –2 | | | | | | | |
| –1 | | | | | | | |
| 0 | | | | | | | |
| 1 | | | | | | | |
| 2 | | | | | | | |
| 3 | | | | | | | |

| + | 15 | –5 | 20 | –10 | 0 | 5 | 10 |
|---|---|---|---|---|---|---|---|
| 10 | | | | | | | |
| 5 | | | | | | | |
| –5 | | | | | | | |
| –15 | | | | | | | |
| 20 | | | | | | | |
| 0 | | | | | | | |
| 15 | | | | | | | |

**Directions:** The following tables are *scrambled* addition tables with the additional challenge of missing numbers in the shaded rows and columns. Fill in the missing squares with the appropriate numbers.

| + | 2 | | | 3 | |
|---|---|---|---|---|---|
| –5 | | | | | –5 |
| –2 | | | –3 | | |
| | | | | 7 | |
| | 8 | | | | |
| | | –4 | | 3 | |

| + | | –10 | 15 | | –5 |
|---|---|---|---|---|---|
| –20 | –15 | | | –20 | |
| | | | 5 | | |
| | | | | 0 | |
| 10 | | | | | |
| | | 10 | | | |

# Student Activity
**Signed Numbers Magic Puzzles**

**Directions:** In these "magic" puzzles, each row and column adds to be the same "magic" number. Fill in the missing squares in each puzzle so that the rows and columns each add up to be the given magic number.

### Magic Puzzle #1

| −2 | 8 | |
| | −9 | 4 |
| | | |

Magic Number = 5

### Magic Puzzle #2

| | | 4 |
| −5 | | |
| 8 | | −6 |

Magic Number = 0

### Magic Puzzle #3

| | | $2\frac{1}{4}$ |
| | $2\frac{1}{2}$ | $-2\frac{1}{2}$ |
| $\frac{1}{4}$ | | |

Magic Number = $\frac{1}{2}$

### Magic Puzzle #4

| −2 | | 9 | −7 |
| | −9 | | 4 |
| 8 | −4 | | −5 |
| | | −6 | |

Magic Number = 1

### Magic Puzzle #5

| −8 | −3 | | 7 |
| | | | |
| | −11 | 2 | 6 |
| 8 | −2 | | −7 |

Magic Number = −2

# Guided Learning Activity
## Language of Subtraction

**How do you interpret the − sign?** It is a minus sign if it is **between** two numbers as a mathematical operation. Otherwise, it is a negative.

Other ways to signify minus: difference, less than, subtract … from …
Other ways to signify negative: opposite

**How do you tell if *less* means < or −?** Look for the distinction between "**is** less than" and "less than." See the two examples in the table below.

| | Expression? Equation? Or Inequality? | Equivalent statement or phrase in words |
|---|---|---|
| $-(-4)$ | Expression | the opposite of the opposite of 4<br>the opposite of negative 4 |
| $8-3$ | Expression | the difference of 8 and 3<br>subtract 3 from 8<br>8 minus 3<br>3 less than 8 |
| $8-3=5$ | Equation | The difference of 8 and 3 is 5.<br>Subtract 3 from 8 to get 5.<br>8 minus 3 is 5.<br>3 less than 8 is 5. |
| $-2-4$ | Expression | the difference of negative 2 and 4<br>subtract 4 from negative 2<br>negative 2 minus 4<br>4 less than negative 2 |
| $-5<-2$ | Inequality | Negative 5 is less than negative 2. |
| $9-(-2)$ | Expression | 9 minus negative 2<br>subtract negative 2 from 9<br>the difference of 9 and negative 2 |

Note that expressions are represented in words by phrases (no verb) and equations and inequalities are represented by sentences (with verbs).

Now try these!  For any problem with subtraction, find at least two ways to write it in words.

| | | Expression? Equation? Or Inequality? | Equivalent statement or phrase in words |
|---|---|---|---|
| a. | | | Zero is less than 8. |
| b. | $10 - 2 = 8$ | | |
| c. | | | the difference of 2 and negative 5 |
| d. | $-(-10)$ | | |
| e. | $5 - (-10) = 15$ | | |
| f. | | | Negative 6 is less than negative 3. |
| g. | | | The opposite of negative 5 is 5. |
| h. | $3 < -(-6)$ | | |
| i. | $\dfrac{1}{2} - \dfrac{3}{4}$ | | |
| j. | $-19 < -18$ | | |

# Student Activity
## Match Up on Subtraction of Real Numbers

**Match-up:** Match each of the expressions in the squares of the table below with its simplified value at the top. If the solution is not found among the choices A through D, then choose E (none of these).

**A** 3 **B** −6 **C** −1 **D** 1 **E** None of these

| | | | |
|---|---|---|---|
| $-1-(-4)$ | $0-(-1)$ | $-30-(-25)+(-1)$ | $-9-(-8)$ |
| $2-(-1)$ | $-3+(-3)$ | $3-(-3)$ | $-6-(-3)$ |
| $0-(-3)$ | $12-(-6)$ | $0-(-1)$ | $3-6$ |
| $-\dfrac{9}{4}-\left(-\dfrac{1}{2}\right)+\dfrac{3}{4}$ | $\dfrac{6}{4}-\dfrac{1}{2}$ | $\dfrac{5}{2}-4-2\dfrac{1}{2}+3$ | $5\dfrac{2}{3}-\left(-\dfrac{1}{3}\right)+(-3)$ |
| $\dfrac{1}{2}-\left(-\dfrac{1}{2}\right)$ | | | |

Watch the SIGNS!

# Student Activity
## Signed Numbers Using a Calculator

When you input expressions with signed numbers into a calculator, you must be careful to tell the calculator which "−" signs represent a minus, and which represent a negative.

1.  The minus button on your calculator looks like this: $\boxed{-}$. It is found with the addition, multiplication, and division functions. The button on your calculator that is used to denote a negative may look like $\boxed{+/-}$ or $\boxed{(-)}$, or it may be above one of the keys, accessed with a $2^{nd}$ function, $\boxed{2nd}$. Locate where your calculator input for a negative is, and draw it here:

2.  On some calculators, the negative is typed before the number, and on some it is typed after the number. We need to figure out which type you have. We'll calculate $-2+5$ (which should be ___). Try it both ways. Write down exactly how to do $-2+5$ on your calculator here:

3. Let's try something more complicated now. How would we write $-8-3$ in words using the word minus? _____
What should the answer be? ___ Now write down the keystrokes for inputting this expression into your calculator here:

4. Work out each of these expressions by hand, then write down how to express them in words, and finally, write down how to input the keystrokes properly into your calculator.

| Expression | Answer | In words (using negative and/or minus) | Keystrokes |
|---|---|---|---|
| $-6+3$ | | | |
| $5-(-4)$ | | | |
| $-19-6$ | | | |
| $-1.25-0.25$ | | | |
| $10-(-8)$ | | | |
| $-3-(-3)$ | | | |

# Student Activity
## Pick Your Property

**Match-up:** Match each of the equations in the squares of the table below with the proper property of the real numbers.

A  Associative Property of Addition
B  Associative Property of Multiplication
C  Commutative Property of Addition
D  Commutative Property of Multiplication

E  Inverse Property of Addition
F  Inverse Property of Multiplication
G  Identity Property of Addition
H  Identity Property of Multiplication
I  Multiplication Property of Zero

| | | |
|---|---|---|
| $0+(-2)=-2$ | $77(-2)=(-2)77$ | $\left(\dfrac{1}{4}+\dfrac{2}{3}\right)+\dfrac{1}{3}=\dfrac{1}{4}+\left(\dfrac{2}{3}+\dfrac{1}{3}\right)$ |
| $2\left(\dfrac{3}{100}\right)\left(\dfrac{100}{3}\right)=2(1)$ | $8(-9+9)=8(0)$ | $\left(\dfrac{1}{4}+\dfrac{2}{3}\right)+\dfrac{1}{3}=\left(\dfrac{2}{3}+\dfrac{1}{4}\right)+\dfrac{1}{3}$ |
| $\dfrac{4}{5}\left(\dfrac{5}{4}\right)=\dfrac{5}{4}\left(\dfrac{4}{5}\right)$ | $5+(6+(-6))=(5+6)+(-6)$ | $\left(2\cdot\dfrac{3}{100}\right)\left(\dfrac{100}{3}\right)=2\cdot\left(\dfrac{3}{100}\cdot\dfrac{100}{3}\right)$ |
| $5\left(\dfrac{1}{5}\cdot 1\right)=5\left(\dfrac{1}{5}\right)$ | $-5+(-8+2)=(-8+2)+(-5)$ | $5+(6+(-6))=5+0$ |
| $\dfrac{4}{5}\left(\dfrac{5}{4}\right)=1$ | $5+9(0)=5+0$ | $24+6(1)=24+6$ |

# Student Activity
## What's Wrong with Division by Zero?

Let's spend some time investigating why division by zero is undefined. You will need a calculator for this activity.

**1.** First, let's see what your calculator thinks. Try the following division problems on your calculator and write down the results:

$$0 \div 5 \qquad\qquad 5 \div 0 \qquad\qquad \frac{12}{0} \qquad\qquad \frac{0}{12}$$

**2.** Even your calculator will reject the idea of division by zero, so let's try dividing by numbers *close* to zero. Looking at the number line below, name and label some numbers that are really close to zero (on both sides of zero).

**3.** Find the following quotients using your calculator.

$$5 \div 0.1 \qquad\qquad 5 \div 0.01 \qquad\qquad 5 \div 0.001 \qquad\qquad 5 \div 0.0001$$

**4.** What is happening to the quotients in #3 as the divisor gets closer to zero?

**5.** Find the following quotients using your calculator.

$$5 \div (-0.1) \qquad\qquad 5 \div (-0.01) \qquad\qquad 5 \div (-0.001) \qquad\qquad 5 \div (-0.0001)$$

**6.** What is happening to the quotients in problem **5** as the divisor gets closer to zero?

**7.** Using the results from problems **4** and **6**, why do you think that division by zero is undefined?

# Student Activity

## Match Up on Multiplying and Dividing Real Numbers

**Match-up:** Match each of the expressions in the squares of the table below with its simplified value at the top. If the solution is not found among the choices A through D, then choose E (none of these). Note that some of the expressions involve other operations besides multiplication and division, so be careful!

**A**  2          **B**  0          **C**  −8          **D**  12          **E**  None of these

| | | | |
|---|---|---|---|
| $\dfrac{-6}{-3}$ | $-\dfrac{4}{3}\left(-\dfrac{3}{2}\right)$ | $(-4)(-2)(-1)$ | $-2 \div 0$ |
| $(-0.5)(-24)$ | $(-0.25)(-8)$ | $(0)(-2)(-1)$ | $32 \div (-4)$ |
| $36 \div (-3)$ | $-3\left(\dfrac{4}{6}\right)$ | $2 \div \left(-\dfrac{1}{2}\right)$ | $-36\left(-\dfrac{2}{6}\right)$ |
| $4 - (-3) + 5$ | $5 \div \left(-\dfrac{5}{8}\right)$ | $\left(-\dfrac{4}{3}\right)(6)$ | $0(-8) - (-2)$ |
| $-2(-2)(-3)(-1)$ | $-\dfrac{1}{3}(-6)$ | $-\dfrac{3}{2} \div \left(-\dfrac{1}{8}\right)$ | $\dfrac{0}{-5}$ |

# Student Activity
## The Exponent Trio

**Directions:** In each of the "trios" below, place three equivalent expressions of the following format:

| Expanded Expression using multiplication | |
|---|---|
| Compact exponential expression | Simplified expression |

The first one has been done for you. Sometimes there are two possible trios for a simplified exponential expression, so you will see some of these listed twice.

| $(-3)(-3)$ | |
|---|---|
| $(-3)^2$ | 9 |

| $3 \cdot 3$ | |
|---|---|
| | |

| $2 \cdot 2 \cdot 2 \cdot 2$ | |
|---|---|
| | |

| | |
|---|---|
| $-3^2$ | |

| | |
|---|---|
| $(-2)^4$ | |

| $-(2 \cdot 2 \cdot 2 \cdot 2)$ | |
|---|---|
| | |

| | |
|---|---|
| | 8 |

| $(-2)(-2)(-2)$ | |
|---|---|
| | |

| | |
|---|---|
| $-2^3$ | |

| $-(4 \cdot 4)$ | |
|---|---|
| | |

| | |
|---|---|
| $(-4)^2$ | |

| | |
|---|---|
| $-10^3$ | |

| | |
|---|---|
| $\left(-\dfrac{1}{2}\right)^2$ | |

| | |
|---|---|
| | $\dfrac{1}{4}$ |

The Exponent Trio

| $2 \cdot 2 \cdot 2 \cdot 2 \cdot 2 \cdot 2 \cdot 2 \cdot 2 \cdot 2$ | |
|---|---|
| | |

| | |
|---|---|
| $(-9)^2$ | |

| | |
|---|---|
| | 125 |

# Student Activity
## Order Operation

**Directions:** With a highlighter, shade the numbers and operation that comes first in the order of operations in each problem. For example, for $7-3\cdot2$, you would highlight $3\cdot2$. If more than one operation could be done first (at the same time), shade both sets. Once you are certain that you have chosen the **first** steps correctly, then simplify each expression.

**1.** $5+3\cdot5-2$

**2.** $12\div2\cdot6+1$

**3.** $4-3^2+6$

**4.** $6\cdot2-8\div2+4$

**5.** $-12-(4+3)$

**6.** $5-2(4\cdot3)-5$

**7.** $9\cdot2\div6-5(2)^2$

**8.** $\dfrac{2-4}{(-3)^2+1}$

**9.** $\left[4+2(2-5)^2\right]-3$

**10.** $\dfrac{3}{4}-\left(\dfrac{2}{3}\right)\left(\dfrac{1}{2}\right)^2$

# Assess Your Understanding

**Real Numbers**

For each of the following, describe the strategies or key steps that will help you **start** the problem. You do **not** have to complete the problems.

| | | **What will help you to start this problem?** |
|---|---|---|
| **1.** | Add: $-3 + (-7)$ | |
| **2.** | Use $<$ or $>$ to make this statement true: $-3 \boxed{\phantom{x}} -4$ | |
| **3.** | Divide: $\dfrac{3}{4} \div \dfrac{1}{6}$ | |
| **4.** | Simplify: $7 - 2 \cdot 3 + 4$ | |
| **5.** | Subtract: $-2 - (-4)$ | |
| **6.** | Is 1.5 rational or irrational? | |
| **7.** | Add: $\dfrac{1}{3} + \dfrac{4}{5}$ | |
| **8.** | Fill in the missing value to make the statement true: $\dfrac{2}{5} = \dfrac{\boxed{\phantom{x}}}{20}$ | |
| **9.** | Is $-3$ a natural number? A whole number? An Integer? | |
| **10.** | Simplify: $(-4)^4$ | |

# Metacognitive Skills
## Real Numbers

Metacognitive skills refer to the ability to judge how well you have learned something and to effectively direct your own learning and studying. This is a self-evaluation tool designed to help you focus your studying and to improve your metacognitive skills with regards to this math class.

Fill the 1ˢᵗ column out **before** you begin studying. Fill the 2ⁿᵈ column out after you study for your test.

Go back to this assessment after your test and circle any of the ratings that you would change – this identifies the "disconnects" between what you **thought** you knew well and what you **actually** knew well.

**Use the scale below to assign a number to each topic.**

**5** *I am confident I can do any problems in this category correctly.*
**4** *I am confident I can do most of the problems in this category correctly.*
**3** *I understand how to do the problems in this category, but I still make a lot of mistakes.*
**2** *I feel unsure about how to do these problems.*
**1** *I know I don't understand how to do these problems.*

| Topic or Skill | Before Studying | After Studying |
|---|---|---|
| Finding the prime-factored form of a number; finding a factor-pair for a number. | | |
| Simplifying a fraction to write it in lowest terms. | | |
| Multiplying or dividing (signed) fractions and simplifying the result. | | |
| Building equivalent fractions. | | |
| Adding and subtracting (signed) fractions and simplifying the result. | | |
| Working with mixed numbers in mathematical expressions. | | |
| Knowing what opposite, inverse and reciprocal mean in terms of real numbers. | | |
| Categorizing numbers in different number sets (Real, rational, natural, etc.) | | |
| Evaluating expressions involving absolute value. | | |
| Using an inequality symbol ($<$ or $>$) to determine the order of real numbers (like they would be found on a number line). | | |
| Adding and subtracting real numbers (signed numbers). | | |
| Identifying which addition or multiplication property has been used in a statement (commutative, associative, identity, inverse, etc.). | | |
| Solving application problems that involve addition or subtraction of signed numbers. | | |
| Multiplying or dividing real numbers (signed numbers). | | |
| Identifying division by zero and understanding why division by zero is undefined. | | |
| Evaluating or rewriting exponential expressions. | | |
| Distinguishing between exponential expressions that involve parentheses, for example: $(-3)^2$ and $-3^2$. | | |
| Knowing the rules of the order of operations. | | |
| Applying the order of operations to a numerical expression. | | |
| Applying the order of operations when it involves absolute value or a fraction bar. | | |
| Finding the average of a set of data. | | |

# ALG: Algebraic Expressions

# Student Activity
## Translating Mathematical Operations

**Directions:** For each line, fill in the missing boxes with the proper words or notation.

| ADDITION: Written in words. | Using the word *sum*. | Using the + sign. |
|---|---|---|
| 5 plus 9 is 14. | | |
| | | $3 + 8 = 11$ |
| | The sum of 2 and $x$ is 9. | |

| SUBTRACTION: Written in words. | Using the word *difference*. | Using the − sign. |
|---|---|---|
| 12 minus 4 is 8. | | |
| | | $11 - 7 = 4$ |
| | The difference of 9 and 3 is 6. | |

| MULTIPLICATION: Written in words. | Using the word *product*. | Using a raised dot. | Using parentheses. |
|---|---|---|---|
| 3 multiplied by 5 is 15. | | | |
| | | $8 \cdot 5 = 40$ | |
| | The product of 2 and $n$ is 14. | | |
| | | | $(6)(7) = 42$ or $6(7) = 42$ |

| DIVISION: Written in words. | Using the division symbol. | Using long division notation. | Using the fraction bar. |
|---|---|---|---|
| The quotient of 16 and 2 is 8. | | | |
| | | $5\overline{)35}\ \ ^7$ | |
| | $27 \div 3 = 9$ | | |
| | | | $\dfrac{32}{4} = 8$ |

# Student Activity
## Translating Expressions and Equations

**Directions:** For each line, fill in the missing boxes with the proper words or notation. The first one has been done for you.

| | Phrase or sentence | Expression or Equation? | Write the expression or equation. |
|---|---|---|---|
| **a.** | the number of feet, $n$, times 12 | Expression | $12n$ |
| **b.** | | | $17 - x = 3$ |
| **c.** | The sum of the measures of angles $x$ and $y$ is 180°. | | |
| **d.** | the quotient of $m$ and 100 | | |
| **e.** | | | $z + 100$ |
| **f.** | The product of $y$ and 10 is 1000. | | |
| **g.** | | | $20 \div 4$ |
| **h.** | | | $u + v = 90°$ |
| **i.** | The number of inches, $n$, divided by 12 is the number of feet, $f$. | | |
| **j.** | 100 less than $x$ | | |
| **k.** | the difference of 100 and $x$ | | |

# Guided Learning Activity

## Learning the Language of Algebraic Expressions

The **terms** of an algebraic expression are separated by addition or subtraction. We rewrite subtraction as addition, like this: $3x - 5$ becomes $3x + (-5)$ and we call each part that is separated by addition a **term**. Thus, $3x$ is a term and $-5$ is a term.

If a term involves both a variable and a numerical factor, then the number part is called the **coefficient**. Thus, for the term $3x$, 3 is the coefficient of the term. A term that consists of a single number is called the **constant term**.

| Expression | How many terms? | Coefficient of... (or constant term) | | | |
|---|---|---|---|---|---|
| | | 1$^{st}$ term | 2$^{nd}$ term | 3$^{rd}$ term | 4$^{th}$ term |
| $5x^2 - 2x + 7$ <br> $5x^2 + (-2x) + 7$ | 3 | 5 | $-2$ | 7 | None |
| $x - 2.5$ | | | | | |
| $a^2 + \frac{1}{2}a - 1$ | | | | | |
| $b^3 + \frac{1}{3}b^2 - \frac{1}{2}b + 12$ | | | | | |
| $x^2 - 3xy - 10y^2$ | | | | | |
| $2\ell + 2w$ | | | | | |

If an algebraic expression is held together by multiplication, then we call the multiplied parts **factors**. If you have colored highlighters, highlight each factor in a separate color.

| Expression | How many factors? | 1$^{st}$ factor | 2$^{nd}$ factor | 3$^{rd}$ factor |
|---|---|---|---|---|
| $7x$ | 2 | 7 | $x$ | None |
| $(x+3)(x-3)$ | | | | |
| $-3ab$ | | | | |
| $h(b_1 + b_2)$ | | | | |
| $x(x+2)(x-3)$ | | | | |

## Mathematical operations in words:

| Addition | Subtraction |
|----------|-------------|
| add | subtract |
| sum | minus |
| plus | difference |
| more than | less than |
| greater than | decreased by |
| increased by | reduced by |
| exceeds | less |

| Multiplication | Division |
|----------------|----------|
| multiply | divide |
| product | quotient |
| times | ratio |
| twice, thrice, … | split into equal |
| double, triple, … | parts |
| fraction words | |

This is an expression of my love:
a = holding hands
b = walks at sunset
c = dinner with your mom

Now try these!  If indicated, use the desired variable. The first one has been done for you.

| | Phrase | Indicated Operation | Algebraic expression |
|---|--------|---------------------|----------------------|
| **a.** | the ratio of $a$ to $h$ | division | $\dfrac{a}{h}$ |
| **b.** | 10 added to $x$ | | |
| **c.** | $x$ less than $y$ | | |
| **d.** | one-half of $t$ | | |
| **e.** | $j$ increased by 25 | | |
| **f.** | triple the length, $l$ | | |
| **g.** | the product of $x$ and 27 | | |
| **h.** | $T$ reduced by 50 | | |
| **i.** | the sum of $u$ and $v$ | | |
| **j.** | the quotient of $c$ and 100 | | |
| **k.** | twice $g$ | | |
| **l.** | exceeds $R$ by $t$ | | |
| **m.** | $x$ greater than 32 | | |

# Student Activity
## Cleaning Out the Math Closet

**Skeleton Method:** To evaluate an expression, it is helpful to "see" the expression with empty spaces where the variables are. To make the "skeleton" of the expression, replace each variable with an empty parentheses "skeleton." Here are two examples:

| Evaluate $x^2 + 4x + 3$ at $x = -2$. |
|---|
| Skeleton: $(\quad)^2 + 4(\quad) + 3$ |
| $\quad\quad\quad (-2)^2 + 4(-2) + 3$ |
| $\quad\quad\quad\quad 4 - 8 + 3$ |
| $\quad\quad\quad\quad\quad -1$ |

| Evaluate $x^2 - 2xy$ at $x = 5$ and $y = -3$. |
|---|
| Skeleton: $(\quad)^2 - 2(\quad)(\quad)$ |
| $\quad\quad\quad (5)^2 - 2(5)(-3)$ |
| $\quad\quad\quad\quad 25 + 30$ |
| $\quad\quad\quad\quad 55$ |

MATH CLOSET

**Now try these!** For each problem, first write a "skeleton," then evaluate.

| **1.** Evaluate $6x - 3x^2$ at $x = -1$. | **2.** Evaluate $b^2 - 4ac$ at $a = 2$, $b = -5$, and $c = -8$. |
|---|---|
| Skeleton: | Skeleton: |

| **3.** Evaluate $\dfrac{3xy + 4}{x + y^2}$ for $x = 3$ and $y = -2$. | **4.** Evaluate: $y^2 - 2y + 3$ for $y = \dfrac{2}{3}$. |
|---|---|
| Skeleton: | Skeleton: |

# Student Activity
## Algebraic Expressions Through Patterns

**Directions:** Fill in the tables. The last row of each table will be an algebraic expression.

| Number of days | Number of hours |
|----------------|-----------------|
| 1 | |
| 2 | |
| 3 | |
| D | |

| Number of dimes | Value in cents |
|-----------------|----------------|
| 2 | |
| 5 | |
| 25 | |
| d | |

| Number of inches | Number of feet |
|------------------|----------------|
| 72 | |
| 36 | |
| 24 | |
| x | |

Can I get a double cheeseburger, minus tomato, twice the mayo, and add a side of fries?

Man, I hate story problems!

| Number of $8 tickets | Number of $6 tickets | Total cost |
|----------------------|----------------------|------------|
| 2 | 2 | |
| 3 | 5 | |
| 8 | 4 | |
| A | C | |

| Sodas ($2 each) | Burgers ($3.50 each) | Total cost |
|-----------------|----------------------|------------|
| 4 | 6 | |
| 2 | 5 | |
| 10 | 10 | |
| S | H | |

# Student Activity
## Count Consecutive Integers

**1.** Fill in the table below:

| Let $x = \dots$ | $x+1$ | $x+2$ | $x+3$ |
|---|---|---|---|
| 5 | | | |
| 10 | | | |
| 22 | | | |
| $a$ | | | |
| $2n$ | | | |

$n, ah, ah, ah, ah, ah \dots$
$n+1, ah, ah, ah, ah, ah \dots$
$n+2 \dots$

**a.** How would you represent the two integers **after** $x$: _____ and _____

**b.** How would you represent the two integers **before** $x$: _____ and _____

Billy didn't care! He loved algebra more than candy!

**2.** Fill in the table below:

| Let $x = \dots$ | $2x$ | $2x+1$ | $2x+2$ | $2x+3$ | $2x+4$ |
|---|---|---|---|---|---|
| 1 | | | | | |
| 2 | | | | | |
| 3 | | | | | |
| 18 | | | | | |
| 25 | | | | | |
| $a$ | | | | | |
| Odd or even? | | | | | |

**a.** How would you represent an unknown even integer? _____

**b.** How would you represent an unknown odd integer? _____

**c.** How would you represent the two integers that follow $2x$? _____ and _____

**d.** How would you represent the two **even** integers after $2x$? _____ and _____

**e.** How would you represent the two **odd** integers after $2x+1$? _____ and _____

# Student Activity
## Follow the Multiplication Road

**Match-up:** Find your way from start to finish along the multiplication road by shading in matching pairs of algebraic expressions. The first pair has been shaded for you!

| START | | | | | |
|---|---|---|---|---|---|
| $3(4a)$ | $7a$ | $\frac{4}{3}(4x)$ | $\frac{x}{3}$ | $\frac{4x}{3}(4)$ | $\frac{z}{5}$ |
| $12a$ | $\frac{2}{3}(3x)$ | $2x$ | $-x(-9)$ | $-9x$ | $5\left(\frac{2z}{5}\right)$ |
| $\frac{1}{12}(2a)$ | $6x$ | $\frac{4x^2}{5x^2}$ | $9x$ | $9-x$ | $5\left(\frac{5z}{2}\right)$ |
| $(0.8x)(-2)$ | $1.6x$ | $\frac{4x^2}{5}$ | $\frac{4}{5}(x^2)$ | $8x$ | $5z$ |
| $(5t)(-4)$ | $12a$ | $-3(-4a)$ | $-7a$ | $4(4x)$ | $4x^2$ |
| $-20t$ | $(-5t)(-4)$ | $-9t$ | $\frac{2}{3}(12t^2)$ | $9t^2$ | $\frac{x^2}{7x^2}$ |
| $5(-2x)$ | $-10x$ | $\frac{3}{4}(8t^2)$ | $6t^2$ | $\frac{1}{7}(x^2)$ | $\frac{x^2}{7}$ |
| | | | | | FINISH |

# Student Activity
## Match Up on Like Terms and Distribution

**Match-up:** Match each of the expressions in the squares of the lower table with an equivalent expression from the top. If the solution is not found among the choices A through D, then choose E (none of these).

**A** $4x+16-12y$    **B** $5+4x+6y$    **C** $x+y-z$    **D** $3x+3y+3z$    **E** None of these

| | | | |
|---|---|---|---|
| $3(x+y)+z$ | $5+2x+6y+2x$ | $3y+3(x+z)$ | $2(8+2x-6y)$ |
| $(x-z)+y$ | $4(x-3y+4)$ | $4(x+3y)+16$ | $6y+5+4x$ |
| $4(x+4)-12y$ | $2(2x+8-6y)$ | $(5+6y)+4x$ | $\frac{1}{3}(-3z+3y+3x)$ |
| $5+2(3y+2x)$ | $2y-(z+y)+x$ | $4x+12y-16$ | $(4x+16)-12y$ |
| $2(2x+8)+6y$ | $3x+3(z+y)$ | | |
| $5x+3y+3z-2x$ | $6y+5+4x$ | | |
| $6x+6y-3(x+y-z)$ | $x-(z-y)$ | | |
| $(z+y)-z$ | $4x-(12y-16)$ | | |

Please! We realize you are all in the same Parentheses! Hang on and you will each get one!

ALGEBRA RELIEF

Distribution Point

2a   3x   z   5y   7m

# Student Activity
## Language of Parentheses

For each problem below either write out the mathematical expression in words, or write the missing expression. Then simplify the expressions. Remember to use words like "the quantity of" or "the sum of" or "the difference of" when you describe more than one term in parentheses.

| | Expression | Write the expression in words |
|---|---|---|
| **1.** | | 8 times the difference of $5y$ and 3 |
| **2.** | $8 - 5(y - 3)$ | |
| **3.** | | 8 minus the quantity of $5y$ minus 3 |
| **4.** | $8 - 5y - 3$ | |
| **5.** | | the difference of 8 and $5y$, times $-3$ |
| **6.** | $-6 - 2(x + 7)$ | |
| **7.** | | negative 6 times the sum of $-2x$ and 7 |
| **8.** | $-6 - (2x + 7)$ | |
| **9.** | | negative 6 minus $2x$ plus 7 |
| **10.** | $(-6 - 2x) \cdot 7$ | |

# Assess Your Understanding

**Algebraic Expressions**

For each of the following, describe the strategies or key steps that will help you **start** the problem. You do **not** have to complete the problems.

| | | What will help you to start this problem? |
|---|---|---|
| **1.** | Write an expression to represent: $y$ increased by 10 | |
| **2.** | Simplify: $9y + 3z - 3y - z$ | |
| **3.** | How many terms are in the expression $7t^2 - 8t + 5$? | |
| **4.** | Evaluate $4a^2 - 2$ when $a = -5$. | |
| **5.** | Simplify: $-2(4c + 5) - 3c$ | |
| **6.** | Write an algebraic expression to represent the value in cents of $q$ quarters. | |
| **7.** | Write $4(x - 3)$ in words. | |
| **8.** | Simplify: $9 - (3x + 6)$ | |
| **9.** | Write using algebra: *Five more than twice a number is 10.* | |
| **10.** | What is the coefficient of the $x$-term in $4x^3 - 2x + 5$? | |

# Metacognitive Skills
## Algebraic Expressions

Metacognitive skills refer to the ability to judge how well you have learned something and to effectively direct your own learning and studying. This is a self-evaluation tool designed to help you focus your studying and to improve your metacognitive skills with regards to this math class.

Fill the 1$^{st}$ column out **before** you begin studying. Fill the 2$^{nd}$ column out after you study for your test.

Go back to this assessment after your test and circle any of the ratings that you would change – this identifies the "disconnects" between what you **thought** you knew well and what you **actually** knew well.

**Use the scale below to assign a number to each topic.**
5   *I am confident I can do any problems in this category correctly.*
4   *I am confident I can do most of the problems in this category correctly.*
3   *I understand how to do the problems in this category, but I still make a lot of mistakes.*
2   *I feel unsure about how to do these problems.*
1   *I know I don't understand how to do these problems.*

| Topic or Skill | Before Studying | After Studying |
|---|---|---|
| Identifying the number of terms in an expression, its coefficients, or its factors. | | |
| Identifying words that translate into mathematical operations. | | |
| Translating a phrase or sentence into a mathematical expression or equation and vice versa. | | |
| Evaluating an algebraic expression for a given number or numbers. | | |
| Distinguish between equations and expressions. | | |
| Using algebra to represent a varying quantity in an application problem. | | |
| Simplifying expressions involving like and unlike terms. | | |
| Multiplying an algebraic term by a constant. | | |
| Applying the distributive property. | | |
| Applying a negative that is in front of a set of parentheses. | | |

# EQN: Solving Linear Equations and Inequalities

# Student Activity

## Is it a Solution?

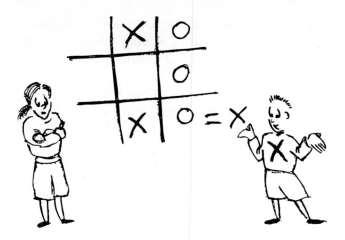

**Tic-tac-toe Directions:** If the number in the square **IS** a solution of the equation, then put an **O** on the square. If it **IS NOT** a solution, then put an **X** on the square.

| | | |
|---|---|---|
| $x + 5 = 9$<br><br>4 | $\lvert y - 3 \rvert = 5$<br><br>2 | $4.3 + x = 7.7$<br><br>3.4 |
| $10 - \dfrac{x}{2} = 4$<br><br>6 | $3z + 7 = -1$<br><br>$-2$ | $x^2 - 3x - 4 = 0$<br><br>4 |
| $0.2x = 3$<br><br>0.6 | $\lvert 6 - a \rvert = 9$<br><br>15 | $x^2 - 3x - 4 = 0$<br><br>$-1$ |

# Student Activity
## Checking Solutions with a Calculator

When you check the solution to an equation, you hope to see an equality (a true statement) when you're finished. However, when the solution is a rounded decimal value, you may only see an approximate equality. To denote this, use the symbol $\approx$ instead of $=$.

For example, we check two solutions in the equation $7x - 2 = 4$.

| | |
|---|---|
| Check $x = \dfrac{6}{7}$: $\quad 7x - 2 = 4$ <br><br> $7\left(\dfrac{6}{7}\right) - 2 \overset{?}{=} 4$ <br><br> $6 - 2 \overset{?}{=} 4$ <br><br> $4 = 4$ <br><br> True. | Check $x = 0.86$: $\quad 7x - 2 = 4$ <br><br> $7(0.86) - 2 \overset{?}{=} 4$ <br><br> $6.02 - 2 \overset{?}{=} 4$ <br><br> $4.02 \approx 4$ <br><br> True. |

Note that $\dfrac{6}{7} = 0.8571428... \approx 0.86$, so both answers should work, but in the second check, we get only an approximate equality. These numbers are close enough for us to believe that $6/7$ is a solution of the equation.

**Directions:** In each box, determine if the given number creates an equality, an approximate equality, or is not a solution. Use the symbols $=$, $\approx$, and $\neq$ where appropriate.

| | | | |
|---|---|---|---|
| $8x = 4 - x$ <br><br> 0.44 | $2y + 1 = 4$ <br><br> 1.5 | $3u^2 + 5u - 2 = 0$ <br><br> 0.33 | $\lvert 2s + 3 \rvert = 3.5$ <br><br> $-0.25$ |
| $w^2 = \dfrac{9}{16}$ <br><br> 0.75 | $x = \dfrac{x+2}{4}$ <br><br> 0.67 | $3z + 3 = 2 - z$ <br><br> $-0.12$ | $a^3 = 7$ <br><br> 1.91 |

# Student Activity

## Paint by the Addition Property

**Directions:** Solve each equation using the Addition Property of Equality.

| | | | |
|---|---|---|---|
| $x - 3 = 2$ | $x + 50 = 25$ | $\dfrac{5}{7} = a - \dfrac{11}{14}$ | $x - 4 = 0$ |
| $-7 + z = -16$ | $3 = 24 + y$ | $1.23 + w = 4.56$ | $a + 42 = 86$ |
| $-1.6 = w - 3.2$ | $1.25 + x = -1.26$ | $-19 = d - 20$ | $u - 6 = -19$ |
| $3 + y = 3$ | $x + \dfrac{1}{4} = -\dfrac{3}{4}$ | $\dfrac{5}{4} = \dfrac{21}{16} + k$ | $r + 2 = -9$ |
| $2.05 = 2 + t$ | $2.72 + r - 3.14 = 0$ | $q + 4 = 12$ | $6006 = 6000 + q$ |

Then shade the solutions here.

| | | | | | | |
|---|---|---|---|---|---|---|
| 33.3 | 9 | $-9$ | $-8$ | 8 | 44 | 0 |
| 0.333 | 3.33 | 5 | $-5$ | $-44$ | $-4$ | 4 |
| $-1$ | 25 | $-25$ | $-1.6$ | $-2$ | 2 | $-21$ |
| $-2.51$ | $\dfrac{3}{2}$ | 1.6 | 16 | 13 | 1 | 21 |
| 2.51 | $-2.49$ | 6 | 0.5 | $-13$ | 11 | $\dfrac{1}{16}$ |
| $\dfrac{21}{28}$ | $\dfrac{2}{3}$ | 0.42 | $-0.05$ | 0.05 | $-11$ | $-\dfrac{1}{16}$ |

# Student Activity
## Crack the Eggs with the Multiplication Property

**Directions:** Solve each equation using the Multiplication Property of Equality.  All the solutions can be found somewhere in the eggs below.

| | | | |
|---|---|---|---|
| $3x = 2$ | $5x = \dfrac{5}{3}$ | $\dfrac{x}{2} = 6$ | $42x = 6$ |
| $\dfrac{4x}{3} = -1$ | $-8x = 8$ | $6x = 42$ | $100x = -25$ |
| $\dfrac{3}{2}x = \dfrac{3}{2}$ | $-\dfrac{x}{5} = -\dfrac{6}{10}$ | $-x = \dfrac{1}{9}$ | $\dfrac{15x}{6} = 20$ |

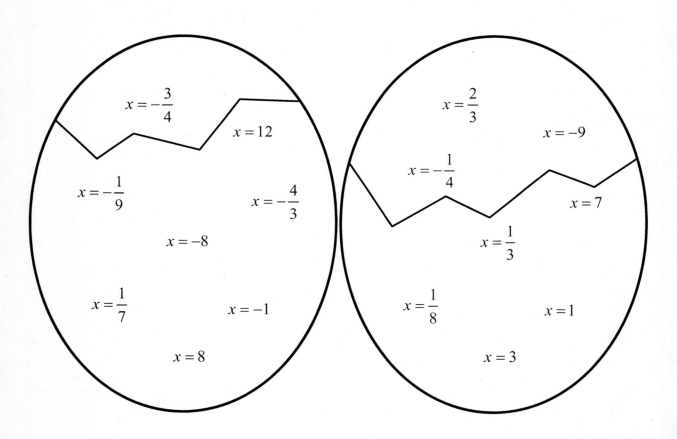

# Student Activity
## Match Up on One-step Equations

**Match-up:** Match each of the equations in the squares in the lower table with its solution from the top. If the solution is not found among the choices A through D, then choose E (none of these).

**A** $-2$

**B** $8$

**C** $-1$

**D** $0$

**E** None of these

There's going to be some give and take in this process, and some things may need to be divided up, but I'm here to make sure each side gets treated the same. Hopefully we can all arrive at a solution.

| | | | |
|---|---|---|---|
| $2 = x + 4$ | $-5a = -40$ | $\dfrac{m}{4} = 0$ | $-3 + t = 5$ |
| $\dfrac{5r}{2} = 5$ | $\ell - 5 = -7$ | $-\dfrac{2}{3}u = \dfrac{2}{3}$ | $1 = h + 3$ |
| $8 = -8 + g$ | $-x = -8$ | $b - 4 = -4$ | $32 = 4d$ |
| $-7 = n - 6$ | $\dfrac{3}{4}f = -\dfrac{3}{2}$ | $-2 = \dfrac{w}{-4}$ | $z - \dfrac{1}{6} = -\dfrac{7}{6}$ |

47

# Student Activity
## Match Up on Solving Equations

**Match-up:** Match each of the equations in the squares of the table below with its solution from the top. If the solution is not found among the choices A through E, then choose F (none of these).

A  $x = -2$

B  $x = 3$

C  $x = 2$

D  $x = 0$

E  $\varnothing$

F  None of these

No, "admitting you have a problem" is not the first step..

Solve

$3 = 2(4x - 2)$

| | | | |
|---|---|---|---|
| $5x - 4 = 3x + 2$ | $-2(1 - 4x) = 3x + 8$ | $5x - 3x = 0$ | $5x + 4x = x - 4$ |
| $5 - 4x = -(1 + x)$ | $5 - (3x + 2) = -6$ | $5 = 5 - 3x$ | $2(x + 3) = -x$ |
| $2(x - 1) - x = x + 5$ | $3x = 4 - 3x$ | $2x - 4 = -8$ | $3x - 1 = 2x + x$ |

# Student Activity
## Stepping Stones

**Directions:** Shade pairs of equivalent expressions to create a path of stepping stones from Start to Finish.

| | | | | | |
|---|---|---|---|---|---|
| **Start** $10\left(\dfrac{1}{2}\right)$ | $5$ | $8\left(\dfrac{3x}{2}\right)$ | $12x$ | $6\left(-\dfrac{z}{3}\right)$ | $3z$ |
| $8x$ | $15x$ | $-12x$ | $-3z$ | $-2z$ | $2z$ |
| $4\left(\dfrac{3x}{2}\right)$ | $-15x$ | $24x\left(-\dfrac{5}{8}\right)$ | $12$ | $9\left(\dfrac{4}{3}\right)$ | $36$ |
| $6x$ | $\dfrac{3}{2}\left(\dfrac{1}{x}\right)$ | $-4u$ | $3\left(\dfrac{1}{4}\right)$ | $\dfrac{1}{12}$ | $x\left(-\dfrac{1}{7}\right)$ |
| $5\left(\dfrac{3x}{10}\right)$ | $\dfrac{3x}{2}$ | $-6u\left(-\dfrac{2}{3}\right)$ | $4u$ | $4\left(\dfrac{2}{u}\right)$ | $\dfrac{x}{7}$ |
| $\dfrac{x}{2}$ | $-9u$ | $9u$ | $\dfrac{2x}{3}\left(\dfrac{3}{2}\right)$ | $x$ | **Finish** |

# Student Activity
**Clear the Way!**

**Directions:** Use the LCD to clear each equation of fractions and shade in the corresponding square in the grid at the bottom. The first one has been done for you. There's a surprise when you're finished!

**1.** $10\left(\dfrac{1}{10}x - \dfrac{1}{5}y\right) = 10(1)$

$\boxed{x - 2y = 10}$

**2.** $\dfrac{1}{10}x + \dfrac{1}{15}y = \dfrac{1}{30}$

**3.** $\dfrac{1}{100}(3x - 5y) = 3$

**4.** $0.1x + 2 = 0.2y$

**5.** $0.04(x - 50) = 0.02y$

**6.** $0.33(x + 1) = 7y$

**7.** $\dfrac{x}{2} - \dfrac{y}{4} = \dfrac{1}{16}$

**8.** $0.001x + 0.0002 = 0.003y$

**9.** $\dfrac{2}{3}y + \dfrac{4}{9} = \dfrac{1}{18}x$

**10.** $\dfrac{x}{10} + \dfrac{y}{100} = \dfrac{1}{1000}$

| $4x - 50 = 20y$ | $x - 2y = 10$ | $3x - 5y = 30$ |
|---|---|---|
| $4x - 200 = 2y$ | $3x - 5y = 300$ | $3x - 5y = 3$ |
| $2x + 3y = 1$ | $3x + 2y = 1$ | $3x + y = 1$ |
| $8x - 2y = 1$ | $8x - 4y = 1$ | $8x - 4y = 2$ |
| $x + 20 = y$ | $x + 20 = 2y$ | $x + 20 = 0.2y$ |
| $10x + y = 1$ | $33x + 33 = 700y$ | $33x + 33 = 7y$ |
| $10x + 2 = 30y$ | $100x + 10y = 1$ | $12y + 8 = x$ |

# Student Activity
## Equations vs. Expressions

**Tic-Tac-Toe:** In each box below, decide whether you are being given an expression or an equation – if it's an expression, simplify the expression and put an **X** in the box. If it's an equation, then solve the equation and circle the solution (thus making an **O**).

| | | |
|---|---|---|
| $4x - (6x + 2)$ | $5x + 3 = 7$ | $3 - x = 9$ |
| $3(x - 2) - 2x$ | $5x + 3 - 7$ | $\dfrac{2}{5}x - \dfrac{1}{3} + \dfrac{x}{5} = \dfrac{2}{3}$ |
| $3(x - 2) = 2x$ | $4x - (2x - 4) = 8$ | $\dfrac{2}{5}x - \dfrac{1}{3} + \dfrac{x}{5} - \dfrac{2}{3}$ |

**QUESTION:** Do the answers to equations look any different than the answers to expressions? Why or why not?

# Student Activity
## Easy Ten Percent

**Directions:** Without using your calculator, find the given percentages by first finding the "easy 10%" of the total amount.

| | 10% of Total | Final Answer |
|---|---|---|
| 1. Your restaurant bill is $22, and you want to leave a 20% tip. How much is the tip? | 10% of $22 is $2.20. | Tip is $4.40. |
| 2. You want to donate 30% of the $400 earned at your yard sale to charity. How much should you donate? | | |
| 3. Your church suggests that you tithe 5% of your annual income. Your annual income is $24,000. How much is the 5% tithe? | | |
| 4. You have a scholarship that covers 40% of tuition. If tuition is $10,000 per year, how much will your scholarship pay? | | |
| 5. At breakfast this morning, you thought the service was excellent and you want to leave a 30% tip on your $12 bill. How much tip should you leave? | | |
| 6. You want to pay 40% of your $2,100 credit card bill, how much should you pay? | | |
| 7. You just sold the first car you ever owned and you want to give your parents 60% of the $2,500 you received. How much should you give them? | | |
| 8. You are splitting the bill for repairs to your apartment, and each roommate will pay 20% of the $6,000 bill. How much do you owe? | | |
| 9. You have a coupon for 15% off the price of your textbooks. If the textbooks cost $460, how much will you save with the coupon? | | |
| 10. The top 5% of the 4,200 students at your college will receive a $200 tuition refund. How many students will receive the refund? | | |

# Student Activity
## Writing and Solving Percent-sentences

**Directions:** Rewrite each sentence as a percent-sentence, translate it into to an equation, then solve. Use $x$ or $p$ to represent the unknown quantity in the equations. Remember percents must be expressed as decimals in equations.

1.  What is 25% of 50?

    _____ is _____ % of _____

    _____ = _____ · _____

    Solution: _____

2.  20 is what percent of 60?

    _____ is _____ % of _____

    _____ = _____ · _____

    Solution: _____

3.  What percent of 80 is 5?

    _____ is _____ % of _____

    _____ = _____ · _____

    Solution: _____

4.  What is 60% of 200?

    _____ is _____ % of _____

    _____ = _____ · _____

    Solution: _____

5.  12 is 6% of what number?

    _____ is _____ % of _____

    _____ = _____ · _____

    Solution: _____

6.  0.5 is what percent of 60?

    _____ is _____ % of _____

    _____ = _____ · _____

    Solution: _____

7.  What is 2% of 50?

    _____ is _____ % of _____

    _____ = _____ · _____

    Solution: _____

8.  8 is 40% of what number?

    _____ is _____ % of _____

    _____ = _____ · _____

    Solution: _____

9.  What is 120% of 200?

    _____ is _____ % of _____

    _____ = _____ · _____

    Solution: _____

10. What percent of 80 is 100?

    _____ is _____ % of _____

    _____ = _____ · _____

    Solution: _____

# Student Activity
## Working with Percent Increase or Decrease

**Directions:** Calculate the amount of increase or decrease for each problem, then find what is being asked.

(Percent of increase as decimal)·(Original amount) = Amount of increase

(Percent of decrease as decimal)·(Original amount) = Amount of decrease

Percent of increase or decrease as a decimal = $\dfrac{\text{Amount of increase (or decrease)}}{\text{Original amount}}$

---

1. The population of a small town is 890. If the population decreases 10% after a factory closing, what is the new population of the town?

By how many people does the population decrease? _____

What is the new population? _____

2. The assessed value of a $120,000 home increases by 5% in one year. What is the new assessed value of the home?

By how much does the home value increase? _____

What is the new assessed value of the home? _____

3. A stock was valued at $34.50 per share. If the stock value decreases by 20% after a poor sales report, what is the new stock price?

By how much does the stock decrease? _____

What is the new stock price? _____

4. An elderly woman receives $582.00 in social security each month. After a recent increase, she began receiving $616.92 per month. What was the percent increase in her social security check?

By how much did the check amount increase? _____

Calculate the percent increase (and show your work):


5. A laptop computer was advertised for $1699. When you call to inquire about purchasing the laptop, the phone salesperson says they will give you the laptop for $1449 instead. What was the percent decrease in the cost of the computer?

By how much did the computer cost decrease? _____

Calculate the percent decrease (and show your work):

# Student Activity

## Pie Charts

**Hints:** Remember that all the pieces of a pie chart must add up to 100%. It may also help to first write a percent-sentence before solving each problem.

On a recent exam, Sarah's algebra class received the grade distribution shown in the pie chart to the right.

**1.** If 10 students scored in the 80-89 range on the test, how many students took the exam?

**2.** What percentage of the class scored a 70 or better?

**3.** How many students scored at least a 70?

### Algebra Exam Scores

### Household Budget

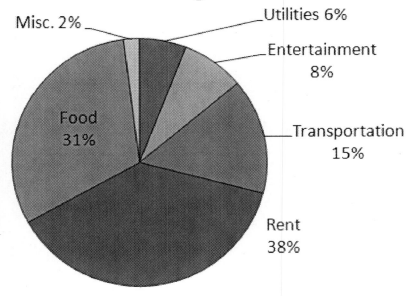

The pie chart shown on the left represents Richard's monthly household budget.

**4.** If Richard's total take home pay is $1500 per month, how much does he spend on food and rent together?

**5.** How much does he spend on entertainment?

**6.** What percentage of his income is not spent on utilities?

# Student Activity
## Circle Scavenger Hunt

**Materials:** Your instructor will provide you with a long piece of string and a ruler.

**Directions:** You have 10 minutes to find five circular objects to measure on your campus. Use the string and ruler to measure the circumference and diameter of each circle. Measure the first two objects in inches and the rest in centimeters. Once you have found five circles, return to your classroom to do the rest of the calculations.

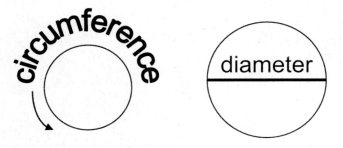

| Object | Circumference | Diameter | Calculate: $\dfrac{circumference}{diameter}$ |
|---|---|---|---|
| **1.** (measure in inches) | | | |
| **2.** (measure in inches) | | | |
| **3.** (measure in centimeters) | | | |
| **4.** (measure in centimeters) | | | |
| **5.** (measure in centimeters) | | | |
| | | | Average: |

**Conclusion:**

# Student Activity
## Is it Length, Area, or Volume?

| Measuring Length | Measuring Area | Measuring Volume |
|---|---|---|
| $P = 2w + 2\ell$ (rectangle) | $A = \ell w$ (rectangle) | $V = \ell w h$ (rectangular solid) |
| $P = 4s$ (square) | $A = s^2$ (square) | $V = s^3$ (cube) |
| $C = 2\pi r = \pi d$ (circle) | $A = \pi r^2$ (circle) | $V = \frac{4}{3}\pi r^3$ (sphere) |
| $P = a + b + c$ (triangle) | $A = \frac{1}{2}bh$ (triangle) | $V = \pi r^2 h$ (cylinder) |
| English units: in, ft, yd, mi | English units: in², ft², yd², mi² | English units: in³, ft³, yd³, mi³ |
| Metric units: mm, cm, m, km | Metric units: mm², cm², m², km² | Metric units: mm³, cm³, m³, km³ |

Some see the glass as half empty, others as half full...
I see it as $V = \frac{\pi r^2 h}{2}$.

**Directions:** Read the problems and decide whether you will have to find a length, area, or volume. Then identify the appropriate formula to use and decide what the resulting units will be. You do not have to solve the problems.

| Problem | Length, Area, or Volume? | Formula | Units of answer |
|---|---|---|---|
| **1.** Mark is going to paint a large rectangular wall in a museum that measures 20 feet by 9 feet. How many square feet will Mark be painting? | | | |
| **2.** Alice wants to build a fence around her circular garden. If the diameter of the garden is 6 meters, how long does the fence need to be? | | | |

| Problem | Length, Area, or Volume? | Formula | Units of answer |
|---|---|---|---|
| **3.** The cylindrical mixing barrel on a cement truck measures 4 yards long and has a radius of 0.8 yards. How much cement can the truck hold in its mixing barrel? | | | |
| **4.** A rectangular fish tank measures 24 inches by 10 inches by 10 inches. How much water can the tank hold? | | | |
| **5.** A real-estate agent is pricing a New York studio apartment for someone. If the apartment is square, with 15 feet on each side, and apartments cost approximately $6.00 per square foot per month, how much will this one rent for? | | | |
| **6.** Stephen wants to buy fabric for a triangular sailboat sail that will measure 20 feet tall by 10 feet wide. How much fabric does he need? | | | |
| **7.** A glass globe hummingbird feeder has a diameter of 12 centimeters. How much sugar-water will it hold? | | | |
| **8.** Sarah is buying a circular cage for her 3 foot long python. She needs one that has a diameter at least twice as long as the snake. Will a cage with a 36 square foot floor be big enough? | | | |
| **9.** Brian is installing an invisible dog fence at his house (a wire that is buried underground). How much wire will it take to fence in his rectangular yard, which measures 20 meters by 30 meters? | | | |
| **10.** How many 8-oz (approximately 236.6 cubic centimeters) servings of iced tea can fit in a jar with diameter 25.4 centimeters and height 38.1 centimeters? | | | |

58

## Whose Turn?

**Directions:** For each formula given below, you will solve for each of the variables.

---

**1.** $r = c + m$

Already solved for: _r_

| Solve for $c$: | Solve for $m$: |
|---|---|
|  |  |

---

**2.** $I = Prt$

Already solved for: ___

| Solve for $P$: | Solve for $r$: | Solve for $t$: |
|---|---|---|
|  |  |  |

---

**3.** $\omega = \dfrac{\theta}{t}$

Already solved for: _____

| Solve for $\theta$: | Solve for $t$: |
|---|---|
|  |  |

---

**4.** $y = mx + b$

Already solved for: _____

| Solve for $m$: | Solve for $x$: | Solve for $b$: |
|---|---|---|
|  |  |  |

---

**5.** $F = ma$

Already solved for: __

| Solve for $m$: | Solve for $a$: |
|---|---|
|  |  |

---

**6.** $E = mgh$

Already solved for: ___

| Solve for $m$: | Solve for $g$: | Solve for $h$: |
|---|---|---|
|  |  |  |

---

**7.** $PV = nRT$

| Solve for $P$: | Solve for $V$: | Solve for $n$: | Solve for $R$: | Solve for $T$: |
|---|---|---|---|---|
|  |  |  |  |  |

---

**8.** $m\left(x_2 - x_1\right) = y_2 - y_1$

| Solve for $m$: | Solve for $x_2$: | Solve for $x_1$: | Solve for $y_2$: | Solve for $y_1$: |
|---|---|---|---|---|
|  |  |  |  |  |

# Guided Learning Activity
## Problem Solving with Equations

You may be tempted to just "guess" at the answers to these problems. However, the point of this exercise is to get used to setting up equations so that when the problems are harder, you know how to do it well!

So please humor your instructor as he or she walks you through the steps to analyzing a problem in order to solve it with algebra.

**Problem 1:** In a *sprint* triathlon, the participants swim 0.75 km, bike 20 km, and then complete the triathlon with a run. If the total distance of the triathlon is 25.75 km, then what is the running distance?

**Analyze the problem:**

| | |
|---|---|
| Swim | |
| Bike | |
| Run | |
| **Total** | |

**Write an equation:**

**Solve the equation:**

**State the conclusion in a complete sentence:**

**Explain why your solution makes sense:**

**Problem 2:** The terminal bus at an airport drives a 4.2 mile loop between the three terminals (A, B, and C). If Terminal A is 1.6 miles from Terminal B, and Terminal B is 0.9 miles from Terminal C, then how far is the drive from Terminal C back to Terminal A?

**Analyze the problem:**

**Write an equation:**

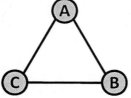

**Solve the equation:**

Total:

**State the conclusion in a complete sentence:**

**Explain why your solution makes sense:**

---

**Problem 3:** Three consecutive <u>even</u> integers sum to 78. What are the three integers?

**Analyze the problem:**

**Write an equation:**

| | |
|---|---|
| First even integer | |
| Second | |
| Third | |
| **Sum** | |

**Solve the equation:**

**State the conclusion in a complete sentence:**

**Explain why your solution makes sense:**

**Problem 4:**  In April, the water level in a mountain reservoir increased 0.6 m as the snow began to melt.  In May, the water level increased another 0.3 m.  In June the water level remained constant, but in July the water level decreased by 0.2 m because of a dry spell.  At the end of July, the water level was 14.2 m; what was the water level at the beginning of April?

**Analyze the problem:**                    **Write an equation:**

| Beg. of April | |
|---|---|
| Change in April | |
| Change in May | |
| Change in June | |
| Change in July | |
| End of July | |

**Solve the equation:**

**State the conclusion in a complete sentence:**

**Explain why your solution makes sense:**

---

**Problem 5:**  The campsite fee at a state park is $27 per day plus $8 to make your campsite reservation online.  Based on a budget of $170 to make the reservations and pay for camping, how many days can a family camp at the state park?

**Analyze the problem:**                    **Write an equation:**

| Cost for 1 day | |
|---|---|
| Cost for 2 days | |
| Cost for 3 days | |
| Cost for 4 days | |
| Cost for $x$ days | |

**Solve the equation:**

**State the conclusion in a complete sentence:**

**Explain why your solution makes sense:**

# Student Activity
## What was the Problem?

For each of the solutions below, **write a problem** that could go with the solution. Luckily for you, each student has done a particularly good job of showing their work!

## Problem 1:

**Student work:**

Let $x =$ the measure of one of the isosceles angles.
Then $2x + 20$ is the measure of the large angle.

$$x + x + (2x + 20) = 180$$
$$4x + 20 = 180$$
$$4x = 160$$
$$x = 40$$

The angles measure $40°$, $40°$, and $100°$.

Check: $40 + 40 + (2 \cdot 40 + 20) \overset{?}{=} 180$
$$180 = 180$$

## Problem 2:

**Student work:**

Let $C =$ commission and $P =$ selling price of house., then $C = 0.07P$.
The owner would receive $P - C$ or $P - 0.07P$.

$$P - 0.07P = 250,000$$
$$0.93P = 250,000$$
$$P \approx \$268,817$$

The selling price of the house should be $268,817.

Check: $268,817 - 0.07(268,817) \overset{?}{=} 250,000$
$$250,000 = 250,000$$

## Problem 3:

**Student work:**

Susan has a budget of $65. For 30 pages, the scrapbook price is $30.
Let $p =$ the number of extra pages in the scrapbook.
Then $1.75p$ is the cost of the extra pages.

$$30 + 1.75p = 65$$
$$1.75p = 35$$
$$p = 20$$

Susan can get 20 extra pages, for a total of 50 pages in the scrapbook.

Check: $30 + 1.75(20) \overset{?}{=} 65$
$$65 = 65$$

# Student Activity
## Working with the Smaller Pieces

**Practice with Distance-Rate-Time Problems**

Distance, rate and time can be related using the formula $d = r\,t$.

I need a ticket on a train leaving west from chicago at 80 mph! And quick! My Algebra grade depends on it...

1. If you travel 8 miles from your house to visit a friend, what is your round-trip distance? _____

2. Will you run faster when you run *with* the wind or *against* it? _____

3. How far can you drive in 90 minutes at a rate of 70 miles per hour? _____

4. If you work 30 miles away from your home, how long will it take you to get there if your average rate is 45 miles per hour?

   _____

5. If you drive for 3 hours and travel 210 miles, what was your average rate?

   _____

6. If you are in a boat that can travel 4 miles per hour in still water, will you be traveling *faster* or *slower* than 4 miles per hour when traveling upstream? _____

**Practice with Value Mixture and Percent Mixture Problems**

7. If a store owner mixes peanuts that cost $3 per pound with cashews that cost $8 per pound, a pound of mixed nuts should cost
   **a.** less than $3    **b.** between $3 and $8    **c.** more than $8

8. Simon has to mix two solutions in his chemistry class. One contains 3% sulfuric acid, and the other contains 10% sulfuric acid. How much sulfuric acid will the resulting mixture contain?
   **a.** less than 3%
   **b.** between 3% and 10%
   **c.** between 10% and 13%
   **d.** more than 13%

**9.** Lucy has to combine plain water with a solution that contains 15% ammonia. The resulting solution will be

    **a.** less than 15% ammonia   **b.** more than 15% ammonia

**10.** If you wanted to obtain 5 liters of a solution that is 4% nitric acid, which solution could you **not** add?

    **a.** 3 liters of a 3% solution
    **b.** 2 liters of a 5% solution
    **c.** 6 liters of a 2% solution

**11.** At Meg's coffee shop, frozen coffees cost $3.50 each. If Meg took in a total of $42 on Tuesday, what is the maximum number of frozen coffees she could have sold?_____

**Practice with Investment Problems**

Interest earned, principle invested, interest rate per year, and time invested (years) can be related using the formula $I = Prt$.

**12.** If you invest $2,500 for one year at an annual rate of 6%, how much interest does the investment earn? _____

**13.** If you invested a total of $15,000 in two different accounts, and you put $6,000 in the first account, how much did you put in the second account?_____

**14.** If you invest $1,000 in an account that pays 5% annual interest, how much money will you have after one year?_____

**15.** Martin invests $2,000 in a bank CD. If the interest after one year is $160, what was the annual interest rate? _____

# Guided Learning Activity
## Organizing Information into Tables

**For each problem:**
a) Fill in each table with numbers in the appropriate spaces (there will be a few unfilled spaces).
b) One of your unfilled spaces should be filled with the unknown variable or expressions involving the unknown variable (like $x$, $2x$, $x+5$, etc.).
c) The last column of each table is calculated using the previous columns. Go ahead and fill this in.

---

1. A man decided to invest the $15,000 inheritance he received so that he could use the annual interest earned to pay the annual taxes on his home, which are $1,200. The highest bank rate that he could find was 6% annual simple interest, but this is not high enough to make the $1,200. So, instead of investing all the money at the bank, he invested some of the money in a riskier, but more profitable, investment offering a 10% return. How much should the man invest in each account to make exactly $1,200 in one year?

| Investment | P | r | t | Interest |
|---|---|---|---|---|
| Bank |  |  |  |  |
| Riskier investment |  |  |  |  |

2. A chemistry experiment calls for a 40% hydrochloric acid solution. If the lab supply room has only 60% and 10% hydrochloric acid solutions, how much of each should be mixed to obtain 12 liters of the 40% acid solution?

| Solution | Solution Amount | Strength | Acid Amount |
|---|---|---|---|
| Strong Acid |  |  |  |
| Weak Acid |  |  |  |
| Mixed Acid |  |  |  |

3. A bulk-foods department purchaser wants to mix up 20 pounds of trail mix consisting of peanuts, chocolate candies, and raisins. The mixture should contain twice as many peanuts as raisins (by weight) and the amount of raisins and chocolate candies should be equal. If peanuts cost $2.40 per pound, chocolate candies cost $1.80 per pound, and raisins cost $3.20 per pound, how many pounds of each ingredient should be used in order to create the trail mix and how much will it cost per pound?

| Ingredient | Amount | Price per pound | Total Value |
|---|---|---|---|
| Peanuts |  |  |  |
| Raisins |  |  |  |
| Chocolate |  |  |  |
| Trail Mix |  |  |  |

4. A feed store sells sunflower seeds for $2.50 per pound and feed corn for $0.60 per pound. The manager wants to create 100 pounds of small animal feed that costs $1.50 per pound by mixing these two ingredients.  How much of each should be used?

| Ingredient | Amount | Price per pound | Value |
|---|---|---|---|
| Sunflower seed | | | |
| Feed corn | | | |
| Small Animal Feed | | | |

5. Two runners leave traveling in opposite directions on a 22-mile loop.  The first runner sets a pace of 5 mph and the second runner sets a pace of 6 mph.  How long will it take them to meet back up on the loop?

| | Rate | Time | Distance |
|---|---|---|---|
| First Runner | | | |
| Second Runner | | | |

6. Two tractor trailer trucks are 540 miles apart and their speeds differ by 5 mph.  Find the speed of each truck if they are traveling toward each other and will meet in 4 hours.

| | Rate | Time | Distance |
|---|---|---|---|
| First truck | | | |
| Second truck | | | |

7. At an electronics store, the number of 37-inch HDTVs that is expected to be sold in a month is double that of 58-inch HDTVs and 50-inch HDTVs combined.  Sales of 58-inch and 50-inch HDTVs are expected to be equal.  58-inch HD-TVs sell for $4,800, 50-inch HD-TVs sell for $2,000, and 37-inch HDTVs sell for $1800.  If total sales of $168,000 are expected this month, how many of each HDTV should be stocked?

| | Number sold | Price | Sales $ |
|---|---|---|---|
| 37-inch HDTV | | | |
| 50-inch HDTV | | | |
| 58-inch HDTV | | | |

# Student Activity
## Tic-Tac-Toe with Inequalities

**Tic-tac-toe #1:** If the inequality in the square is true, then put an **O** on the square. If it is false, then put an **X** on the square.

| | | |
|---|---|---|
| $-5 < -1$ | $8 > 8$ | $-6 > -2$ |
| $2 - 3 < -1$ | $\dfrac{1}{2} < \dfrac{1}{3}$ | $7 \le 7$ |
| $a \ge a$ | $0 < -4$ | $\left| -3 \right| > 1$ |

**Tic-tac-toe #2:** If the number in the square is a solution to the inequality, then put an **O** on the square. If it is false, then put an **X** on the square.

| | | |
|---|---|---|
| $x - 2 < 7$ <br><br> $9$ | $x + 4 \ge -2$ <br><br> $-7$ | $x - 3 \le -1$ <br><br> $1$ |
| $-2x < 4$ <br><br> $-1$ | $\dfrac{1}{2}x \ge 2$ <br><br> $6$ | $-1 \le 2 - x$ <br><br> $3$ |
| $8 > 4 - x$ <br><br> $-5$ | $-x < 4$ <br><br> $-3$ | $2x + 3 < 3x$ <br><br> $2$ |

# Guided Learning Activity
## Graphing and Notation of Inequalities

Think of interval notation as a way to tell someone how to draw the graph, from left to right, giving them only a "begin" value and an "end" value for each interval.

- Always give intervals from LEFT to RIGHT on the number line.
- Use $-\infty$ and $\infty$ to denote the "ends" of the number line (as shown above).
- Use ( or ) to denote an endpoint that is approached, but not included.
- Use [ or ] to denote an endpoint that is included.
- The parenthesis or bracket needs to open in the direction of the true part of the inequality.

**Directions:** Each inequality can be represented in three ways. One is given to you; fill in the missing ones.

|     | Set notation | Graph | Interval Notation |
|-----|--------------|-------|-------------------|
| 1.  | $x > 2$ | -5 -4 -3 -2 -1 0 1 2 3 4 5 | |
| 2.  | $x \le -3$ | -5 -4 -3 -2 -1 0 1 2 3 4 5 | |
| 3.  | $x \ge -1$ | -5 -4 -3 -2 -1 0 1 2 3 4 5 | |
| 4.  | | -5 -4 -3 -2 -1 0 1 2 3 4 5 | |
| 5.  | | -5 -4 -3 -2 -1 0 1 2 3 4 5 | |
| 6.  | $-2 < x \le 3$ | -5 -4 -3 -2 -1 0 1 2 3 4 5 | |
| 7.  | | -5 -4 -3 -2 -1 0 1 2 3 4 5 | |
| 8.  | | -5 -4 -3 -2 -1 0 1 2 3 4 5 | $\left(-\infty, 4\right)$ |
| 9.  | | -5 -4 -3 -2 -1 0 1 2 3 4 5 | $\left(-2, 0\right]$ |
| 10. | | -5 -4 -3 -2 -1 0 1 2 3 4 5 | $\left(-\infty, \infty\right)$ |

## Match Up on Solving Inequalities

**Match-up:** Match each of the inequalities in the squares of the table below with an equivalent inequality. If the solution is not found among the choices A through D, then choose E (none of these).

**A** $x > 2$　　　　**B** $x < 3$　　　　**C** $x > -1$　　　　**D** $x < -3$　　　　**E** None of these

| | | | |
|---|---|---|---|
| $2x < -6$ | $-3x < -6$ | $-5x < 5$ | $x + 2 < 5$ |
| $1 > x - 1$ | $-1 < x$ | $8 < 4x$ | $15 < -5x$ |
| $\dfrac{3}{2}x > 3$ | $6 > -6x$ | $2x + 3 < -3$ | $\dfrac{x}{2} + 1 < \dfrac{5}{2}$ |
| $5 - 2x < 11$ | $2 - 3x < 5$ | $\dfrac{x}{3} < 1$ | $x - 7 > -8$ |
| $2x < -2$ | $\dfrac{x+1}{2} > \dfrac{1}{2}$ | $1 - x > 4$ | $-9 > x - 6$ |

# Student Activity
## Match Up on Equations and Inequalities

**Match-up:** Match each of the equations and inequalities in the squares of the table below with an equivalent equation or inequality. If one is not found among the choices A through D, then choose E (none of these).

**A** $x > -1$     **B** $x = -1$     **C** $x < 4$     **D** $x = 4$     **E** None of these

| | | | |
|---|---|---|---|
| $2x + 4(2x - 1) = -14$ | $-8(x - 1) = 4x + 5$ | $-4x + 1 > -15$ | $\dfrac{x}{3} - \dfrac{x}{4} > -\dfrac{1}{12}$ |
| $3x + 7(2x + 5) = 103$ | $\dfrac{1}{3}x - 2 < \dfrac{1}{4}x - \dfrac{5}{3}$ | $19x - 21 = -(4x - 71)$ | $\dfrac{x + 1}{3} > \dfrac{-3x - 3}{2}$ |
| $\dfrac{5(x + 2)}{2} + 1 = 4x$ | $-\dfrac{5(2 - x)}{3} < 2x - 3$ | $0.03x + 0.08 > 0.2x - 0.6$ | $\dfrac{3}{2}(x + 5) = \dfrac{17 + 5x}{2}$ |
| $\dfrac{11 + 3(2x - 1)}{2} = 20 - x$ | $6 > \dfrac{5}{4}x + 1$ | $-2(x - 3) < -(x - 2)$ | $\dfrac{4x + 1}{3} = \dfrac{2x - 3}{5}$ |
| $-2x - 1 > 5x + 6$ | $-\dfrac{3}{4}x + 1 = -2 + \dfrac{9}{4}x$ | $-2x > 7 - (13x + 18)$ | $1.2x + 7.3 = -5.1x + 1$ |

# Assess Your Understanding
## Solving Linear Equations

For each of the following, describe the strategies or key steps that will help you **start** the problem. You do **not** have to complete the problems.

| | | What will help you to start this problem? |
|---|---|---|
| **1.** | Solve: $3x + 5 = 11$ | |
| **2.** | Solve: $-4x = 12$ | |
| **3.** | Solve: $\dfrac{3x}{4} - 2 = \dfrac{x}{6}$ | |
| **4.** | Is $-3$ a solution to $4x + 6 = -2$? | |
| **5.** | Solve: $3(-2n + 4) = -30$ | |
| **6.** | Write $-2 < x \le 4$ in interval notation | |
| **7.** | Solve: $\dfrac{b}{2} - \dfrac{1}{3} = \dfrac{3}{4}$ | |
| **8.** | Solve: $-4x \le 24$ | |
| **9.** | Solve $A = bh + Bh$ for $B$. | |
| **10.** | Is $x = -2$ a solution of $2x + 1 < -3$? | |

| | | **What will help you to start this problem?** |
|---|---|---|
| **11.** | Change 0.25% to a decimal. | |
| **12.** | Find 2.5% annual simple interest on $500. | |
| **13.** | Find the volume of a sphere if the radius is 4 mm. | |
| **14.** | If the circumference of a circle is $14\pi$ cm, find the radius. | |
| **15.** | 25 is what percent of 60? | |
| **16.** | If the sales tax on an item is 6%, and the price is $129.95, find the total paid by the customer. | |
| **17.** | How many liters of a 2% acid solution must be added to 30 liters of a 10% acid solution to dilute it to an 8% acid solution? | |
| **18.** | A car averaged 60 mph for part of a trip and 70 mph for the remainder. If the 7 ½ -hour drive covered 500 miles, for how long did the car average 60 mph? | |
| **19.** | If the vertex angle of an isosceles triangle is 24°, find the measure of each base angle. | |
| **20.** | The cost of a stereo to an electronics store is $268. If the markup is 20%, what is the sales price of the stereo? | |

73

# Metacognitive Skills
## Solving Linear Equations and Inequalities

Metacognitive skills refer to the ability to judge how well you have learned something and to effectively direct your own learning and studying. This is a self-evaluation tool designed to help you focus your studying and to improve your metacognitive skills with regards to this math class.

Fill the 1$^{st}$ column out **before** you begin studying. Fill the 2$^{nd}$ column out after you study for your test.

Go back to this assessment after your test and circle any of the ratings that you would change – this identifies the "disconnects" between what you **thought** you knew well and what you **actually** knew well.

**Use the scale below to assign a number to each topic.**
**5**  *I am confident I can do any problems in this category correctly.*
**4**  *I am confident I can do most of the problems in this category correctly.*
**3**  *I understand how to do the problems in this category, but I still make a lot of mistakes.*
**2**  *I feel unsure about how to do these problems.*
**1**  *I know I don't understand how to do these problems.*

| Topic or Skill | Before Studying | After Studying |
|---|---|---|
| Checking the solution to an equation. | | |
| Applying the Addition Property of Equality to solve an equation. | | |
| Applying the Multiplication Property of Equality to solve an equation. | | |
| Setting up an equation to represent an application problem. | | |
| Setting up a percent-sentence and then translating it into an equation. | | |
| Setting up and solving application problems that involve percents, including those with a percent increase or decrease. | | |
| Solving linear equations like $ax + b = c$. | | |
| Solving linear equations that involve parentheses and like terms. | | |
| Clearing the fractions from equations involving fractions. | | |
| Understanding the difference between solving an equation and simplifying an expression. | | |
| Solving a formula for a selected variable. | | |
| Choosing a formula and solving a problem involving business or science formulas. | | |
| Choosing the formula and solving a problem involving length, area, or volume. | | |
| Using problem solving skills to solve application problems. | | |
| Placing the correct units on the results to application problems. | | |
| Setting up a table to organize the information in an application problem. | | |
| Deciding whether an inequality is true or false. | | |
| Solving an inequality. | | |
| Graphing an inequality. | | |
| Writing the answer to an inequality in interval notation. | | |
| Solving a compound inequality (like $-3 < x + 1 < 5$). | | |

# ABS: Absolute Value Equations and Inequalities

# Student Activity

## Working with Intersection and Union

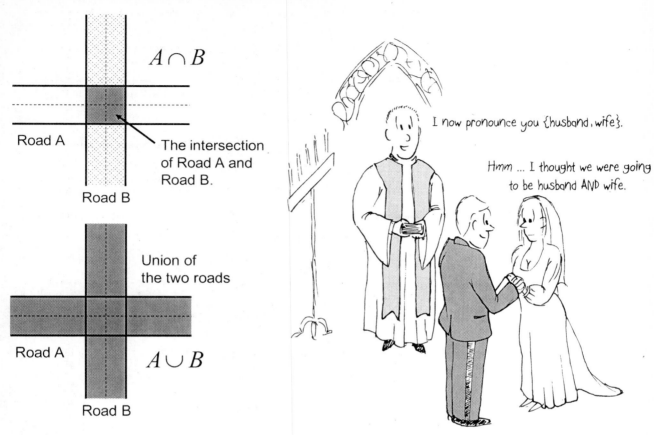

**Section I:** Working with Sets of Numbers

| | **Set A** | **Set B** | **Intersection** $A \cap B$ (*A* and *B*) | **Union** $A \cup B$ (*A* or *B*) |
|---|---|---|---|---|
| **a.** | $\{-2, 0, 2, 4\}$ | $\{0, 1, 2, 3\}$ | | |
| **b.** | $\{-3, -2, -1\}$ | $\{1, 2, 3\}$ | | |
| **c.** | $\{-3, 1, 2, 4\}$ | $\{-2, -1, 2\}$ | | |
| **d.** | $\{1, 2, 3\}$ | $\{1, 2, 3\}$ | | |
| **e.** | $\{0\}$ | $\{-1, 0, 1\}$ | | |
| **f.** | $\mathbb{Z}$ | $\mathbb{N}$ | | |
| **g.** | $\{..., 1, 2, 3\}$ | $\{2, 3, 4, ...\}$ | | |

## Section II: Working with Number-line Graphs

**a.**  A:

B:

$A \cap B$:

**b.**  A:

B:

$A$ or $B$:

**c.**  A: $x < 0$

B: $x \leq -2$

$A \cap B$:

**d.**  A: $x < 0$

B: $x \leq -2$

$A \cup B$:

**e.**  A:

B:

$A$ and $B$:

**f.**  A:

B:

$A \cup B$:

**g.**  A: $[-3,3]$

B: $(-2,2)$

$A$ or $B$:

**h.**  A: $[-3,3]$

B: $(-2,2)$

$A$ and $B$

# Guided Learning Activity
## Solving Compound and Double Inequalities

**Solving a Compound Inequality (AND or OR)**
1. Solve each inequality separately.
2. Graph each inequality together on a number line. (*use different colors*)
3. Graph the desired intersection (and) or union (or) on a number line. (*use a 3ʳᵈ color*)
4. Write the interval solution that represents the graphed solution. (*the 3ʳᵈ color graph*)

**Solving a Double Inequality**
Apply the properties of inequality to all three of its parts to isolate *x* in the middle.

---

**1.** Solve: $3x \le 15$ and $2x > -6$

**3.** Solve: $5 < 2x + 1 \le 9$

Graph the separate pieces:

Graph the solution:

Graph the solution:

Write the solution: _____

Write the solution: _____

**2.** Solve: $3x < 3$ or $-2x < -10$

**4.** Solve: $2x - 5 < -11$ or $5x + 1 \le 6$

Graph the separate pieces:

Graph the separate pieces:

Graph the solution:

Graph the solution:

Write the solution: _____

Write the solution: _____

# Student Activity
## The Match is Out There

**Directions:** Solve each compound or double inequality and then match the problem up with the graph of its solution.

_____ **1.** Solve: $x + 2 > 4$ and $x - 1 < 4$

a.

b.

_____ **2.** Solve: $0 \leq x + 3 \leq 5$

c. Empty set

d.

_____ **3.** Solve: $2 + x \leq 2$ or $2 - 3x \geq -4$

e.

f.

g. All reals

_____ **4.** Solve: $4x \leq -12$ and $\dfrac{x}{2} \leq 1$

h.

i.

_____ **5.** Solve: $\dfrac{x}{2} + 3 \leq 1$ or $\dfrac{5x}{4} - 1 > \dfrac{3}{2}$

j.

_____ **6.** Solve: $2x - 5 < 5$ or $3x + 1 > -2$

# Student Activity
## Matching Up the Different Cases

**Directions:** Begin by isolating the absolute value in each of the equations and inequalities below. Then categorize the problem as one of the four cases or a special case. The first one has been done for you.

Let $k$ be a **positive** constant and let $X$, $X_1$, and $X_2$ be mathematical expressions.

| Case 1: | Case 2: | Case 3: | Case 4: | Special Case: |
|---|---|---|---|---|
| $\mid X \mid = k$ | $\mid X_1 \mid = \mid X_2 \mid$ | $\mid X \mid < k$ <br> $\mid X \mid \leq k$ | $\mid X \mid > k$ <br> $\mid X \mid \geq k$ | If the constant is negative or zero. |

| | | | |
|---|---|---|---|
| $2\mid x-3 \mid < 4$ <br> $\mid x-3 \mid < 2$ <br> **CASE 3** | $\mid x-3 \mid -4 \geq 2$ | $\mid 2a-3 \mid +5 > 3$ | $\dfrac{\mid x \mid}{3} - 2 = 4$ |
| $\mid x+4 \mid = \mid 2x-2 \mid$ | $1 = \mid 2t-4 \mid +4$ | $5 - \mid 2x \mid < 2$ | $4\mid x+3 \mid -3 \leq 2$ |
| $-2\mid 4y \mid < 10$ | $\dfrac{\mid 3x-4 \mid}{-6} \leq -2$ | $1 > \mid 3-b \mid -6$ | $\mid 2x \mid = \mid x+3 \mid$ |
| $5 - \mid x+4 \mid = 2$ | $6 - 2\mid x+4 \mid \leq 0$ | $\mid x-4 \mid +5 = 5$ | $3 < \dfrac{\mid d+12 \mid}{5}$ |

# Guided Learning Activity
## Solving Absolute Value Equations and Inequalities

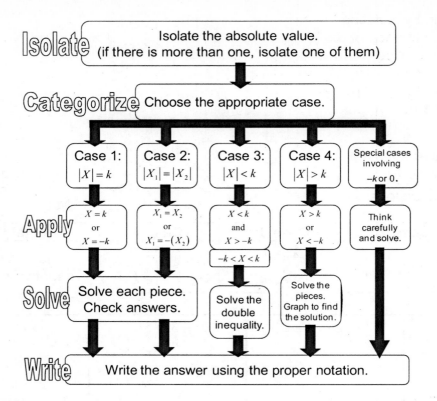

**Directions:** With the guidance of your instructor, work through the strategy for solving an absolute value equation or inequality with each of the problems below.

1. Solve $|2x| - 3 = 11$

2. Solve $3 + |2x + 1| \leq 7$

**3.** Solve $\dfrac{\left|3x+4\right|}{5} > 1$

**5.** Solve $\left|x+4\right| = \left|3x-8\right|$

**4.** Solve $-4\left|x+1\right| < 20$

# Student Activity

## Connecting Absolute Value Problems to Graphs

The graph of $y = |x|$ gives a V-shaped curve. We can interpret why the absolute value equations and inequalities come out the way they do by looking at graphs that correspond to the equations. For example:

$y = |x - 2| - 3$

$y = |x - 2| - 3$

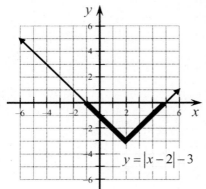

$y = |x - 2| - 3$

Where is $|x - 2| - 3 = 0$?

Where $x = -1$ or $x = 5$.

Where is $|x - 2| - 3 > 0$?

Where $x < -1$ or $x > 5$.

Intervals: $(-\infty, -1) \cup (5, \infty)$

Where is $|x - 2| - 3 < 0$?

Where $-1 < x < 5$.

Interval: $(-1, 5)$

Now solve the corresponding equations and inequalities using the techniques we have learned in this section. Remember to isolate the absolute values first!

**1. a.** Solve: $|x - 2| - 3 = 0$     **b.** Solve: $|x - 2| - 3 > 0$     **c.** Solve: $|x - 2| - 3 < 0$

As you answer the following problems, you might find it helpful to trace each answer on the graph in a different color.

**2. a.** Where is $|x| - 4 = 0$? _____

    **b.** Where is $|x| - 4 < 0$? _____

    **c.** Where is $|x| - 4 \geq 0$? _____

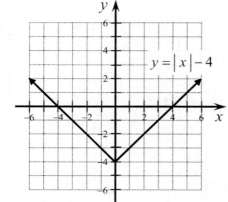

$y = |x| - 4$

**3.** **a.** Where is $|x+1|-4=0$? _____

   Solve: $|x+1|-4=0$

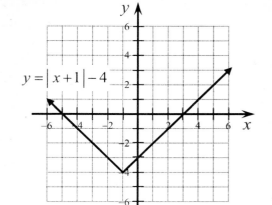

$y=|x+1|-4$

   **b.** Where is $|x+1|-4\le 0$? _____

   Solve: $|x+1|-4\le 0$

---

**4.** **a.** Where is $|x+3|+1<0$? _____

   Solve: $|x+3|+1<0$

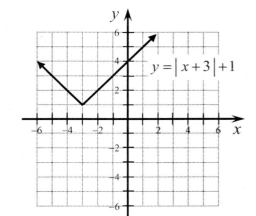

$y=|x+3|+1$

   **b.** Where is $|x+3|+1\ge 0$? _____

   Solve: $|x+3|+1\ge 0$

---

**5.** **a.** Where is $|x-3|\le 0$? _____

   Solve: $|x-3|\le 0$

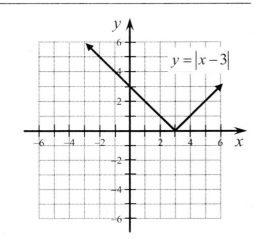

$y=|x-3|$

   **Pay attention to this next one!**

   **b.** Where is $|x-3|\ge 2$? _____

   Solve: $|x-3|\ge 2$

# Assess Your Understanding
## Absolute Value Equations and Inequalities

For each of the following, describe the strategies or key steps that will help you **start** the problem.  You do **not** have to complete the problems.

| | | What will help you to start this problem? |
|---|---|---|
| **1.** | Solve: $|x+4|-1<6$ | |
| **2.** | Solve: $-13 \le 5x+2 \le 17$ | |
| **3.** | Solve the compound inequality: $5x+1<11$ or $3x-2>13$ | |
| **4.** | Solve: $|x+1|=|2x-1|$ | |
| **5.** | Let $A=\{-2,0,2,4\}$ and $B=\{1,3,5\}$. Find $A \cup B$. | |
| **6.** | Let $A=\{1,3,5,7\}$ and $B=\{5,6,7,8\}$. Find the intersection of $A$ and $B$. | |
| **7.** | Solve: $|x+2|=-4$ | |
| **8.** | Solve: $|2x+1| \ge 9$ | |

# Metacognitive Skills
## Absolute Value Equations and Inequalities

Metacognitive skills refer to the ability to judge how well you have learned something and to effectively direct your own learning and studying. This is a self-evaluation tool designed to help you focus your studying and to improve your metacognitive skills with regards to this math class.

Fill the 1$^{st}$ column out **before** you begin studying. Fill the 2$^{nd}$ column out after you study for your test.

Go back to this assessment after your test and circle any of the ratings that you would change – this identifies the "disconnects" between what you **thought** you knew well and what you **actually** knew well.

**Use the scale below to assign a number to each topic.**
5   *I am confident I can do any problems in this category correctly.*
4   *I am confident I can do most of the problems in this category correctly.*
3   *I understand how to do the problems in this category, but I still make a lot of mistakes.*
2   *I feel unsure about how to do these problems.*
1   *I know I don't understand how to do these problems.*

| Topic or Skill | Before Studying | After Studying |
|---|---|---|
| Understanding intersection and union, including their relationship to the words "and" and "or" and the symbols $\cup$ and $\cap$. | | |
| Solving a compound inequality (one that uses **and** or **or**). | | |
| Solving a double inequality (like $-3 < x + 1 < 5$). | | |
| Isolating the absolute value in an equation or inequality. | | |
| Determining the type of absolute value equation or inequality (Case 1, 2, 3, 4, or special case). | | |
| Remembering the proper procedure after you determine the category of absolute value equation or inequality. | | |
| Reasoning out the answer to a "special case" absolute value equation or inequality. | | |

# LINE: Lines and More

# Guided Learning Activity
## Graphing Paired Data

### Example 1: Gas mileage

In 2007, the most fuel-efficient mid-sized car for city driving was the Toyota Prius, which was rated at 60 miles per gallon.  The least efficient mid-sized car for city driving was the Ferrari 612 Scaglietti, rated at 10 miles per gallon. (www.fueleconomy.gov)

In the table, each row of data represents an ordered pair.  Plot the data using two different colors (one for each type of car) and then draw a straight line through each set of plotted points.

Toyota Prius

| gallons | miles |
|---------|-------|
| 1 | 60 |
| 2 | 120 |
| 3 | 180 |
| 4 | 240 |

Ferrari 612

| gallons | miles |
|---------|-------|
| 2 | 20 |
| 4 | 40 |
| 6 | 60 |
| 8 | 80 |

### Example 2: Temperature Scales

Although we still measure temperature in the U.S. using the Fahrenheit scale for non-scientific purposes, the rest of the countries in the world use the Celsius Scale.  In the data table below, we see paired data that show both Fahrenheit and Celsius temperatures.  To read some interesting historical theories about why the Fahrenheit scale is set up the way it is, look at the Wikipedia entry for Fahrenheit under "History." (www.en.wikipedia.org)

| °F | °C |
|-----|------|
| 20 | −6.7 |
| 40 | 4.4 |
| 60 | 15.6 |
| 80 | 26.7 |
| 100 | 37.8 |
| 120 | 48.9 |

# Student Activity
## Winter Severity Index

Deer are adapted to survive most Michigan winter weather conditions. However, research suggests that winter weather can affect the deer population. The Michigan DNR monitors winter conditions by calculating a winter severity index (WSI). The WSI reflects the estimated effect of winter weather conditions on the energy reserves of deer; the higher the index value, the larger the potential impact on the herd. The current WSI System takes advantage of standard weather data available from the National Climate Data Center and a weekly index value is calculated from November through April (for more information visit www.michigan.gov/dnr).

The graph below shows WSI data for the Upper Peninsula, 2006-2007 season. The graph shows the weekly regional WSI value for the current season (dashed line) and the average weekly regional WSI value for the 1996 through 2005 seasons (solid line). Answer the questions using information from the graph. [Graph courtesy of the Michigan Department of Natural Resources.]

**Use the graph to answer the following questions:**

1. How many weeks after November 1 was the highest WSI recorded for the 2006-2007 season? _____

   - Write an ordered pair to represent this point: (_____ , _____)

   - In what month would this peak have occurred? _____

2. When was the first major cold-snap of the season, and what was the associated WSI?

   - Write an ordered pair to represent this point: (_____ , _____)

3. Between what two weekly readings was there the biggest change in the WSI?

   - Write ordered pairs to represent these points: (_____ , _____) & (_____ , _____)

4. Which 6-week period in 2006-07 was harsher than the same period for the 1998-2005 average? _____

# Student Activity
## Take Me to the Point

**Directions:** Capital letters are often used to name specific points. For each point that is graphed below, write down the letter and coordinates of the point in the appropriate category in the table below. Point *A* has been done for you.

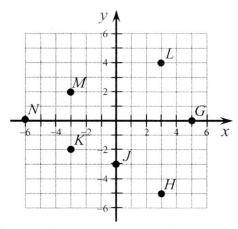

| Quadrant I | Quadrant II | Quadrant III | Quadrant IV | *x*-axis | *y*-axis |
|------------|-------------|--------------|-------------|----------|----------|
| $A(2,3)$   |             |              |             |          |          |

**Directions:** Now plot these points on the axis provided. Indicate each point with its letter on the graph.

$P(2,3)$ $\quad$ $G(-3,4)$ $\quad$ $A(0,0)$ $\quad$ $R(-2,2)$
$U(0,-3)$ $\quad$ $N(3,-4)$ $\quad$ $F(-3,-4)$ $\quad$ $H(4,5)$

# Student Activity
## Tic-Tac-Toe with Ordered Pairs

**Directions for Game #1:** If the ordered pair in the square **is** a solution to the equation $y = 2x - 5$, then circle it (thus placing an **O** on the square). If the ordered pair is **not** a solution, then put an **X** over it.

| | | |
|---|---|---|
| $(-10, -30)$ | $(-10, -25)$ | $\left(\dfrac{2}{3}, -\dfrac{11}{3}\right)$ |
| $(26, 47)$ | $(0, -5)$ | $(0, 0)$ |
| $(5, 0)$ | $\left(-\dfrac{1}{2}, -4\right)$ | $(1, 3)$ |

**Directions for Game #2:** If the ordered pair in the square **is** a solution to the equation $4x - 6y = 12$, then circle it (thus placing an **O** on the square). If the ordered pair is **not** a solution, then put an **X** over it.

| | | |
|---|---|---|
| $(-9, -8)$ | $\left(1, -\dfrac{4}{3}\right)$ | $(3, 0)$ |
| $(-2, 0)$ | $(0, 0)$ | $\left(-\dfrac{3}{2}, -3\right)$ |
| $(1, 1)$ | $\left(\dfrac{2}{3}, -\dfrac{14}{9}\right)$ | $(9, -4)$ |

# Guided Learning Activity
## Graphing Linear Equations

**Example 1:** Fill in the missing values in the table of solutions for the linear equation, $y = -x + 4$, plot the points, and then draw the line through the points.

| $x$ | $y$ |
|-----|-----|
| 3   |     |
|     | 3   |
| 0   |     |
| −2  |     |
|     | 5   |

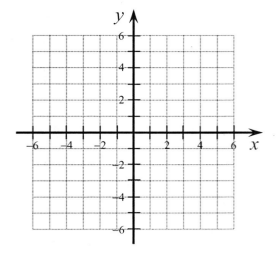

| If $x = 3$, find $y$. | If $y = 3$, find $x$. | If $x = 0$, find $y$. | If $x = -2$, find $y$. | If $y = 5$, find $x$. |
|------|------|------|------|------|
| $y = -( \quad ) + 4$ | $( \quad ) = -x + 4$ | $y = -( \quad ) + 4$ | $y = -( \quad ) + 4$ | $( \quad ) = -x + 4$ |

**Example 2:** Fill in the missing values in the table of solutions for the linear equation, $3x - y = -2$, plot the points, and then draw the line through the points.

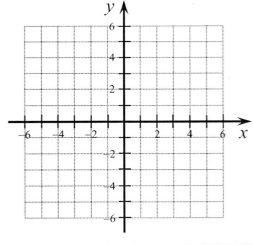

| $x$ | $y$ |
|-----|-----|
|     | 5   |
| 0   |     |
|     | −1  |
| −2  |     |
|     | 3   |

| If $y = 5$, find $x$. | If $x = 0$, find $y$. | If $y = -1$, find $x$. | If $x = -2$, find $y$. | If $y = 3$, find $x$. |
|------|------|------|------|------|
| $3x - ( \quad ) = -2$ | $3( \quad ) - y = -2$ | $3x - ( \quad ) = -2$ | $3( \quad ) - y = -2$ | $3x - ( \quad ) = -2$ |

# Student Activity

## Out-of-line Suspects

**Directions:** A student has created a table of solutions for each **linear** equation below.

- Plot the points in the table.
- Use the graphed points to identify which points (if any) are most likely incorrect.
- Circle the incorrect solution in the table of values.
- Find the correct ordered pairs for any that are circled.
- Plot the corrected solutions.

*Chief, my get gut tells me the second one from the right is a little suspect...*

**1.** Linear equation: $y = 3x - 1$

| x | y |
|---|---|
| −1 | 2 |
| −2 | 3 |
| 0 | −1 |
| $\frac{1}{3}$ | 0 |
| 2 | 5 |

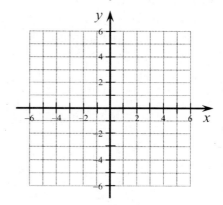

**2.** Linear equation: $y = \frac{1}{2}x + 2$

| x | y |
|---|---|
| −2 | 0 |
| −1 | $\frac{3}{2}$ |
| 2 | 5 |
| 0 | 2 |
| 4 | 4 |

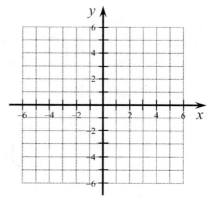

**3.** Linear equation: $y = 4 - x$

| x | y |
|---|---|
| −2 | 6 |
| −6 | 1 |
| 1 | 3 |
| 2 | −4 |
| 6 | −2 |

# Student Activity

## The Cost of College

When you pay for your college education, there are typically three different types of charges:

1) ***Variable cost*** – Tuition, which is paid per credit hour (or unit of study).  There may also be fees that vary depending on the number of credit hours (for example, an additional $6 per credit hour).

2) ***Fixed cost*** – Student fees, which are paid regardless of how many or what types of classes you take, (for example, a $25 Registration Fee).  These are fixed costs.

3) Course fees or lab fees, which are paid for certain courses only (we will ignore these in this model).  Your instructor may tell you to ignore some other fee or tuition oddities to allow a linear model to be used.

Our linear model of college costs will only consider the first two types of charges.  Your instructor will help you locate the information necessary to build the linear model – perhaps they have asked you to bring in a copy of your tuition bill!

How much do you pay in **fixed** costs for your enrollment at your college? _____

**Per credit hour**, how much do you pay in tuition (and **variable** fees)? _____

Fill in the table below for a student at your college:

| Credits (or units) taken by student, $n$ | Fixed student fees | Tuition and other variable costs for this number of credits (or units) | Total cost for this number of credits, $C$ |
|---|---|---|---|
| 4 | | | |
| 8 | | | |
| 12 | | | |
| 16 | | | |

**Create a graph containing the data points above.**

Write a linear equation to model the cost $C$, for taking a certain number of credits $n$.

Show how you could use the linear equation you just wrote to estimate the cost of taking 9 credits (or units).

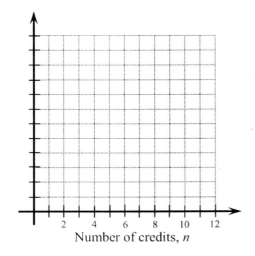

Total Cost

Number of credits, $n$

# Student Activity
## ID the Intercepts

**Directions:** For each graph, identify the $x$- and $y$-intercepts (if they exist) using **both** coordinates of the point.

**1.** $x$-intercept: (   ,   )

    $y$-intercept: (   ,   )

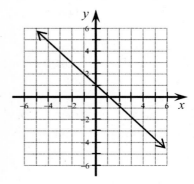

**2.** $x$-intercept: (   ,   )

    $y$-intercept: (   ,   )

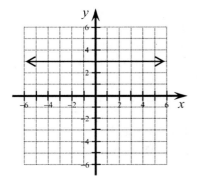

**3.** $x$-intercept: (   ,   )

    $y$-intercept: (   ,   )

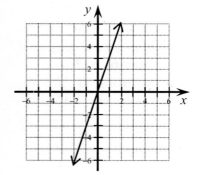

**4.** $x$-intercept: (   ,   )

    $y$-intercept: (   ,   )

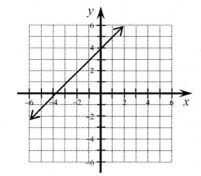

**5.** $x$-intercept: (   ,   )

    $y$-intercept: (   ,   )

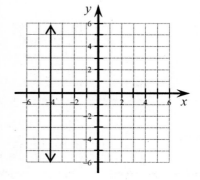

**6.** $x$-intercept: (   ,   )

    $y$-intercept: (   ,   )

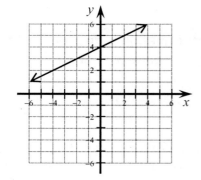

# Student Activity
## A Web of Lines

**Directions:** First, use the method of intercepts to find the $x$- and $y$-intercepts for each equation below. **Then** draw each of the lines on the graph provided using a straight-edge. (Use different colors for each line if you have colored pencils.)

$6x + y = 6$

   $x$-int: (  ,  )

   $y$-int: (  ,  )

$5x + 2y = 10$

   $x$-int: (  ,  )

   $y$-int: (  ,  )

$4x + 3y = 12$

   $x$-int: (  ,  )

   $y$-int: (  ,  )

$3x + 4y = 12$

   $x$-int: (  ,  )

   $y$-int: (  ,  )

$2x + 5y = 10$

   $x$-int: (  ,  )

   $y$-int: (  ,  )

$x + 6y = 6$

   $x$-int: (  ,  )

   $y$-int: (  ,  )

**Challenge:** Now, if you're really a good math "detective" you can construct and graph a set of your own linear equations to make a similar "web" in **Quadrant III**. Look for patterns in the equations above to help you.

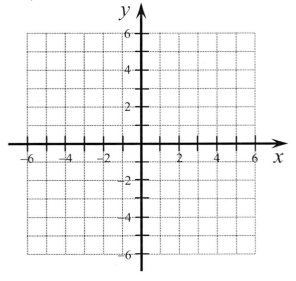

# Guided Learning Activity
## Clues to the Equation

**Directions:** For each graphed line, construct a possible table of values that could accompany the line. Then, together with your class, you will write a linear equation to describe the line mathematically.

**1.**

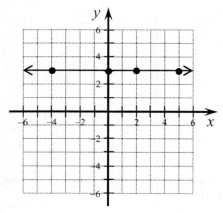

**Clue:** Table of Solutions

| $x$ | $y$ |
|-----|-----|
|     |     |
|     |     |
|     |     |

**Equation:** _____

**2.**

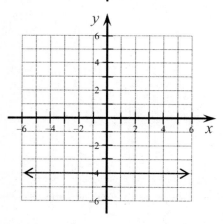

**Clue:** Table of Solutions

| $x$ | $y$ |
|-----|-----|
|     |     |
|     |     |
|     |     |

**Equation:** _____

**3.**

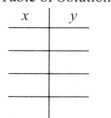

**Clue:** Table of Solutions

| $x$ | $y$ |
|-----|-----|
|     |     |
|     |     |
|     |     |

**Equation:** _____

**4.** What would be the equation of a horizontal line that passes through $(2, 3)$? _____

**5.** What would be the equation of a vertical line that passes through $(-2, 4)$? _____

**6.** What would be the equation of a vertical line with an $x$-intercept of 5? _____

**7.** What would be the equation of a vertical line with an $x$-intercept of 0? _____

# Student Activity
## Graphing Linear Equations with a Calculator

You can use your graphing calculator to graph linear equations. Each model of calculator will have a different set of keystrokes that will allow you to graph these equations. Before you start, it is a good idea to guess what your graph should look like (in case you enter something into your calculator incorrectly). The standard viewing window on most graphing calculators goes from -10 to 10 on the x-axis and from -10 to 10 on the y-axis.

**1.** Sketch a graph of the equation $2x + 3y = 3$ on the axes below.

Man ... if I had just gotten this calculator a little earlier, I could have saved myself a long walk!

Graphing calculators can usually only accept equations beginning with $y =$.

**2.** Solve the equation $2x + 3y = 3$ for $y$.

**3.** Now find the key on your calculator that takes you to the $y =$ screen. It will usually look like $\boxed{Y =}$. It may be a function above another key, so you may have to press the *function* or $\boxed{2^{nd}}$ key first. Below, draw the keys you need to press.

Once you reach the $y =$ screen, you should see your cursor next to $Y_1 =$. **Enter the equation you found in problem 2.** Note that you do not need to type $y =$ since it is already there.

**4.** Next, find your graph key. It may say $\boxed{\text{GRAPH}}$ or it may be a function above another key. Select graph. Draw the keys you need to press.

**5.** Often times, it is convenient to view a graph in a **standard** viewing window. Find your calculator's zoom menu. It may be a key that says $\boxed{\text{ZOOM}}$ or it may be a function above another key. If you have it as an option, select **ZoomStd** or **Zoom Standard**. Draw the keys you need to press to do this.

Hopefully your calculator's graph looks like the one you sketched in problem 1.

**6.** You can change the size of the viewing window by using other options in the zoom menu. You can also change the window to a size of your choice using the $\boxed{\text{WINDOW}}$ key. Note that WINDOW may be found *above* a key on some calculators. Draw the keys you need to press to get into the window menu.

**Now change the window to these settings:**

| | |
|---|---|
| xmin $= 0$ | ymin $= 0$ |
| xmax $= 5$ | ymax $= 5$ |
| xscale (or xscl) $= 1$ | yscale (or yscl) $= 1$ |

Using the $\boxed{\text{GRAPH}}$ key or function, look at the graph again.

**7.** What has happened to the view of the graph on the calculator screen?

**8.** What window settings would show you only the second Quadrant?

| | |
|---|---|
| xmin $=$ | ymin $=$ |
| xmax $=$ | ymax $=$ |
| xscale (or xscl) $=$ | yscale (or yscl) $=$ |

# Guided Learning Activity
## Hit the Slopes

**Directions:** For each line that is graphed below, determine the slope of the line using either a slope triangle or the slope formula.

**1.**

**2.**

**3.**

**4.**

**5.**

**6.**

What ever happened to a nice $y = -x$ ?

# Student Activity
## Match Up on Slopes

**Match-up:** In each box of the grid below, you will find either a pair of points that is on a line or the description of a line in words. Match each line that is described with its slope in the choices A through E. If the slope is not found in A through E, then choose F (none of these).

**A** 2      **B** $\dfrac{1}{2}$      **C** $-1$      **D** 0      **E** undefined      **F** None of these

| | | | |
|---|---|---|---|
| $(0,0)$ and $(3,0)$ | $(4,4)$ and $(-4,0)$ | $\left(\dfrac{3}{2},\dfrac{1}{2}\right)$ and $\left(\dfrac{3}{2},\dfrac{10}{2}\right)$ | $(-3,-6)$ and $(3,-4)$ |
| $(2,5)$ and $(3,4)$ | $\left(\dfrac{7}{6},\dfrac{1}{9}\right)$ and $\left(\dfrac{1}{6},\dfrac{10}{9}\right)$ | $(3.14,\ 2.72)$ and $(1.14,\ 1.72)$ | $(-6,-1)$ and $(-8,0)$ |
| $(-100,300)$ and $(-100,50)$ | $(9,7)$ and $(7,9)$ | $(1,1)$ and $(-5,1)$ | $(8,8)$ and $(7,6)$ |
| The line is a vertical line. | The line goes to the right 1 unit for every 2 units it goes up. | The line is a horizontal line. | The line goes down 1 unit for every 1 unit it goes to the right. |

# Student Activity
## Midpoints as Averages

1. Plot the two given points on each graph provided. Connect the two points with a line **segment** using a straight-edge. Then estimate the midpoint of the line segment.

$(5,0)$ and $(-1,-4)$          $(-6,3)$ and $(3,5)$

          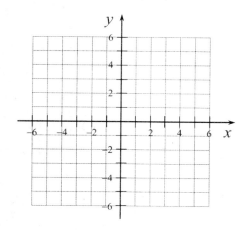

Midpoint: (____, ____)          Midpoint: (____, ____)

2. We find an average (or mean) of several values by finding the sum of the values, and then dividing by the number of values. Find the average of 5, 12, 3, and 4.

   Average: $\dfrac{\Box + \Box + \Box + \Box}{\Box} =$

3. We find the midpoint of two points by averaging the $x$-coordinates and averaging the $y$-coordinates. Find the midpoint for $(5,0)$ and $(-1,-4)$ using this technique.

   $x$-coordinate: $\dfrac{\Box + \Box}{\Box} =$          $y$-coordinate: $\dfrac{\Box + \Box}{\Box} =$

   Midpoint: (____, ____)   Is this consistent with the midpoint you found in #1?

4. Repeat this process for $(-6,3)$ and $(3,5)$

   Midpoint: . $\left( \dfrac{\Box + \Box}{\Box}, \dfrac{\Box + \Box}{\Box} \right) = ($ ____, ____ $)$

5. See if you can find "third-points" for the points $(-2,-6)$ and $(1,6)$. These would be two points that split the line segment into thirds.

# Student Activity
## Entertaining Rates of Change

It is often helpful to summarize raw data with a rate describing the amount of change in one quantity with respect to the amount of change in another. This is called an **average rate of change**, and is often used to describe change that occurs *over time*.

$$\text{Average Rate of Change} = \frac{\text{change in quantity}}{\text{change in time}}$$

**Directions:** Answer the questions below by calculating an average rate of change and using this information to answer the follow-up questions.

1. This table contains box office data from 2000 – 2006. (Source: www.boxofficemojo.com)

| Year | Total Gross (in millions) | Tickets Sold (in millions) | # of Pics | Ticket Price | #1 Picture of the year |
|------|---------------------------|----------------------------|-----------|--------------|------------------------|
| 2006 | $9,209.4 | 1400.0 | 606 | $6.58 | Dead Man's Chest |
| 2005 | $8,840.4 | 1381.3 | 547 | $6.40 | Revenge of the Sith |
| 2004 | $9,418.3 | 1516.6 | 551 | $6.21 | Shrek 2 |
| 2003 | $9,185.9 | 1523.3 | 508 | $6.03 | Return of the King |
| 2002 | $9,167.0 | 1578.0 | 467 | $5.81 | Spider-Man |
| 2001 | $8,412.5 | 1487.3 | 482 | $5.66 | Harry Potter |
| 2000 | $7,661.0 | 1420.8 | 478 | $5.30 | The Grinch |

a. What is the average annual rate of change for the total box office gross between 2000 and 2006? Show your calculation and round your answer to the nearest tenth.

b. Complete the statement: *On average, total box office gross increased _____ dollars per year over the 6-year period between 2000 and 2006. The total increase in box office gross during this period was _____.*

c. What is the average annual rate of change for the total box office gross between 2003 and 2006? Show your calculation and round your answer to the nearest tenth.

d. Complete the statement: *On average, total box office gross increased _____ dollars per year over the 3-year period between 2003 and 2006. The total increase in box office gross during this period was _____.*

e. In general, Is the average annual rate of change for total box office gross increasing or decreasing? _____

The following tables contain U.S. Music purchasing data for 2005 and 2006.

| Units sold (in millions): | 2005 | 2006 |
|---|---|---|
| Overall Music Sales<br>(Albums, singles, music video, digital tracks) | 1,003 | 1,198 |
| Total Album Sales<br>(Includes CD, CS, LP, Digital albums) | 618.9 | 588.2 |
| Digital Track Sales | 352.7 | 581.9 |
| Overall Album Sales<br>(Includes all albums & track equivalent albums) | 654.1 | 646.4 |
| Internet Album Sales<br>(Physical album purchases via e-commerce sites) | 24.7 | 29.4 |
| Digital Album Sales | 16.2 | 32.6 |

(Source: www.businesswire.com)

| Top Ten Selling Albums of 2006 | Units Sold |
|---|---|
| Soundtrack/High School Musical | 3,719,071 |
| Me and My Gang/Rascal Flatts | 3,479,994 |
| Some Hearts/Carrie Underwood | 3,015,950 |
| All the Right Reasons/Nickelback | 2,688,166 |
| Futuresex/Lov…/Justin Timberlake | 2,377,127 |
| Back to Bedlam/James Blunt | 2,137,142 |
| B'day/Beyonce | 2,010,311 |
| Soundtrack/Hannah Montana | 1,987,681 |
| Taking the Long Way/Dixie Chicks | 1,856,284 |
| Extreme Behavior/Hinder | 1,817,350 |

2. Use the tables above to answer the following questions. Show your calculations and round your answer to the nearest hundredth.

   a. What is the average *monthly* rate at which **overall music sales** increased from 2005 to 2006?

   b. What is the average *monthly* rate at which **total album sales** decreased from 2005 to 2006?

   c. What is the average *monthly* rate at which **digital album sales** increased from 2005 to 2006?

3. Based on your results from parts b and c, does it appear that all the people who topped buying physical albums are now purchasing digital albums?  Explain.

4. What is the average *daily* sales rate for the top selling album of 2006?  What is the average daily sales rate for the *tenth* top selling album of 2006?  (Round your answer to the nearest whole number.)

# Student Activity
## Match Up on Slope-Intercept Form

**Match-up:** In each box of the grid below, you will find either the equation of a line or a description of a line. For each, determine the slope **and** the $y$-intercept and match it with the appropriate letters. If the slope or the $y$-intercept is not found among the choices or cannot be determined from the information given, then choose E or N respectively.

**Slope:** **A** 2   **B** $\dfrac{1}{2}$   **C** $-1$   **D** 0   **E** None of these or cannot be determined

$y$-**Intercept:** **J** 3   **K** $-2$   **L** 0   **M** 1   **N** None of these or cannot be determined

| | | | |
|---|---|---|---|
| $y = \dfrac{1}{2}x - 3$ | $y = -x + 3$ | $x + y = 0$ | The line is a vertical line passing through $(-2, 3)$. |
| $6x - 3y = 6$ | $y = 1 - 3x$ | The line passes through $(0, 3)$ and $(2, 1)$. | The line passes through $(0, -2)$ and $(2, -1)$. |
| The line is a vertical line passing through $(2, 1)$. | The line is a horizontal line passing through $(-2, 3)$. | The line has intercepts $(0, 1)$ and $(-2, 0)$. | The line is parallel to a horizontal line and passes through $(0, 1)$. |
| The line is perpendicular to a line with a slope of $-2$ and passes through $(0, 3)$. | $10x - 5y = 10$ | $4y - 2x = 0$ | The line is parallel to a line with a slope of 2 and passes through the origin. |

# Guided Learning Activity
## Graphing with Slope-Intercept Form

**Slope-Intercept Form:** $y = mx + b$ where $m$ is the slope and $(0, b)$ is the $y$-intercept.

**To graph using slope-intercept form:**

**1.** Graph a point (the $y$-intercept).

**2.** Use $m = \dfrac{\text{rise}}{\text{run}}$ to move from that point to locate another point on the line.

**1.** Graph: $y = \dfrac{1}{2}x + 3$

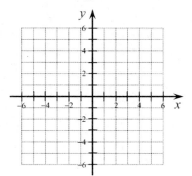

**2.** Graph: $y = 3x - 2$

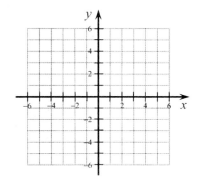

**3.** Graph: $y = -\dfrac{3}{2}x + 3$

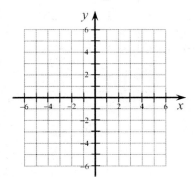

**4.** Graph: $y = -2x$

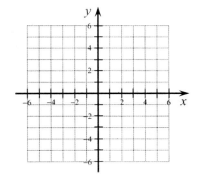

**5.** Graph: $3x - 4y = -4$

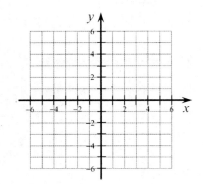

**6.** Graph: $y - x = 2$

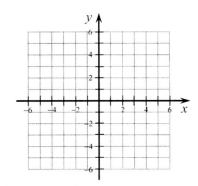

# Student Activity
## Evidence from the Graph

**Directions:** For each line that is graphed below, determine the equation of the line and write it in slope-intercept form.

**1.**

**2.**

**3.**

**4.**

**5.**

**6.**

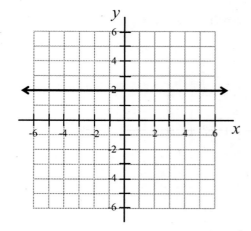

# Student Activity

## What's the Verdict, Parallel or Perpendicular?

**Directions:** Each of the "accused" pairs of lines is going to trial. Your task is to graph each pair of lines, and then assign each pair a verdict: parallel or perpendicular. If the pair of lines is neither parallel nor perpendicular, you may declare a mistrial.

**1.** $y = \frac{1}{2}x + 3$; $y = \frac{1}{2}x + 1$

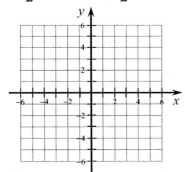

**2.** $y = \frac{1}{3}x - 1$; $y = 3x$

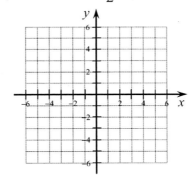

**3.** $y = \frac{1}{2}x + 1$; $y = -\frac{1}{2}x + 2$

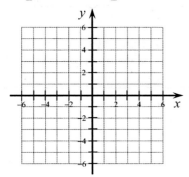

**4.** $y = 2x - 3$; $y = -\frac{1}{2}x + 4$

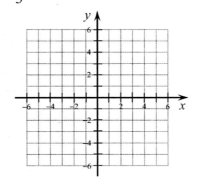

**5.** $y = 3$; $y = -1$

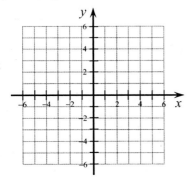

**6.** $y = -x + 4$; $y = x - 2$

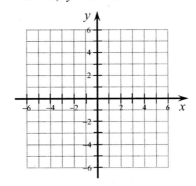

# Student Activity

## Modeling Data with Linear Equations

**Directions:** Fill in the missing information in the table below. The first one has been done for you.

| Mathematical model | Description | Slope and meaning of the slope | $y$-intercept and meaning of the $y$-intercept |
|---|---|---|---|
| $C = 4.5x + 250$ | The cost $C$ of producing $x$ items. | $m = 4.5$ It costs $4.50 to produce each item. | $(0, 250)$ Regardless of the number of items produced, the fixed costs are $250. |
| $C = 0.10n + 25$ | The cost $C$ of renting a car for a day and driving it for $n$ miles. | | |
| $R = 8.5t$ | The revenue $R$ from selling $t$ movie tickets. | | |
| $P = 10h$ | The weekly pay $P$ from working $h$ hours. | | |
| $P = 25t - 500$ | The profit $P$ from selling $t$ tickets to a benefit concert. | | |
| $v = 32t$ | The velocity $v$ of a falling object (in feet per second) $t$ seconds after it is dropped. | | |
| | The distance $d$ a Toyota Prius hybrid car can travel on $g$ gallons of gasoline. | $m = 55$ A Toyota Prius gets 55 miles per gallon of gasoline. (Source: www.toyota.com/ prius/specs.html) | $(0, 0)$ The car can travel zero miles with zero gallons of gasoline in the tank. |
| | The cost $C$ of taking $n$ credit hours at a community college. | $m = 66$ The cost is $66 per credit hour. | $(0, 25)$ There is a $25 registration fee regardless of the number of credit hours taken. |

# Student Activity
## Double Trouble: Two Ways to Find an Equation of a Line

| 1. | Write an equation for the line that passes through $(2,5)$ and $(-1,-1)$. | |
|---|---|---|
| a. | Find the slope between the two points: | |
| b. | Use point slope form $y - y_1 = m(x - x_1)$, the slope from **a** and the point $(2,5)$ to find an equation for this line. | Use point slope form $y - y_1 = m(x - x_1)$, the slope from **a** and the point $(-1,-1)$ to find an equation for this line. |
| c. | What do you notice about the two equations in part **b**? Why does this work? | |

| 2. | Write an equation for the line that passes through $(1,2)$ and $(-3,-1)$. | |
|---|---|---|
| a. | Find the slope between the two points: | |
| b. | In the point $(1,2)$, we know that $x = 1$ and $y = 2$. Use this information, together with the slope from part **a** in $y = mx + b$ and solve for $b$.<br><br>Replace $m$ and $b$ in $y = mx + b$ with the values you now know. | Use point slope form $y - y_1 = m(x - x_1)$, the slope from **a** and the point $(1,2)$ to find an equation for this line.<br><br>Write the line in slope-intercept form: $y = mx + b$. |
| c. | What do you notice about the two equations in part **b**? Why does this work? | |

# Student Activity
## Aliases to a Line Problem

**1.** Graph the line $2x - 4y = 12$ by completing the table of solutions and plotting points..

| $x$ | $y$ |
|-----|-----|
| $-2$ | |
| $-1$ | |
| $0$ | |
| $1$ | |
| $2$ | |

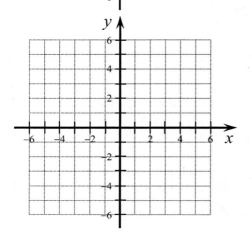

**2.** Find the $x$- and $y$-intercepts of the line $2x - 4y = 12$ and use the intercepts to graph.

$x\text{-int} : (\quad , \quad)$

$y\text{-int} : (\quad , \quad)$

**3.** Write the equation $2x - 4y = 12$ in slope-intercept form:

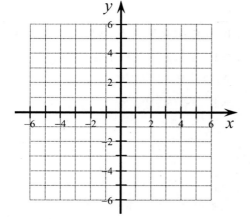

What is the slope of the line? ____

What is the $y$-intercept of the line? ____

Use this information to graph the line.

**4.** If a line were parallel to $2x - 4y = 12$, what would its slope be? ____

**5.** If a line were perpendicular to $2x - 4y = 12$, what would its slope be? ____

# Guided Learning Activity
## Writing Equations that Model Data

**Example 1:** In the first year of a research study, the population of flamingoes at a zoo was 25. In the fifth year of the study the population of flamingoes was 45. Find a linear equation to model the population of flamingoes, $P$, in the year $t$.

**Declare the variables:** _$t =$ year of study, $P =$ number of flamingoes_

**In terms of these variables, what form do ordered pairs have?** _$(t, P)$_

**Rewrite the point-slope form using the new variables:** _$P - P_1 = m(t - t_1)$_

**What information are you given in the problem?**

      **Ordered pairs(s):** _$(1, 25)$ and $(5, 45)$_       **Slope:** _(not given)_

**Strategy to find an equation of the line:** _Find the slope and then use a point and the slope in the point-slope form._

**Finish:**

**Example 2:** A factory produces ethanol at a rate of 4000 liters every 5 hours. After the factory had run for 8 hours, the factory manager noted that there were 10,000 liters of ethanol in the storage tank. Write a linear equation relating the time $t$ in hours since the factory began production and the number of liters $L$ of ethanol in the storage tank.

**Declare the variables:** $t =$ _____

                $L =$ _____

**In terms of these variables, what form do ordered pairs have?** (   ,   )

**Rewrite the point-slope form using the new variables:** _____

**What information are you given in the problem?**

      **Ordered pair(s):** _____       **Slope:** _____

**Strategy to find an equation of the line:** _____

_____

**Finish:**

# Student Activity
## Tic-Tac-Toe on Inequality Solutions

**Directions for Game #1:** If the ordered pair in the square **IS** a solution of the inequality, then circle the ordered pair (thus putting an **O** on the square). If it **IS NOT** a solution, then put an **X** over the ordered pair in the square.

| | | |
|---|---|---|
| $x + y < 5$<br><br>$(2,3)$ | $y < x + 6$<br><br>$(9,3)$ | $y > 0$<br><br>$(2,-1)$ |
| $2x - 3y \le 0$<br><br>$\left(\dfrac{1}{2}, \dfrac{1}{3}\right)$ | $x > y + 4$<br><br>$(4,0)$ | $x \le 5$<br><br>$(4,9)$ |
| $3x + 3 \ge 4y + 4$<br><br>$(2,1)$ | $3 - x < -4$<br><br>$(8,1)$ | $y - 2x \le 3x - y$<br><br>$(1,5)$ |

**Directions for Game #2:** If the ordered pair in the square **IS** a solution of the graphed inequality, then circle the ordered pair (thus putting an **O** on the square). If it **IS NOT** a solution, then put an **X** over the ordered pair in the square.

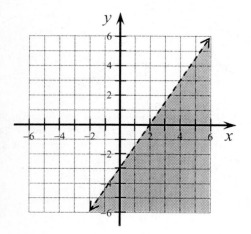

| | | |
|---|---|---|
| $(4,1)$ | $(4,3)$ | $(-2,-4)$ |
| $(1,-3)$ | $(-0.5,-6)$ | $(0,0)$ |
| $(-1,-4)$ | $(-3,-6)$ | $(7,-7)$ |

# Guided Learning Activity

**Shady Inequalities**

**Graphing an Inequality in two variables:**
1. Draw the boundary line (remember to draw it either dashed or solid).
2. Determine which side of the boundary line to shade.
3. Shade the half-plane that shows the solution of the inequality.

---

**1.** Graph: $y \leq \frac{1}{2}x - 1$

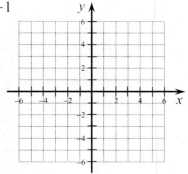

**2.** Graph: $x + y > 3$

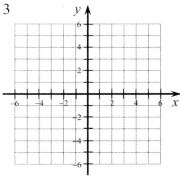

**3.** Graph: $2y + 3x < 6$

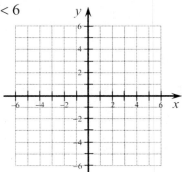

**4.** Graph: $y \geq 4$

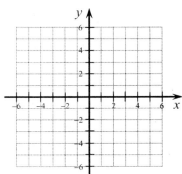

**5.** Graph: $x < -1$

**6.** Graph: $x - 2y > 4$

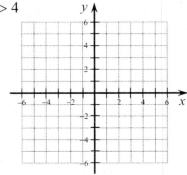

Psssst....Buddy!  Check these out!
All the cool kids got one...

# Student Activity
## Following the Clues Back to the Inequality

**Directions:** In each "crime-scene" below, you are shown the graph of an inequality. Use your mathematical powers of reasoning (and detective skills) to determine what the inequality must have been to result in this graph.

**1.**

**3.**

**2.**

looks like a trail of clues...

(1,2) (3,7) (2,5) (6,4) (9,3)

# Assess Your Understanding

## Lines and More

For each of the following, describe the strategies or key steps that will help you **start** the problem. You do **not** have to complete the problems.

| | | What will help you to start this problem? |
|---|---|---|
| **1.** | Find the intercepts of the graph of $2x - 8y = 8$. | |
| **2.** | Find the slope of a line perpendicular to the one that passes through $(0,0)$ and $(-3,4)$. | |
| **3.** | Find the equation of a vertical line that passes through $(3,7)$. | |
| **4.** | Graph the solution of $y < 2x - 3$. | |
| **5.** | Find the slope between $(-2,3)$ and $(1,-5)$. | |
| **6.** | Graph $2x + 4y = 8$ by creating a table of values. | |
| **7.** | Graph $3x - y = 6$ by using its intercepts. | |
| **8.** | Is $(6,1)$ a solution of $x - 2y \leq 4$? | |
| **9.** | Graph the line that has slope 1 and passes through $(-2,-3)$. | |
| **10.** | Find the intercept(s) of the horizontal line $y = 2$. | |

# Metacognitive Skills
## Lines and More

Metacognitive skills refer to the ability to judge how well you have learned something and to effectively direct your own learning and studying. This is a self-evaluation tool designed to help you focus your studying and to improve your metacognitive skills with regards to this math class.

Fill the 1$^{st}$ column out **before** you begin studying. Fill the 2$^{nd}$ column out after you study for your test.

Go back to this assessment after your test and circle any of the ratings that you would change – this identifies the "disconnects" between what you **thought** you knew well and what you **actually** knew well.

**Use the scale below to assign a number to each topic.**
5  *I am confident I can do any problems in this category correctly.*
4  *I am confident I can do most of the problems in this category correctly.*
3  *I understand how to do the problems in this category, but I still make a lot of mistakes.*
2  *I feel unsure about how to do these problems.*
1  *I know I don't understand how to do these problems.*

| Topic or Skill | Before Studying | After Studying |
|---|---|---|
| Plotting points or identifying points on a graph using an ordered pair. | | |
| Identifying the quadrant or axis that a point is located in/on. | | |
| Creating a line graph given a set of data. | | |
| Deciding whether an ordered pair is a solution of a linear equation. | | |
| Deciding whether an ordered pair is a solution of a linear inequality. | | |
| Constructing a table of solutions for a linear equation. | | |
| Graphing a linear equation by plotting points. | | |
| Finding the *x*- and *y*-intercepts of a linear equation (if they exist). | | |
| Graphing a linear equation by using the *x*- and *y*-intercepts. | | |
| Interpreting a graph or answering questions based on the graph of a linear equation. | | |
| Understanding the equations and graphs of vertical and horizontal lines. | | |
| Finding the slope of a line by looking at its graph. | | |
| Finding the slope of a line if you are given two points. | | |
| Understanding the slopes of horizontal and vertical lines. | | |
| Calculating a rate of change (including the units). | | |
| Understanding the special properties of slope for parallel and perpendicular lines. | | |
| Finding the slope and *y*-intercept of a line given in slope-intercept form. | | |
| Rearranging a linear equation so that it is written in slope-intercept form. | | |
| Using the slope and any point on the line to graph a linear equation. | | |

*Continued on next page.*

**Use the scale below to assign a number to each topic.**

5 *I am confident I can do any problems in this category correctly.*

4 *I am confident I can do most of the problems in this category correctly.*

3 *I understand how to do the problems in this category, but I still make a lot of mistakes.*

2 *I feel unsure about how to do these problems.*

1 *I know I don't understand how to do these problems.*

| Topic or Skill | Before Studying | After Studying |
|---|---|---|
| Writing an equation of a line if you are given the slope and $y$-intercept. | | |
| Using the slope and $y$-intercept to graph a linear equation. | | |
| Using the point-slope form to write an equation of a line. | | |
| Writing an equation for a line if you are given two points on the line. | | |
| Writing an equation of a horizontal or a vertical line. | | |
| Knowing the formulas for slope, slope-intercept form, and point-slope form. | | |
| Applying the formulas for linear equations to write an equation to model data. | | |
| Solving application problems that involve linear equations. | | |
| Graphing a linear inequality. | | |
| Understanding whether the boundary line in a linear inequality is dashed or solid and which side of the boundary line to shade. | | |
| Solving for $y$ in a linear inequality (including the special rule that applies to dividing or multiplying by a negative number when solving inequalities) | | |
| Solving application problems that involve inequalities. | | |

# FUNC: Functions and More

# Student Activity
## Everyday Functions

**Case #1:** During an eight hour shift, hospital patients in the ICU are assigned to a nurse at a ratio of 2:1. If the ICU has eight patients, the nursing assignments would be as follows:

| | |
|---|---|
| Patient 1 | Nurse A |
| Patient 2 | |
| Patient 3 | Nurse B |
| Patient 4 | |
| Patient 5 | Nurse C |
| Patient 6 | |
| Patient 7 | Nurse D |
| Patient 8 | |

**1.** Consider the set of patients to be the domain and the set of nurses to be the range. The situation described here **is** a function. Why?

**2.** Describe the practical implications if this relation were **not** a function.

**3.** If the domain was the set of nurses and the range was the set of patients, would the relation still be a function? Why or why not?

**Case #2:** Suppose the exchange rate for the US Dollar to Mexican Peso is $1=10.880 pesos. The table below shows the conversion of several dollar amounts to pesos.

| US Dollars | Mexican Pesos |
|---|---|
| 5 | 54.4 |
| 24 | 261.12 |
| 78 | 848.64 |
| 110 | 1,196.8 |
| 500 | 5,440 |
| 1250 | 1,360 |

**4.** Let the domain be US Dollars and the range be Mexican Pesos. This relation **is** a function. Why?

**5.** Describe the practical implications if this relation were **not** a function.

**6.** If the domain and range were reversed, would it still be a function? Why or why not?

**Case #3:** These ordered pairs represent the identification numbers of commercial fishing boats and the profit (or loss) per boat in thousands of dollars.

$(1, 37)$     $(2, 43)$     $(5, 37)$     $(7, 13)$     $(8, -5)$     $(10, 37)$     $(12, -117)$

**7.** Let the domain be _____ and the range be _____ .

**8.** Is this a function? Why or why not?

**9.** Describe the practical implications if this relation were **not** a function.

**10.** If the domain and range were reversed, would it still be a function? Why or why not?

# Student Activity
## Skeleton Crew

**Skeleton Method:** To evaluate a function that uses function notation, it is helpful to "see" the function without the variables. To make the "skeleton" of the problem, replace the variables with an empty parentheses "skeleton".

For example, $f(x) = x^2 - 5x + 1$ becomes $f(\ ) = (\ )^2 - 5(\ ) + 1$.

Now if you have to evaluate the expression for $x = -2$, write the $-2$ inside each parentheses skeleton and evaluate.

So $f(\ ) = (\ )^2 - 5(\ ) + 1$ becomes $f(-2) = (-2)^2 - 5(-2) + 1$.

**Directions:** For each row of the table, a function is described in words, in function notation, or as a parentheses skeleton. Complete the table and then evaluate the function for the given value. The first one has been done for you.

| Word Description | $f(x)$ | Skeleton | $f(-2)$ |
|---|---|---|---|
| Multiply by 2 and then subtract 3 | $f(x) = 2x - 3$ | $f(\ ) = 2(\ ) - 3$ | $f(-2) = 2(-2) - 3$ $= -4 - 3 = -7$ |
| Subtract 3 and then multiply by 2 | | | |
| | $f(x) = x^3 - 5$ | | |
| Add 3, then square the result | | | |
| | | $f(\ ) = |(\ ) - 6|$ | |
| | | | $f(-2) = \dfrac{(-2) + 3}{4} = \dfrac{1}{4}$ |
| | $f(x) = 9 - x$ | | |
| Take the reciprocal of it (assume it is not zero) | | | |

# Guided Learning Activity
## Investigating Functions

**Definition of a function:** a relation in which each **first component** corresponds to exactly one **second component**.

**Vertical Line Test:** If a vertical line intersects a graph in more than one point, the graph is **not** the graph of a function.

---

**1.** Graph: $f(x) = |x|$

| $x$ | $f(x)$ |
|-----|--------|
|     |        |
|     |        |
|     |        |
|     |        |
|     |        |

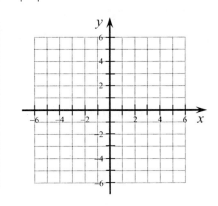

Is it a function? _____

**2.** Graph: $f(x) = x - 3$

| $x$ | $f(x)$ |
|-----|--------|
|     |        |
|     |        |
|     |        |
|     |        |
|     |        |

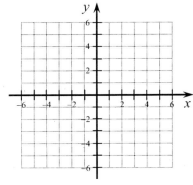

Is it a function? _____

**3.** Graph: $x = |y|$

| $x$ | $y$ |
|-----|-----|
|     | $-2$ |
|     | $-1$ |
|     | $0$ |
|     | $1$ |
|     | $2$ |

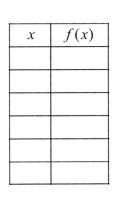

Is it a function? _____

Matt fails the Vertical Line Test

Talk about lack of function

# Student Activity
## Gleaning Information from Graphs

In general, the value of $f(a)$ is given by the $y$-coordinate of a point on the graph of $f$ with $x$-coordinate $a$.

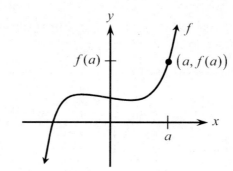

**Directions:** Use the graph for each problem to answer the questions.

1. **a.** $f(-2) =$ _____

   **b.** Find the value(s) of $x$ where $f(x) = 1$. _____

   **c.** $f(5) =$ _____

   **d.** $f(\_\_\_) = -3$

   **e.** Find the value(s) of $x$ where $f(x) = 2$. _____

   **f.** Write the domain of $f$ : _____

   **g.** Write the range of $f$ : _____

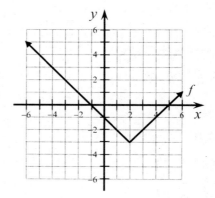

2. **a.** $g(2) =$ _____

   **b.** Find the value(s) of $x$ where $g(x) = 4$. _____

   **c.** $g(0) =$ _____

   **d.** $g(\_\_\_) = 2$

   **e.** What is the slope, $m$, of this line? _____

   **f.** Write the domain of $g$ : _____

   **g.** Write the range of $g$ : _____

Relationship counseling for function properties.

3. **a.** $h(-1) =$ _____

   **b.** Find the value(s) of $x$ where $h(x) = 0$. _____

   **c.** $h(-4) =$ _____

   **d.** $h(\_\_\_) = 2$

   **e.** Find the value(s) of $x$ where $h(x) = -2$. _____

   **f.** Write the domain of $h$: _____

   **g.** Write the range of $h$: _____

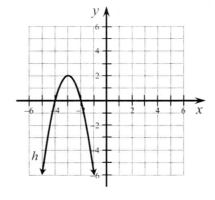

4. **a.** $p(4) =$ _____

   **b.** Find the value(s) of $x$ where $p(x) = 1$. _____

   **c.** $p(-2) =$ _____

   **d.** Find the value(s) of $x$ where $p(x) = 4$. _____

   **e.** What is the slope of $p$ where $x > 2$? _____

   **f.** Write the domain of $p$: _____

   **g.** Write the range of $p$: _____

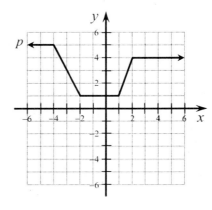

129

# Student Activity
## Shapeshifters

**Directions:** For problems 1-4, a function is drawn using a dashed line. Follow the directions for each problem to draw the graph that is described using a **solid line**.

1. Shift the graph of *f* down 4 units.

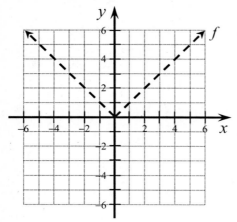

2. Reflect the graph of *g* over the *x*-axis.

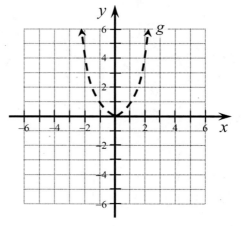

3. Shift the graph of *h* to the right 2 units.

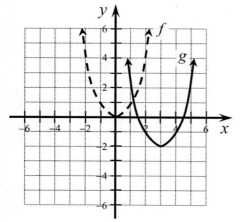

4. Shift the graph of *p* up 3 units and to the left 5 units.

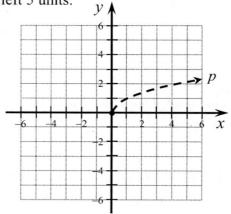

5. Describe the translations and/or reflections of *f* that must occur to get the graph of *g*.

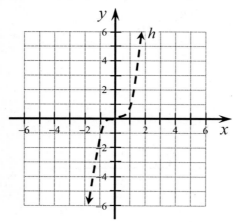

6. Describe the translations and/or reflections of *f* that must occur to get the graph of *g*.

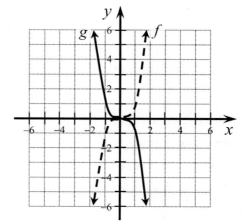

# Student Activity
## Reflecting on Graph Translations

1. In the graphs below, examine the functions $f$ and $g$ in each picture. How is the change in the mathematical function related to the change between the graph of $f$ and the graph of $g$?

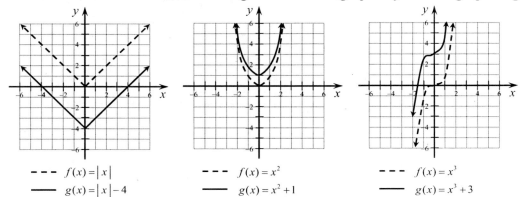

- - - $f(x) = |x|$
——— $g(x) = |x| - 4$

- - - $f(x) = x^2$
——— $g(x) = x^2 + 1$

- - - $f(x) = x^3$
——— $g(x) = x^3 + 3$

If $f$ is a function and $k$ is a positive number then
a. The graph of $y = f(x) + k$ is identical to the graph of $y = f(x)$ except that it is translated $k$ units _____.
b. The graph of $y = f(x) - k$ is identical to the graph of $y = f(x)$ except that it is translated $k$ units _____.

2. In the graphs below, examine the functions $f$ and $g$ in each picture. How is the change in the mathematical function related to the change between the graph of $f$ and the graph of $g$?

If $f$ is a function and $h$ is a positive number then
a. The graph of $y = f(x - h)$ is identical to the graph of $y = f(x)$ except that it is translated $h$ units _____.
b. The graph of $y = f(x + h)$ is identical to the graph of $y = f(x)$ except that it is translated $h$ units _____.

I'm sorry, that's not quite what we had in mind.

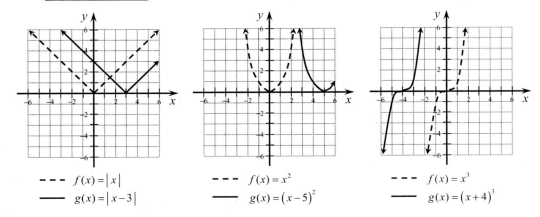

- - - $f(x) = |x|$
——— $g(x) = |x - 3|$

- - - $f(x) = x^2$
——— $g(x) = (x - 5)^2$

- - - $f(x) = x^3$
——— $g(x) = (x + 4)^3$

**3.** In the graphs below, examine the functions *f* and *g* in each picture.

A vertical reflection is a reflection about the *x*-axis. These graphs show vertical reflections:

  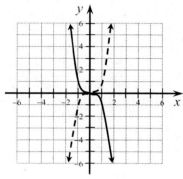

$$--- \quad f(x) = |x|$$
$$\underline{\qquad} \quad g(x) = -|x|$$

$$--- \quad f(x) = x^2$$
$$\underline{\qquad} \quad g(x) = -x^2$$

$$--- \quad f(x) = x^3$$
$$\underline{\qquad} \quad g(x) = -x^3$$

A horizontal reflection is about the *y*-axis. These graphs show horizontal reflections:

  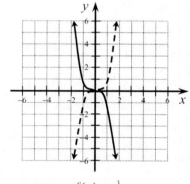

$$--- \quad f(x) = |x|$$
$$\underline{\qquad} \quad g(x) = |-x|$$

$$--- \quad f(x) = x^2$$
$$\underline{\qquad} \quad g(x) = (-x)^2$$

$$--- \quad f(x) = x^3$$
$$\underline{\qquad} \quad g(x) = (-x)^3$$

**a.** The graph of $y = -f(x)$ is the graph of $y = f(x)$ reflected \underline{\qquad\qquad\qquad}.

**b.** The graph of $y = f(-x)$ is the graph of $y = f(x)$ reflected \underline{\qquad\qquad\qquad}.

**c.** Why is the graph of $g(x) = -x^3$ the same as the graph of $g(x) = (-x)^3$?

**d.** Why is the graph of $f(x) = |x|$ the same as the graph of $g(x) = |-x|$?

**4.** Let's recap. Let $k$ and $h$ represent positive numbers.

   **a.** The graph of $y = f(x+h)$ is a translation of $h$ units _____ from $f(x)$.

   **b.** The graph of $y = f(x-h)$ is a translation of $h$ units _____ from $f(x)$.

   **c.** The graph of $y = f(-x)$ is a _____ reflection of $f(x)$.

   **d.** It seems that when $x$ is replaced by another expression involving $x$, the general effect on the graph is _____ (*horizontal or vertical?*).

   **e.** The graph of $y = f(x)+k$ is a translation of $h$ units _____ from $f(x)$.

   **f.** The graph of $y = f(x)-k$ is a translation of $h$ units _____ from $f(x)$.

   **g.** The graph of $y = -f(x)$ is a _____ reflection of $f(x)$.

   **h.** It seems that when something is added to, subtracted from, or multiplied by $f(x)$, the general effect on the graph is _____ (*horizontal or vertical?*).

**5. a.** $g(x) = (x-1)^2$ is the graph of $f(x) = $ ____ shifted ____ units _____.

   **b.** $g(x) = |x| + 6$ is the graph of $f(x) = $ ____ shifted ____ units _____.

   **c.** $g(x) = (x+2)^3$ is the graph of $f(x) = $ ____ shifted ____ units _____.

   **d.** $g(x) = (x+5)^2 - 2$ is the graph of $f(x) = $ ____ shifted 5 units _____ and 2 units _____.

   **e.** $g(x) = -|x|$ is the graph of $f(x) = $ ____ reflected about the _____.

   **f.** $g(x) = \sqrt{x} + 3$ is the graph of $f(x) = \sqrt{x}$ shifted ____ units _____.

   **g.** $g(x) = \sqrt{-x}$ is the graph of $f(x) = \sqrt{x}$ reflected about the _____.

**6.** Suppose we define a new function called the "box" function, mathematically, it is written like this: $f(x) = \boxed{x}$. Based on the graphing rules we just developed, describe how each of the following functions would be graphed, in comparison with the graph of $f$. Think carefully about whether $x$ has been replaced with another expression or whether the $f(x)$ portion of the graph has remained intact.

   **a.** $g(x) = \boxed{x+2}$  _____

   **b.** $g(x) = \boxed{x} - 4$  _____

   **c.** $g(x) = 2 + \boxed{x}$  _____

   **d.** $g(x) = -\boxed{x}$  _____

   **e.** $g(x) = \boxed{x-3}$  _____

   **f.** $g(x) = \boxed{-x}$  _____

   **g.** $g(x) = \boxed{x+3} - 1$  _____

# Student Activity
## Algebra of Domains

When we create a new function using the algebra of functions $f$ and $g$, the domain of the new function is the set of real numbers that are in the domain of both $f$ **and** $g$ (as in, the intersection of the two domains). For a quotient $f / g$, there is the additional restriction that the new denominator, $g$, not be equal to zero.

**Directions:** Begin by writing the domain for the functions $f$, $g$, and $h$. Then complete the table below.

$$f(x) = 2x - 1 \qquad\qquad g(x) = \sqrt{x+4} \qquad\qquad h(x) = x + 5$$

Domain: _____        Domain: _____        Domain: _____

|     | Find the new function and simplify it. | What is the domain of new function? |
| --- | --- | --- |
| 1. | $(f + g)(x)$ | |
| 2. | $(f - h)(x)$ | |
| 3. | $(f / h)(x)$ | |
| 4. | $(f \cdot h)(x)$ | |
| 5. | $(h / g)(x)$ | |
| 6. | $g(h(x))$ | |
| 7. | $g(f(x))$ | |

# Student Activity
## Graphical Algebra of Functions

When we perform an algebraic operation on functions (addition, subtraction, multiplication, division), we are performing the operation on the *output* of the functions.

1.  Below you see the graphs of two functions, $f$ and $g$. A table of values has been given for each of the functions. Fill in the table of values for $f+g$ and $f-g$.

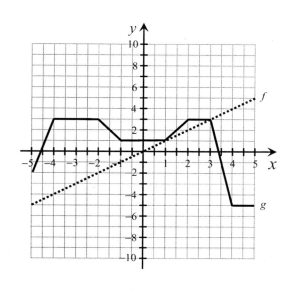

| $x$ | $f(x)$ | $g(x)$ | $(f+g)(x)$ $f(x)+g(x)$ | $(f-g)(x)$ $f(x)-g(x)$ |
|---|---|---|---|---|
| −5 | −5 | −2 | | |
| −4 | −4 | 3 | | |
| −3 | −3 | 3 | | |
| −2 | −2 | 3 | | |
| −1 | −1 | 1 | | |
| 0 | 0 | 1 | | |
| 1 | 1 | 1 | | |
| 2 | 2 | 3 | 5 | −1 |
| 3 | 3 | 3 | | |
| 4 | 4 | −5 | | |
| 5 | 5 | −5 | | |

2.  Now use the table or the graph to evaluate the expressions in the tables below.

| | |
|---|---|
| $f(4)$ | 4 |
| $g(4)$ | −5 |
| $f(4)+g(4)$ | −1 |
| $(f+g)(4)$ | −1 |
| $g(1)$ | 1 |
| $g(3)$ | 3 |
| $g(1+3)$ | −5 |
| $g(1)+g(3)$ | 4 |

| | |
|---|---|
| $f(-3)$ | −3 |
| $g(-3)$ | 3 |
| $f(-3)-g(-3)$ | −6 |
| $(f-g)(-3)$ | −6 |
| $g(1)$ | 1 |
| $g(4)$ | −5 |
| $g(1-4)$ | 3 |
| $g(1)-g(4)$ | 6 |

3.  Why aren't $g(1+3)$ and $g(1)+g(3)$ equivalent?

**4.** Now we will graphically perform the addition and subtraction of the functions. For example, at $x = 2$, $f(2) = 2$ and $g(2) = 3$. Adding the values of $f$ and $g$ gives 5, so we plot a point at $(2,5)$ for the addition graph. Subtracting the values at two, $f - g$ gives $-1$, so we plot a point at $(2,-1)$ for the subtraction graph.

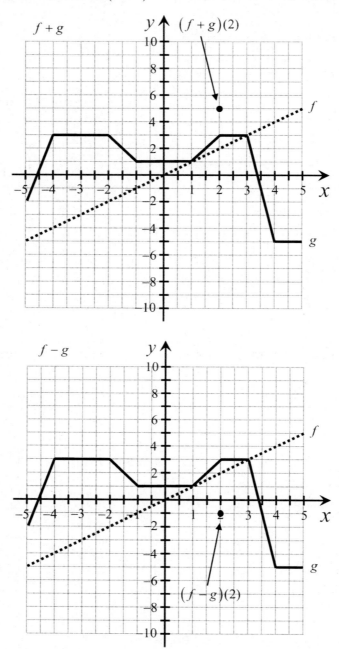

**5.** Now that you see the graphs of $f + g$ and $f - g$, try to explain what's happened to make the new curve for each one.

# Student Activity
## Composing Skeletons

**Skeleton Method:** To evaluate a function that uses function notation, it is helpful to "see" the function without the variables. To make the "skeleton" of the problem, replace the variables with an empty parentheses "skeleton".

For example, $f(x) = x^2 + 4x + 3$ becomes $f(\ ) = (\ )^2 + 4(\ ) + 3$.

For this activity, we define: $f(x) = 3x - 5$ $\qquad$ $g(x) = x - 5$ $\qquad$ $h(x) = x^2 - 5x$

| | Evaluate this expression | Write with parentheses notation | Make a parentheses skeleton of the outer function | Substitute the inner function and simplify |
|---|---|---|---|---|
| **1.** | $(f \circ g)(x)$ | $f(g(x))$ | $f(\ ) = 3(\ ) - 5$ | $f(g(x)) = 3(x-5) - 5$ $= 3x - 15 - 5 = 3x - 20$ |
| **2.** | $(g \circ f)(x)$ | | | |
| **3.** | | $h(g(x))$ | | |
| **4.** | | | | $g(h(x)) = (x^2 - 5x) - 5$ $= x^2 - 5x - 5$ |
| **5.** | $(f \circ f)(x)$ | | | |
| **6.** | | | | $g(g(x)) = (x-5) - 5$ $= x - 10$ |
| **7.** | | $h(f(x))$ | | |

# Student Activity
## From Tables and Graphs

The functions $f$, $g$, and $h$ are evaluated for the values of $x$ in the table below.

For example, we can use the table to see that $f(0) = 2$ or $g(-2) = 7$.

| $x$ | $f(x)$ | $g(x)$ | $h(x)$ |
|-----|--------|--------|--------|
| $-2$ | 0 | 7 | 4 |
| $-1$ | 1 | 5 | 1 |
| 0 | 2 | 3 | 0 |
| 1 | 3 | 1 | 1 |
| 2 | 4 | $-1$ | 4 |

| $x$ | $f(x)$ | $g(x)$ | $h(x)$ |
|-----|--------|--------|--------|
| $-2$ | 0 | 7 | 4 |
| $-1$ | 1 | 5 | 1 |
| 0 | 2 | 3 | 0 |
| 1 | 3 | 1 | 1 |
| 2 | 4 | $-1$ | 4 |

$g(-2) = 7$

$f(0) = 2$

**Directions:** Evaluate each expression using the table of values.

1. $h(-1) =$

2. $(f + g)(2) = f(2) + g(2) =$

3. $(h - g)(1) =$

4. $(g / f)(-1) =$

5. $(f \cdot g)(0) =$

6. $g(3 - 1) =$

7. $h(f(0)) =$

8. $g(h(-1)) =$

9. $(g \circ f)(-2) =$

10. $f(f(-2)) =$

**Directions:** Evaluate each expression below using the graphs of $f$ and $g$.

11. $(f + g)(2) = f(2) + g(2) =$

12. $(g - f)(-2) =$

13. $(f \cdot g)(0) =$

14. $g(f(-1)) =$

15. $(f \circ f)(-1) =$

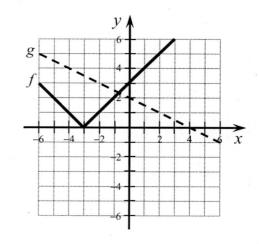

# Student Activity

## Why is 1-1 Important?

1. Below, you see a graph that represents $f$. Is $f$ a function? _____
2. Is $f$ one-to-one? _____
3. Make a table of solutions for $f$.
4. Swap the $x$- and $y$- coordinates of each solution to make a second table.
5. Graph the solutions from the second table on the same set of axes as $f$.
6. Does the new graph represent a function? _____

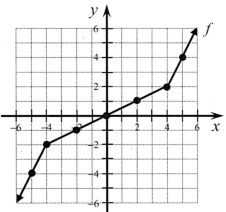

| $x$ | $f(x)$ |
|-----|--------|
| $-5$ | $-4$ |
| $-4$ | |
| $-2$ | |
| $0$ | |
| $2$ | |
| $4$ | |
| $5$ | |

| $x$ | $y$ |
|-----|-----|
| $-4$ | $-5$ |
| | |
| | |
| | |
| | |
| | |
| | |

7. Below, you see a graph that represents $g$. Is $g$ a function? _____
8. Is $g$ one-to-one? _____
9. Make a table of solutions for $g$.
10. Swap the $x$- and $y$- coordinates of each solution to make a second table.
11. Graph the solutions from the second table on the same set of axes as $g$.
12. Does this new graph represent a function? _____

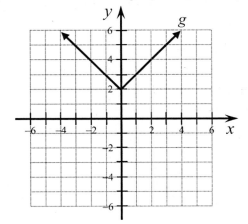

| $x$ | $g(x)$ |
|-----|--------|
| $-3$ | $5$ |
| $-2$ | |
| $-1$ | |
| $0$ | |
| $1$ | |
| $2$ | |
| $3$ | |

| $x$ | $y$ |
|-----|-----|
| $5$ | $-3$ |
| | |
| | |
| | |
| | |
| | |
| | |

13. If the original function is one-to-one, what does that mean for the inverse?

14. If the original function is NOT one-to-one, what does that mean for the inverse?

# Guided Learning Activity
## The Magic of Inverse Functions

| Operation | Inverse Operation |
|---|---|
| Add 5. | Subtract 5. |
| Multiply by 3. | |
| Subtract 7. | |
| Divide by $-2$. | |
| Cube it. | |
| Take a square root. | |

OK, YOU give ME the wallet

And now, Abracadabra... I wave the inverse wand.

I give YOU back the wallet

Huh?huh? Like that?

OK. Now you're got your fists up.

so I wave the inverse wand again ...

*The Mathmagic of Inverse Functions*

We reverse any process by walking through the steps completely backwards.

| $f$ : **Putting on socks & shoes.** | $f$ says: Put on a sock, put on a shoe, tie the shoe. |
|---|---|
| $f^{-1}$ : **Removing socks & shoes.** | $f^{-1}$ says: Untie the shoe, take off the shoe, take off the sock. |

We can write inverse mathematical functions by reversing the function process, in the same way.

| $f(x) = 2x + 3$ | $f$ says: Multiply by 2, then add 3. |
|---|---|
| $f^{-1}(x) = \dfrac{x-3}{2}$ | $f^{-1}$ says: Subtract 3, then divide by 2. |

| $f(x) = \dfrac{x-5}{3}$ | $f$ says: |
|---|---|
| $f^{-1}(x) =$ | $f^{-1}$ says: |

| $f(x) = \sqrt[3]{x+4}$ | $f$ says: |
|---|---|
| $f^{-1}(x) =$ | $f^{-1}$ says: |

140

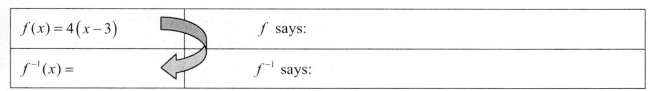

| $f(x) = 4(x-3)$ | $f$ says: |
|---|---|
| $f^{-1}(x) =$ | $f^{-1}$ says: |

Inverse functions have a special property that $f\left(f^{-1}(x)\right) = x$ and $f^{-1}\left(f(x)\right) = x$.
Why is that? Let's input a "value" into our Socks & Shoes function and see.

This represents the composition $f^{-1}\left(f(\text{bare foot})\right) = \text{bare foot}$.

Let's try a few mathematical compositions with numerical inputs.

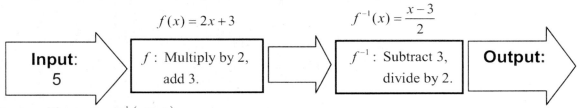

Here we illustrate $f^{-1}\left(f(x)\right) = x$:

| Input | Apply the function | Output of function = Input to inverse | Apply the inverse function | Output |
|---|---|---|---|---|
| 4 | $f(x) = 5x - 2$ | | $f^{-1}(x) = \dfrac{x+2}{5}$ | |
| 3 | $f(x) = \sqrt[3]{x+5}$ | | $f^{-1}(x) = x^3 - 5$ | |
| 8 | $f(x) = \dfrac{x+4}{3}$ | | $f^{-1}(x) = 3x - 4$ | |

And here we illustrate $f\left(f^{-1}(x)\right) = x$:

| Input | Apply the inverse function | Output of inverse = Input to function | Apply the function | Output |
|---|---|---|---|---|
| 13 | $f^{-1}(x) = \dfrac{x+2}{5}$ | | $f(x) = 5x - 2$ | |
| 2 | $f^{-1}(x) = x^3 - 4$ | | $f(x) = \sqrt[3]{x+4}$ | |
| 5 | $f^{-1}(x) = 3x - 4$ | | $f(x) = \dfrac{x+4}{3}$ | |

# Student Activity
## Mirror Images

**Directions:** Recall that if $f$ and $f^{-1}$ are inverse functions, then they should be symmetric over the line $y = x$. Also, if $(a,b)$ is a point on the graph of $f$, then $(b,a)$ will be a point on the graph of $f^{-1}$. On each graph, draw a dashed line for $y = x$ and then use the graph of $f$ to graph $f^{-1}$ on the same axes. The first one has been started for you.

**1.**

**2.**

**3.**

**4.**

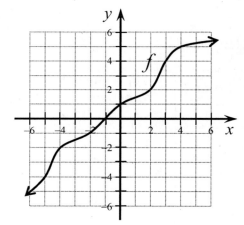

**5.** Now draw your own one-to-one function, and graph its inverse.

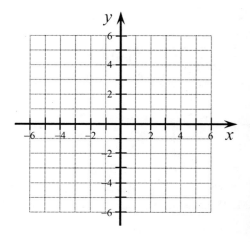

# Student Activity
## Put it All Together

If $f$ and $f^{-1}$ are inverse functions of each other then the following are true:

- The domain of $f$ is the range of $f^{-1}$ and the range of $f$ is the domain of $f^{-1}$.
- Let $(a,b)$ be a point on the graph of $f$, then $(b,a)$ is a point on $f^{-1}$.
- The graphs of $f$ and $f^{-1}$ are symmetric over the line $y = x$.
- $f\left(f^{-1}(x)\right) = x$ and $f^{-1}\left(f(x)\right) = x$

---

**Problem A:** Let $f(x) = \dfrac{3}{2}x + 3$.

1. The point $(2,6)$ is on the graph of $f$. What point must be on the graph of $f^{-1}$? $(\_\_,\_\_)$

2. Find $f^{-1}(x)$. Check this step with someone else before you go on.

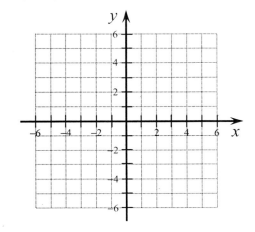

3. Graph $f$ and $f^{-1}$ on the axes to the right. Label which is which.

4. Draw the line $y = x$ on the same graph using a dashed line.

5. Fill in the domain and range for each:

   $f(x)$ Domain: _____      $f^{-1}(x)$ Domain: _____

   Range: _____          Range: _____

6. Find and simplify $f\left(f^{-1}(x)\right)$.     7. Find and simplify $f^{-1}\left(f(x)\right)$.

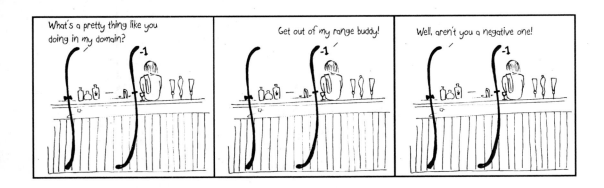

**Problem B:** Let $g(x) = \sqrt{x+3} + 1$.

1. Graph $g$ on the axes below. Use the table of solutions to assist you. Label the graph with $g$.

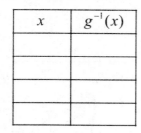

| $x$ | $g(x)$ |
|-----|--------|
| $-5$ | |
| $-4$ | |
| $-3$ | |
| $-2$ | |
| $1$ | |
| $6$ | |

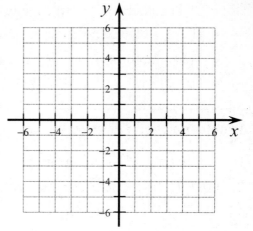

2. What is the domain of $g$? _____ What is the range? _____

3. Draw the line $y = x$ on the same graph using a dashed line. Draw a reflection of the graph of $g$ over the line of symmetry. Label this $g^{-1}$.

4. Find the formula for $g^{-1}(x)$ using the formula for $g(x)$. Check this step with someone else before you go on.

5. Based on the domain and range of $g$, what is the domain of $g^{-1}$? _____ What is the range? _____

6. Fill in the table of solutions for $g^{-1}$ using the formula you just found **and** the domain for $g^{-1}$. Then graph this function on the new axes provided.

| $x$ | $g^{-1}(x)$ |
|-----|-------------|
| | |
| | |
| | |
| | |

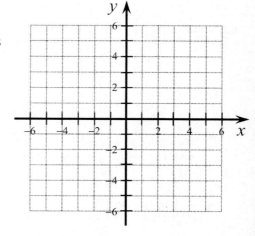

7. Include a condition with the formula for $g^{-1}$ so that $g^{-1}(x)$ gives the true inverse of $g$. $g^{-1}(x) = $ _____ , $x \geq$ _____ .

# Assess Your Understanding
## Functions and More

For each of the following, describe the strategies or key steps that will help you **start** the problem. You do **not** have to complete the problems.

|  |  | **What will help you to start this problem?** |
|---|---|---|
| **1.** | If $f(x) = -3x + 1$, find $f(4)$. | |
| **2.** | Is the relation $\{(2,3),(4,1),(3,4)\}$ a function? | |
| **3.** | Let $f(x) = 2x - 7$. Find $f^{-1}(x)$. | |
| **4.** | Graph $g(x) = (x-1)^2$ and write the domain and range. | |
| **5.** | Describe how to graph $f(x) = |x - 2|$ given the graph of $y = |x|$. | |
| **6.** | Describe how to graph $g(x) = -x^3 + 4$ given the graph of $y = x^3$. | |
| **7.** | Let $f(x) = x^2 - 4x$ and $g(x) = 2x - 1$. Find $(f \circ g)(x)$. | |
| **8.** | Let $f(x) = 2x + 3$ and $g(x) = x - 4$. Find $(f / g)(x)$ and state the domain of the new function. | |
| **9.** | Is $y = x^2$ a function? Is it one-to-one? | |
| **10.** | Let $f(x) = \dfrac{x+1}{x-3}$. Find $f^{-1}(x)$. | |

# Metacognitive Skills
## Functions and More

Metacognitive skills refer to the ability to judge how well you have learned something and to effectively direct your own learning and studying. This is a self-evaluation tool designed to help you focus your studying and to improve your metacognitive skills with regards to this math class.

Fill the 1st column out **before** you begin studying. Fill the 2nd column out after you study for your test.

Go back to this assessment after your test and circle any of the ratings that you would change – this identifies the "disconnects" between what you **thought** you knew well and what you **actually** knew well.

**Use the scale below to assign a number to each topic.**
5  *I am confident I can do any problems in this category correctly.*
4  *I am confident I can do most of the problems in this category correctly.*
3  *I understand how to do the problems in this category, but I still make a lot of mistakes.*
2  *I feel unsure about how to do these problems.*
1  *I know I don't understand how to do these problems.*

| Topic or Skill | Before Studying | After Studying |
|---|---|---|
| Understand the difference between a relation and a function. | | |
| Be able to use function notation and evaluate for a specific value. | | |
| Identify whether a relation written as a set is a function. | | |
| Apply the vertical line test to decide if a graph represents a function. | | |
| Write the description of a function using function notation. | | |
| Create a table of solutions and sketch the graph of an equation. | | |
| Identify information on graphs using function notation. | | |
| Find the domain and range of a graph. | | |
| Graph the basic functions (identity, squaring, cubing, absolute value). | | |
| Given a graph, draw various translations, and transformations. | | |
| Identify how a change to an equation will change its graph. | | |
| Determine the domain of a function. | | |
| Perform operations on functions (sum, difference, product, quotient). | | |
| Find the domain of a function created from a sum, difference, product, or quotient of two functions. | | |
| Understand what it means graphically to add or subtract functions. | | |
| Interpret and evaluate of function compositions. | | |
| Perform algebra or composition of functions using algebra. | | |
| Perform algebra or composition of functions using a table of values. | | |
| Perform algebra or composition of functions using a graph. | | |
| Identify when a relation, function, or graph is one-to-one. | | |
| Apply the horizontal line test to determine whether a graph is 1-1. | | |
| Know the four properties of functions and their inverses. | | |
| Know how the property of 1-1 relates to finding an inverse function. | | |
| Given the equation for a 1-1 function, find its inverse. | | |
| Given the graph of a relation, draw its inverse. | | |
| Explain (in words) what an inverse is. | | |

# SYS1: Solving Systems

# Student Activity
## Targeting Solutions

**Directions:** In each problem below, test the ordered pairs that are provided to see if they are possible solutions to either of the equations in the system of equations. After testing, place the ordered pairs into the appropriate region of the "target" for each problem.

| | |
|---|---|
| **Outer target region** | The ordered pair IS a solution of the first equation, but not the second equation. |
| **Middle target region** | The ordered pair is not a solution of the first equation, but IS a solution of the second equation. |
| **Bulls-eye (center target region)** | The ordered pair is a solution of both equations. |

**1.** $\begin{cases} y < 5x + 2 \\ y > -x + 8 \end{cases}$

$(1, 6)$

$(4, 25)$

$(2, 10)$

$(-10, 20)$

$(-3, -20)$

$(\frac{1}{2}, 9)$

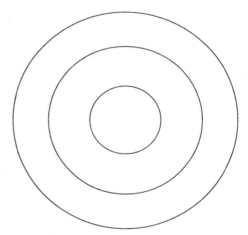

**2.** $\begin{cases} 4x + 2y = 8 \\ -2x + 4y = 1 \end{cases}$

$(5, -6)$

$(7, \frac{15}{4})$

$(\frac{3}{2}, 1)$

$(-4, -\frac{7}{4})$

$(0, 4)$

$(-3, 10)$

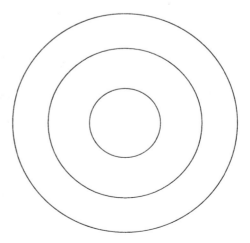

# Guided Learning Activity

**Graphing to Solve Systems of Equations**

**Example 1:** Write some ordered-pair solutions of the equation $x + y = 5$.

( , )　( , )　( , )　( , )　( , )　( , )

Write some ordered-pair solutions of the equation $x - y = -3$.

( , )　( , )　( , )　( , )　( , )　( , )

When we write the two equations like this: $\begin{cases} x + y = 5 \\ x - y = -3 \end{cases}$

we call this a *system of equations*. When we *solve* a system of two equations, we are looking for all of the ordered pairs $(x, y)$ that satisfy *both* equations.

Maybe you found an ordered pair that was a solution of both equations? If so, circle it.

**Example 2:** What does it mean to be a solution to a system?

Consider the system of equations given by $\begin{cases} y = 2x + 2 \\ y = x - 1 \end{cases}$ and the ordered pair $(-3, -4)$.

You can see by evaluating each equation for $x = -3$ and $y = -4$ that this ordered pair is a solution of both equations. This means it is a solution of the system of equations.

| $y = 2x + 2$ | $y = x - 1$ |
|---|---|
| $y = 2( \ ) + 2$ | $y = ( \ ) - 1$ |
| $-4 \overset{?}{=} 2(-3) + 2$ | $-4 \overset{?}{=} (-3) - 1$ |
| $-4 = -4$ | $-4 = -4$ |

So what does the solution of a system of equations look like? Graph $y = 2x + 2$ and $y = x - 1$ on the graphing grid provided to the right.

What is special about the point $(-3, -4)$ on the graph?

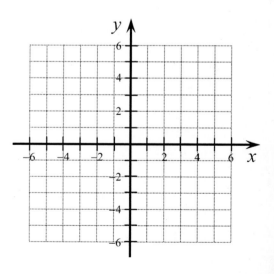

# Now solve these systems of equations by graphing.

**1.** $\begin{cases} y = -3x + 6 \\ y = 2x - 4 \end{cases}$

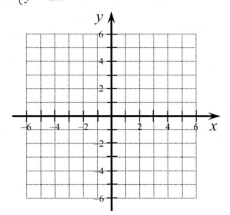

**4.** $\begin{cases} y = -3x - 2 \\ y = 2x + 3 \end{cases}$

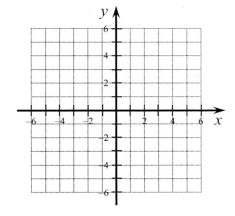

**2.** $\begin{cases} 5x - y = 6 \\ 3x - 3y = 6 \end{cases}$

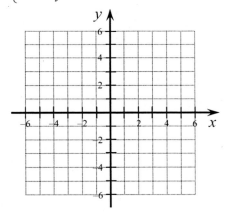

**5.** $\begin{cases} -2x - 4y = -2 \\ 4x + 8y = 4 \end{cases}$

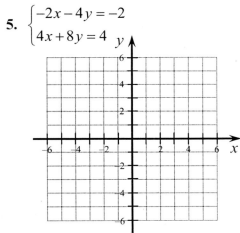

**3.** $\begin{cases} -4x + 2y = 8 \\ 6x - 3y = 3 \end{cases}$

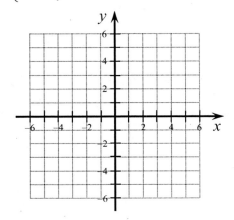

**6.** $\begin{cases} y = \frac{3}{4}x - 1 \\ y = -\frac{4}{3}x - 1 \end{cases}$

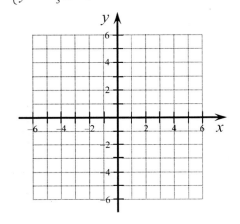

# Student Activity
## Following the Clues Back to the System of Equations

SYS1-4

**Directions:** In each "crime-scene" below, you are shown the graph of a system of equations. Use your mathematical powers of reasoning (and detective skills) to determine what the system of equations must have been to result in this graph.

1.

2.

3.

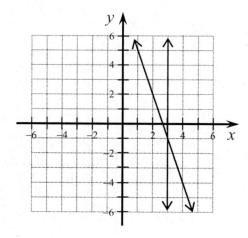

152
© 2011 Cengage Learning. All Rights Reserved. May not be scanned, copied or duplicated, or posted to a publicly accessible website, in whole or in part.

# Student Activity

## Graphing Systems of Equations with a Calculator

You can use your graphing calculator to solve systems of linear equations by graphing. Each model of calculator will use a different set of keystrokes to graph these equations. Before you start to graph any system of equations, it is a good idea to guess what your graph should look like (in case you type something in your calculator incorrectly). The standard zoom on most graphing calculators goes from -10 to 10 on the x-axis and from -10 to 10 on the y-axis.

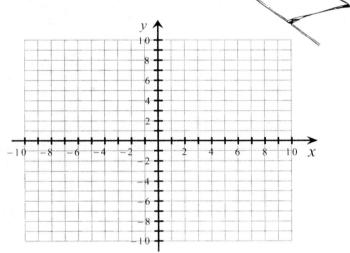

1. Without using a calculator, sketch a graph of the system of equations $\begin{cases} y = 5x - 5 \\ y = -3x + 3 \end{cases}$ on the axes provided, and label the solution point.

2. Find the $y =$ screen on your calculator. Place your cursor next to $Y_1 =$ and enter the first equation in your system. Then move your cursor to $Y_2 =$ and enter the second equation. Make sure you are in the standard viewing window and then view the graph. Hopefully this looks like the graph you produced above.

3. In order to find the solution to this system of equations on the calculator, we need to find the point of intersection. **From your graph screen**, find the *Math* or *Calculate* menu. You'll know you have found it when you see a menu that contains functions like *zero* (or *root*), *max*, *min*, and *intersect* (it will also have a few others).

   Draw the keys or write the steps that were necessary to find this menu.

4. Select *intersect* (or *intersection*) from the menu you just found. Your calculator will give you a series of prompts. At the end of all of the prompts, your calculator should display numerical values labeled $x =$ and $y =$ . This should match the solution you found at the beginning of the activity.

| Calculator Prompt | What does it mean? What should you do? | Possible Errors |
|---|---|---|
| First Curve? | The calculator is asking you to tell it which curve is the first curve in the system of equations. Place your cursor on one of the lines you graphed using the arrow keys and press *enter*. You will then be asked for the second curve. Move the cursor to the other curve and press *enter*. *Why does it ask this? Well, if you had graphed four lines, it would want to know which lines (curves) were the two that you want to find the intersection of.* | If you give the calculator the **same** curve for both the first and second prompts, you will likely get an error. |
| Lower Bound? | The calculator is asking you to give it a smaller region to look in to find the intersection point. It is asking you for a value to the left of the point of intersection (lower bound) first. Use your arrow keys to place your cursor to the left of the point of intersection and press *enter* or use the number keypad to enter a value that you know is to the left of the intersection point, and then press *enter*. You will then be asked for the Upper Bound. Repeat the process for a value to the right of the intersection point. | The upper bound must be **above** (to the right of) the lower bound. If you try to give the calculator an upper bound value that is below the lower bound value, you will get an error. |
| Guess? | It is easier for the calculator to find the exact intersection point if you give it a guess point near the intersection point. Just position your cursor near the intersection point and press enter. *Why does it ask for a guess? If you were graphing more complicated curves (for example polar curves), there might be more than one intersection point in the same lower and upper boundary. In this case, the calculator would want to know which point you wanted to find out exactly.* | You must give the calculator a guess that is **between** the lower and upper bounds. If your guess is outside the region where you told the calculator to look, it will be (understandably) unhappy with you. |

**Note:** If you do not see the intersection values, it could be that you pressed enter one too many times when going through the prompts (inadvertently skipping the solutions). If this happens, repeat the steps a little more slowly, making sure to only press *enter* once for each prompt.

# Student Activity
## Which Egg is Easier to Crack?

When solving systems of equations by substitution, we begin by solving one of the equations for $x$ or $y$; this is called the *substitution equation*. Sometimes one of the variables is easier to solve for than the other (that is, it will take less steps to solve for one than the other).

**Directions:** Look at the given equations and find the substitution equation where solving for the variable requires **the least number of steps**. Then place the substitution equation in the egg that corresponds to the variable that was solved for. For example, in the first equation, it would be much easier to solve for $x$, resulting in the equation $x = 2 - 3y$.

$$x + 3y = 2 \qquad 7x + y = 14 \qquad 3y = 2x + 2 \qquad -2y + x = 18$$

$$x - 2y = 9 \qquad 15 = -5y + x \qquad 16x = 16y + 16 \qquad \frac{2}{3}x + y = \frac{1}{3}$$

$$25x = 225y + 225 \qquad 12x - y = 0 \qquad 3y = 3x + 6 \qquad 5x + y = 10$$

$$18x = 9y + 36 \qquad \frac{1}{10}x + y = 1 \qquad 12x + y = 2 \qquad 4x = 2y + 6$$

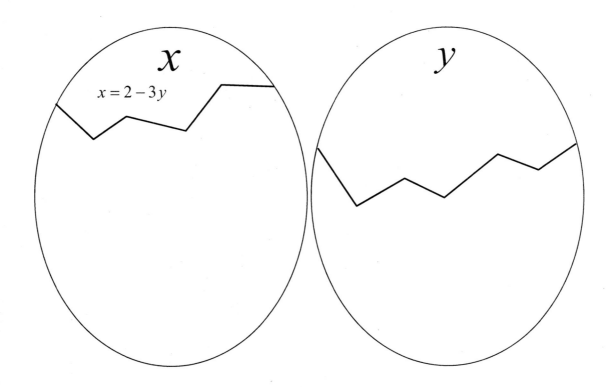

*x*

$x = 2 - 3y$

*y*

# Student Activity
## Double Trouble on Substitution

**Directions:** When you solve a system of equations by substitution, there are *at least* two ways to tackle the problem. For example, in the first system, you could easily solve for $y$ in the first equation, or for $x$ in the second equation to find your substitution equation. **Some choices are easier than others!** For each of the systems below, solve the system of equations two ways as indicated. Your ordered pair solution should be the same for both methods. If they are not, you'll have to go back and look for a mistake. The first one has been started for you.

| | System of Equations | Solve for a variable in the FIRST equation, then substitute and solve. | Solve for a variable in the SECOND equation, then substitute and solve. |
|---|---|---|---|
| 1. | $\begin{cases} 2x + y = 5 \\ x - 4y = 7 \end{cases}$ | Solve for $y$ in the 1$^{st}$ equation: $y = -2x + 5$ Substitute into 2$^{nd}$ equation and solve: $x - 4(-2x + 5) = 7$ $x + 8x - 20 = 7$ $9x - 20 = 7$ $9x = 27$ $x = 3$ Find the missing coordinate of the ordered pair: $y = -2(3) + 5$ $= -6 + 5 = -1$ $(3, -1)$ | |
| 2. | $\begin{cases} x + 8y = 20 \\ 4x - y = 14 \end{cases}$ | | |

| | System of Equations | Solve for a variable in the FIRST equation, then substitute and solve. | Solve for a variable in the SECOND equation, then substitute and solve. |
|---|---|---|---|
| **3.** | $\begin{cases} 3x - y = -2 \\ -12x + 4y = 16 \end{cases}$ | | |
| **4.** | $\begin{cases} 3x + 4y = 0 \\ \dfrac{3}{4}x + \dfrac{8}{3}y = -\dfrac{5}{12} \end{cases}$ | | |
| **5.** | $\begin{cases} x - 7y = -1 \\ -\dfrac{1}{7}x + y = \dfrac{1}{7} \end{cases}$ | | |

# Student Activity
## Forced Elimination

**Directions:** In order to develop some intuition about which variable is easier to eliminate, we will **start** a few problems by trying both of the possible eliminations. Use the elimination method to solve for a variable. **Then circle the elimination that was easier.** The first one has been started for you.

1.

| Eliminate $x$ | Eliminate $y$ |
|---|---|
| $\begin{cases} x + 3y = 7 \\ 2x - 3y = -4 \end{cases}$ <br> Multiply the first equation by $-2$. <br> $\begin{cases} -2x - 6y = -14 \\ 2x - 3y = -4 \end{cases}$ <br> $-9y = -18$ <br> $y = 2$ | $\begin{cases} x + 3y = 7 \\ 2x - 3y = -4 \end{cases}$ |

2.

| Eliminate $x$ | Eliminate $y$ |
|---|---|
| $\begin{cases} -10x + y = -5 \\ 10x - 2y = 0 \end{cases}$ | $\begin{cases} -10x + y = -5 \\ 10x - 2y = 0 \end{cases}$ |

3.

| Eliminate $x$ | Eliminate $y$ |
|---|---|
| $\begin{cases} 5x + 7y = 12 \\ 10x - 3y = 7 \end{cases}$ | $\begin{cases} 5x + 7y = 12 \\ 10x - 3y = 7 \end{cases}$ |

**4.**

| Eliminate $x$ | Eliminate $y$ |
|---|---|
| $\begin{cases} 3x + y = 2 \\ 4x - 2y = 6 \end{cases}$ | $\begin{cases} 3x + y = 2 \\ 4x - 2y = 6 \end{cases}$ |

Now inspect these systems of equations. For each system, decide which variable will be easier to eliminate and describe the steps that are necessary to achieve this. Problem 5 has been done for you.

**5.** $\begin{cases} 3x + y = 1 \\ 2x + 3y = -11 \end{cases}$  Eliminate $y$.

Multiply the first equation by $-3$ and then add the equations.

**6.** $\begin{cases} 5x - y = -5 \\ x + y = 5 \end{cases}$

**7.** $\begin{cases} x + y = 12 \\ -x + y = 0 \end{cases}$

**8.** $\begin{cases} -4x + 2y = 2 \\ 4x + 6y = 22 \end{cases}$

**9.** $\begin{cases} 3x + 14y = -7 \\ 2x + 2y = 12 \end{cases}$

**10.** $\begin{cases} 3x + 8y = 2 \\ -6x - 7y = 5 \end{cases}$

# Student Activity
## Triple the Fun on Systems of Equations

**Directions:** For each of the problems below, solve the system of equations by graphing, by substitution, and by elimination – your solution should be the same for all three methods. If they are not, go back and look for a mistake in your work.

**1.**

**System of Equations:**
$$\begin{cases} 10x + 5y = 15 \\ x + 2y = -6 \end{cases}$$

**Solve by Graphing:**

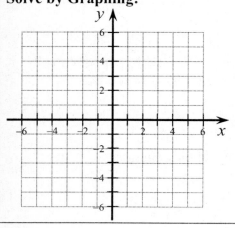

**Solve by Substitution**

**Solve by Elimination**

**2.**

**System of Equations:**
$$\begin{cases} \dfrac{1}{4}x - y = -\dfrac{1}{3} \\ 3x - 12y = -12 \end{cases}$$

**Solve by Graphing:**

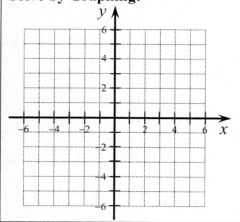

**Solve by Substitution**

**Solve by Elimination**

**3.**

**System of Equations:**
$$\begin{cases} 3y - 4x = 2 \\ 3x - 4y = -5 \end{cases}$$

**Solve by Graphing:**

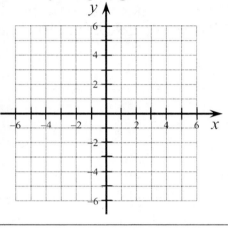

**Solve by Substitution**

**Solve by Elimination**

**4.**

**System of Equations:**
$$\begin{cases} -4x + y = -5 \\ 12x - 3y = 15 \end{cases}$$

**Solve by Graphing:**

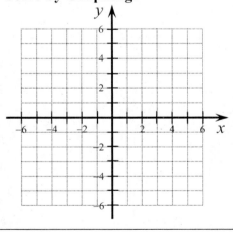

**Solve by Substitution**

**Solve by Elimination**

# Student Activity
## Choose Your Tactic

**Directions:** For each system of equations below, decide whether it will be easier to solve using substitution or elimination (choose a tactic). Then describe your tactical plan: For **substitution,** which variable in which equation will you solve for? For **elimination** – which variable will you eliminate and how?

| | System of Equations | Tactic<br>Substitution or Elimination | Tactical Plan<br>Substitution: Which variable will you solve for?<br>Elimination: Which variable will you eliminate? |
|---|---|---|---|
| **1.** | $\begin{cases} 5x + 3y = -9 \\ y = 2x + 8 \end{cases}$ | | |
| **2.** | $\begin{cases} 2x + y = 9 \\ 5x + 3y = 26 \end{cases}$ | | |
| **3.** | $\begin{cases} y = -x \\ 6x + 6y = 0 \end{cases}$ | | |
| **4.** | $\begin{cases} 4x + 11y = 7 \\ 4x + 3y = -1 \end{cases}$ | | |
| **5.** | $\begin{cases} 16x - 2y = 16 \\ 4x = 2y - 8 \end{cases}$ | | |
| **6.** | $\begin{cases} 0.02x + 0.01y = 0.1 \\ -2x + 3y = -18 \end{cases}$ | | |
| **7.** | $\begin{cases} -\frac{x}{5} - \frac{y}{3} = 2 \\ -\frac{3x}{10} + \frac{2y}{10} = \frac{9}{10} \end{cases}$ | | |
| **8.** | $\begin{cases} x = 12y - 7 \\ 12y - x = 12 \end{cases}$ | | |

Hold the line men, they're bringing in the -1 .

# Student Activity
## The Last Sentence

**Directions:** In each example below you are given **the last sentence** of an application problem. In many cases, it is possible to declare the variables in the problem just by looking at this sentence. Take your best guess about what the variables should be based on the sentences you are given. The first one has been done for you.

1. Find the length and width of the painting.

> Let $L$ = the length of the painting.
> Let $W$ = the width of the painting.

2. Determine much Maria spent on the rental car and on the hotel.

3. How much money was invested in each stock?

4. How many pounds of peanuts and cashews are required to make the snack mix?

5. Find the speed of the plane and the speed of the wind.

6. Determine how many liters of acid and how many liters of water are required to make the required batch.

7. What base and height measurements give the triangle an area of 12 ft$^2$?

8. Find the cost of each adult and child ticket.

# Guided Learning Activity
## Using the Problem Solving Strategy

**Problem A:** At a scrapbooking party, customers can purchase two special packages, the vacation scrapbook package and the baby album package. Shaude purchases 2 vacation packages and one baby album package for a total of $93.50. Stephan purchases one vacation package and three baby album packages for a total of $138. Find the cost of each of the special packages that were offered at the party

1. **Analyze.** Draw diagrams, tables, and charts to help make sense of the problem.
2. **Declare.** Declare the variables and the quantitative givens in the problem.
3. **Formulate.** Write equations to represent the problem.
4. **Solve.** Solve the equation or system of equations.
5. **Validate.** Check the result. Does it make sense? Is it reasonable?
6. **Conclude.** State the conclusion. Don't forget units.

**Problem B:** Jasper invests a total of 250,000 L$ (Linden dollars). He splits his investment between a 30-day Linden Bank CD that pays 7% interest at the end of the term and a 30-day Second Life real estate investment scheme organized by his brother that ends up losing 4% of its original value at the end of 30 days. At the end of the 30 days, Jasper makes 2,100 L$. How much did Jasper invest in the CD and how much did he invest in the real estate scheme?

---

1. **Analyze.** Draw diagrams, tables, and charts to help make sense of the problem.
2. **Declare.** Declare the variables and the quantitative givens in the problem.
3. **Formulate.** Write equations to represent the problem.
4. **Solve.** Solve the equation or system of equations.
5. **Validate.** Check the result. Does it make sense? Is it reasonable?
6. **Conclude.** State the conclusion. Don't forget units

**Problem C:** Technical grade hydrogen peroxide is 35% $H_2O_2$. Common household hydrogen peroxide is only 3% $H_2O_2$. Imogen wants to make a 10% $H_2O_2$ solution to use in removing a toxic mold in a city building. What amount of the technical grade solution and what amount of the household solution (to the nearest 0.1 mL) should she mix to create the 500 mL of solution she desires?

---

1. **Analyze.** Draw diagrams, tables, and charts to help make sense of the problem.
2. **Declare.** Declare the variables and the quantitative givens in the problem.
3. **Formulate.** Write equations to represent the problem.
4. **Solve.** Solve the equation or system of equations.
5. **Validate.** Check the result. Does it make sense? Is it reasonable?
6. **Conclude.** State the conclusion. Don't forget units

---

# Student Activity
## Setting Up the System

**Directions:** Read each problem carefully.
a) **Declare** the variables you would use to solve the problem.
b) **Formulate** a system of two equations that could be used to solve each problem. You do not need to solve the application problems.

1.  The elevation in Denver, Colorado, is 13 times greater than that of Waikapu, Hawaii. The sum of their elevations is 5,600 feet. Find the elevation of each city.

2.  The owner of a pet food store wants to mix birdseed that costs $1.25 per pound with sunflower seeds that cost $0.75 per pound to make 50 pounds of a mixture that costs $1.00 per pound. How many pounds of each type of seed should he use?

3.  An airplane can fly 500 miles in 2 hours when flying with the wind. The same journey takes 4 hours when flying against the wind. Find the speed of the wind and the speed of the plane.

4.  Barb invested a total of $25,000 in two accounts. One account has an annual interest rate of 3.5%, while the other has an annual rate of 6%. She earned $1,257 in interest in one year. How much did she invest in each account?

5.  Two angles are supplementary. The measure of the first angle is 10 degrees more than three times the second angle. Find the measure of each angle.

# Student Activity
## The Moving Walkway Problem

A moving walkway is like an escalator (transporting you from one place to another), only it is a flat belt that you can walk on. Moving walkways can

Normal walking
Rate = 3 mph

Rate of moving walkway = 1.4 mph

commonly be found now in airports, museums, amusement parks, or zoos.  When you walk onto a moving walkway going in your direction and continue to walk, your walking speed is increased by the speed of the walkway.

**1.** Using the information from the diagram above, fill in the missing rates.

Person on moving walkway
(walking with the flow)

Rate = _____ mph

Person on moving walkway
(walking against the flow)

Rate = _____ mph

**2.** Often, you don't know one of the rates in a problem.  In these cases, you have to express the rate as an algebraic expression.  Assume the person is jogging at a rate of 4.7 mph and the rate of the moving walkway is $x$ mph.  Fill in the missing rates.

Person on moving walkway
(walking with the flow)

Rate = _____ mph

Person on moving walkway
(walking against the flow)

Rate = _____ mph

**3.** Now the rate of the person is $R$ mph and the rate of the walkway is 2.3 mph.  Fill in the missing rates.

Person on moving walkway
(walking with the flow)

Rate = _____ mph

Person on moving walkway
(walking against the flow)

Rate = _____ mph

**4.** At the Montparnasse station in Paris, there is a high-speed moving walkway that was designed to help commuters get through the station faster. This moving walkway, which is 0.18 km long, moves at a speed of 9 km/hr (originally the walkway was designed to go 11 km/hr, but there were too many accidents). David walks at his normal pace in the direction of flow on the walkway – it takes 54 seconds to go the full distance of the walkway. How fast was David walking?

**5.** While on a layover at the Detroit International Airport, a bored math teacher decides to calculate the speed of one of the many moving walkways in the airport. He knows (from years of treadmill walking) that her normal walking speed is 2.5 mph (or 2.2 ft/sec). Walking with the flow of the walkway, it takes 43 seconds to go the distance of the walkway. Walking against the flow of the walkway, it takes 2 minutes and 47 seconds. What is the approximate rate of the moving walkway in feet per second and what is the length of the walkway in feet?

|  | Rate | Time | Distance |
|---|---|---|---|
| With Flow of Walkway |  |  |  |
| Against Flow of Walkway |  |  |  |

**6.** The Central-Mid-Levels Escalator in Hong Kong is the world's longest covered escalator, stretching 800m (2,625 ft). A trip on the escalator, from top to bottom, takes about 20 minutes when a person stands still. What is the approximate rate of the escalator in feet per second? How long would the trip take if the person walked downward at a rate of 2 ft/sec?

# Student Activity
## Tic-tac-toe on Inequalities

**Directions for Tic-tac-toe #1:** If the ordered pair in the square **IS** a solution of the system of inequalities shown below, then circle the ordered pair (thus putting an **O** on the square).  If it **IS NOT** a solution, then put an **X** over the ordered pair.

System of Inequalities: $\begin{cases} y < 4x + 2 \\ 3x - 3y \geq 6 \end{cases}$

| | | |
|---|---|---|
| $(-2,-6)$ | $(1,1)$ | $(4,1)$ |
| $(6,2)$ | $(8,6)$ | $(4,4)$ |
| $(2,0)$ | $(0,2)$ | $(0,-2)$ |

**Directions for Tic-tac-toe #2:** If the ordered pair in the square **IS** a solution of the graphed system of inequalities, then circle the ordered pair (thus putting an **O** on the square).  If it **IS NOT** a solution, then put an **X** over the ordered pair.

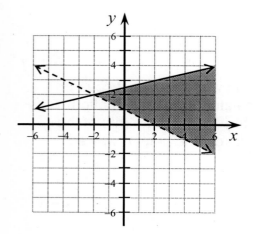

| | | |
|---|---|---|
| $(-2,2)$ | $(2,0)$ | $(2,3)$ |
| $(8,3)$ | $(3,4)$ | $(0,4)$ |
| $(6,4)$ | $(6,-2)$ | $(4,0)$ |

# Student Activity
## Testing Test Points

**Directions:** In each problem below you are given a system of inequalities, and the graphed inequality boundary lines. In each region separated by the boundary lines, a test point is provided. Test this point in both inequalities to determine if it is true (T) or false (F). Shade the region of the graph where the ordered pair **is** a solution to both inequalities. The first test point has been done for you.

**1.**

|  | $y < \frac{1}{2}x$ | $y > 2x - 8$ |
|---|---|---|
| $(0,3)$ | F | T |
| $(2,-2)$ |  |  |
| $(-5,0)$ |  |  |
| $(-4,-4)$ |  |  |

**3.**

|  | $y < \frac{2}{3}x + 4$ | $y > \frac{2}{3}x$ |
|---|---|---|
| $(-6,3)$ |  |  |
| $(-3,1)$ |  |  |
| $(3,-2)$ |  |  |

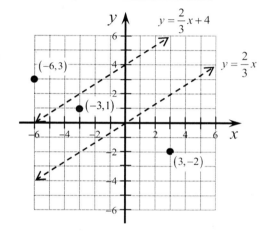

**2.**

|  | $y \leq -\frac{1}{3}x + 2$ | $y \geq -\frac{3}{2}x - \frac{3}{2}$ |
|---|---|---|
| $(-6,5)$ |  |  |
| $(-6,1)$ |  |  |
| $(6,4)$ |  |  |
| $(6,-2)$ |  |  |

**4.**

|  | $y \geq -\frac{5}{4}x + \frac{7}{2}$ | $y \leq -\frac{5}{4}x - \frac{5}{2}$ |
|---|---|---|
| $(-2,-3)$ |  |  |
| $(0,0)$ |  |  |
| $(2,4)$ |  |  |

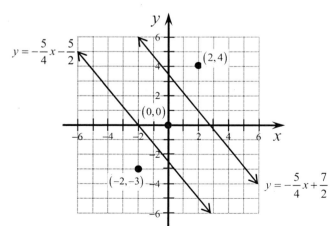

# Guided Learning Activity
## Graphing Systems of Linear Inequalities

**Example:** Consider the system of inequalities given by $y \geq -\dfrac{4}{3}x + 2$ and $y > \dfrac{5}{6}x + 1$.

What does the solution of a system of inequalities look like? Separately, the graphs of our inequalities look like this:

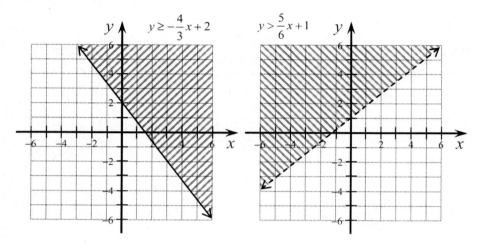

When you graph the inequalities on the same set of axes, the solution is the region where the two graphs intersect. This region of intersection (shown in solid gray) with the appropriate boundary lines is the solution to the system of inequalities.

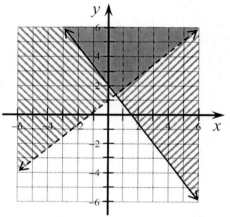

The final solution to this system of inequalities is shown below.

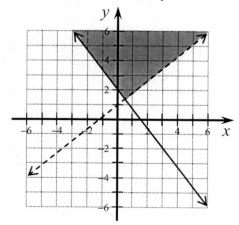

# Now try your hand at graphing the solutions to these systems of inequalities.

**1.** $\begin{cases} y < -2x + 3 \\ y < 2x - 3 \end{cases}$

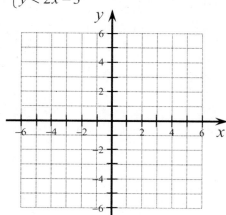

**4.** $\begin{cases} 3x + y \geq -5 \\ -3x - y \geq -5 \end{cases}$

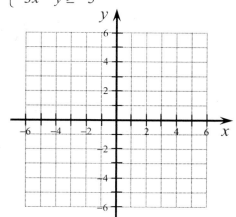

**2.** $\begin{cases} 2 - y \leq 0 \\ y - \frac{1}{2}x \leq 4 \end{cases}$

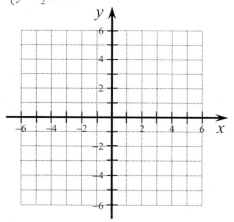

**5.** $\begin{cases} y \leq 0 \\ x \leq 0 \end{cases}$

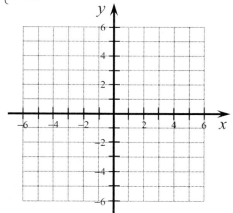

**3.** $\begin{cases} y > \frac{1}{3}x + 1 \\ y < \frac{1}{3}x - 3 \end{cases}$

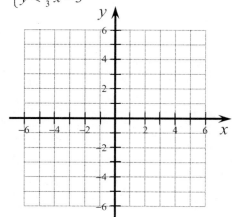

**6.** $\begin{cases} y \leq -\frac{1}{2}x + 2 \\ y > 2x + 2 \end{cases}$

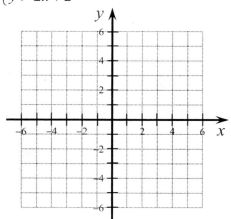

# Student Activity
## Following the Clues Back to the System of Inequalities

**Directions:** In each "crime-scene" below, you are shown the graph of a system of inequalities. Use your mathematical powers of reasoning (and detective skills) to determine what the inequality must have been to result in this graph.

1.

2.

3.

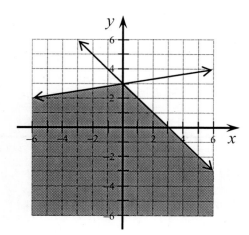

# Assess Your Understanding
## Solving Systems

For each of the following, describe the strategies or key steps that will help you **start** the problem.  You do **not** have to complete the problems.

| | | **What will help you to start this problem?** |
|---|---|---|
| **1.** | Use substitution to solve the system: $$\begin{cases} x - 3y = -1 \\ 3x - 4y = 7 \end{cases}$$ | |
| **2.** | Is the system below consistent or inconsistent? $$\begin{cases} x - 4y = 6 \\ 2x = 8y + 10 \end{cases}$$ | |
| **3.** | Is $(1, -2)$ a solution to the system? $$\begin{cases} 3x + y = 1 \\ x - y = -1 \end{cases}$$ | |
| **4.** | Use elimination to solve the system: $$\begin{cases} x + 3y = -2 \\ x + 4y = 0 \end{cases}$$ | |
| **5.** | Solve the system. $$\begin{cases} x = y - 7 \\ x - 2y = 14 \end{cases}$$ | |
| **6.** | Is the pair of equations dependent or independent? $$\begin{cases} x - 4y = 6 \\ 2x = 8y + 10 \end{cases}$$ | |
| **7.** | Solve the system by graphing: $$\begin{cases} y = -2x + 4 \\ y = 2x \end{cases}$$ | |
| **8.** | Solve the system using the method of your choice: $$\begin{cases} 3x - 4y = 7 \\ 5x - 8y = 8 \end{cases}$$ | |

| | | **What will help you to start this problem?** |
|---|---|---|
| **9.** | Is $(1,-3)$ a solution to the system? $$\begin{cases} 3x+y<1 \\ x-y>-1 \end{cases}$$ | |
| **10.** | Rafi paddles his canoe upstream 24 miles in 4 hours. If he can make the downstream trip in 3 hours, find the rate that Rafi can paddle and the rate of the current. | |
| **11.** | Solve the system by the method of your choice: $$\begin{cases} y-3x=2 \\ 4x-y=-1 \end{cases}$$ | |
| **12.** | Two angles are supplementary and one angle measures $30°$ more than the other angle. Find the measure of each angle. | |
| **13.** | Graph the solution to the system: $$\begin{cases} x\leq 3 \\ y\leq \frac{1}{2}x-1 \end{cases}$$ | |
| **14.** | Is the pair of equations dependent or independent? $$\begin{cases} 3x+y=8 \\ 6x+2y=8 \end{cases}$$ | |
| **15.** | Graph the solution to the system: $$\begin{cases} 3x+y<4 \\ 2y-x>-2 \end{cases}$$ | |

# Metacognitive Skills
## Solving Systems

Metacognitive skills refer to the ability to judge how well you have learned something and to effectively direct your own learning and studying. This is a self-evaluation tool designed to help you focus your studying and to improve your metacognitive skills with regards to this math class.

Fill the 1ˢᵗ column out **before** you begin studying. Fill the 2ⁿᵈ column out after you study for your test.

Go back to this assessment after your test and circle any of the ratings that you would change – this identifies the "disconnects" between what you **thought** you knew well and what you **actually** knew well.

**Use the scale below to assign a number to each topic.**
5   *I am confident I can do any problems in this category correctly.*
4   *I am confident I can do most of the problems in this category correctly.*
3   *I understand how to do the problems in this category, but I still make a lot of mistakes.*
2   *I feel unsure about how to do these problems.*
1   *I know I don't understand how to do these problems.*

| Topic or Skill | Before Studying | After Studying |
|---|---|---|
| Checking whether an ordered pair is a solution of a system of equations. | | |
| Classifying a system of equations as consistent or inconsistent. | | |
| Classifying a system of equations as dependent or independent. | | |
| Graphing a system of equations to find the solution. | | |
| Understanding what the "solution" to a system of equations is. | | |
| Finding or identifying a substitution equation for a system of equations. | | |
| Solving a system of equations by substitution. | | |
| Clearing the fractions from an equation or inequality. | | |
| Identifying a variable to eliminate and performing the steps to set up that elimination in a system of equations. | | |
| Solving a system of equations by elimination. | | |
| Deciding whether substitution or elimination would be an easier method for solving a system of equations. | | |
| Clearing the decimals from an equation or inequality. | | |
| Declaring the variables for an application involving a system of equations. | | |
| Writing the system of equations for a problem involving $d = rt$. | | |
| Writing the system of equations for a value-mixture problem or a percent-solution problem. | | |
| Writing the system of equations for an investment problem. | | |
| Solving a system of equations from an application problem. | | |
| Writing the conclusion to an application problem involving a system of equations. | | |
| Checking whether an ordered pair is a solution of a system of inequalities. | | |
| Determining whether the boundary lines for a system of inequalities are dashed or solid. | | |
| Graphing the boundary lines for a system of inequalities. | | |
| Determining the proper region to shade in a system of inequalities. | | |

# SYS2: More on Solving Systems

# Guided Learning Activity

## Systems Island

### Example 1:

**HOW TO SOLVE:**
A system of three equations
with three variables.

⟹

**GOAL:** A system of
two equations
with two variables.

$$\begin{cases} x + 3y - z = -6 \\ 2x + y + z = 8 \\ 3x - y - 2z = -3 \end{cases}$$

Choose a variable
to eliminate.

$$\begin{cases} 3x + 4y = 2 \\ 7x + y = 13 \end{cases}$$

Pick a different pair of
equations and eliminate
the **same** variable.

$$\begin{cases} x + 3y - z = -6 \\ 2x + y + z = 8 \\ 3x - y - 2z = -3 \end{cases}$$

Let's eliminate z.

Pick any two equations
and eliminate the
chosen variable.

$$\begin{cases} x + 3y - z = -6 \quad A \\ 2x + y + z = 8 \quad B \\ 3x - y - 2z = -3 \quad C \end{cases}$$

Add 2 times
equation B
and equation C.

$$7x + y = 13$$

We eliminated z.

$$\begin{cases} x + 3y - z = -6 \quad A \\ 2x + y + z = 8 \quad B \\ 3x - y - 2z = -3 \quad C \end{cases}$$

Add equations
A and B.

$$3x + 4y = 2$$

We eliminated z.

*There you have it viewers! Equations one and two will be combined and the z's will be eliminated from the competition!*

**Example 2:**

HOW TO SOLVE:
A system of three equations with three variables when there are missing variables.

GOAL: A system of two equations with two variables.

$$\begin{cases} x & - & 2y & & & = & 1 \\ 2x & + & y & + & z & = & 1 \\ & & 3y & + & 2z & = & -9 \end{cases}$$

$$\begin{cases} 3y & + & 2z & = & -9 \\ 5y & + & z & = & -1 \end{cases}$$

Choose a variable to eliminate. If the system is missing a variable in one of the equations, that would be an easy choice.

Pick a pair of equations and eliminate the **same** variable.

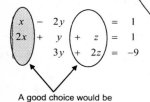

$$\begin{cases} x & - & 2y & & & = & 1 \\ 2x & + & y & + & z & = & 1 \\ & & 3y & + & 2z & = & -9 \end{cases}$$

A good choice would be choosing either x or z to eliminate. Let's choose x.

Notice that one of the equations already has the chosen variable eliminated.

$$\begin{cases} x & - & 2y & & & = & 1 \\ 2x & + & y & + & z & = & 1 \\ & & 3y & + & 2z & = & -9 \end{cases}\begin{matrix} A \\ B \\ C \end{matrix}$$

Add -2 times equation A and equation B.

$$5y + z = -1$$

↑
We eliminated x.

$$\begin{cases} x & - & 2y & & & = & 1 \\ 2x & + & y & + & z & = & 1 \\ & & 3y & + & 2z & = & -9 \end{cases}$$

x was already eliminated.

$$3y + 2z = -9$$

**Directions:** For each system of three equations, use linear combinations to get a system of two equations by eliminating the chosen variable. **Write the system of two equations in the box for each problem.**

**1.** Solve by eliminating **z**:

$$\begin{cases} x & - & 3y & + & z & = & 2 \\ 2x & + & 4y & - & z & = & 10 \\ 3x & - & 4y & + & 2z & = & 8 \end{cases}\begin{matrix} A \\ B \\ C \end{matrix}$$

**a.** Use a combination of A and B to eliminate z.

System of Two Equations:

**b.** Use a combination of B and C to eliminate z.

**2.** Solve by eliminating **y:**
$$\begin{cases} 2x + y + z = -1 & A \\ x + 2y + 3z = 5 & B \\ x - y - 4z = -2 & C \end{cases}$$

**a.** Use a combination of A and C to eliminate y.

System of Two Equations:

**b.** Use a combination of B and C to eliminate y.

**3.** Solve by eliminating **x:**
$$\begin{cases} 2x + y + z = -1 & A \\ 4x - y - z = 6 & B \\ y + 3z = -2 & C \end{cases}$$

**a.** Write the equation that already has x eliminated.

System of Two Equations:

**b.** Use a combination of A and B to eliminate x.

**Still have time?** Finish solving problems **2** and **3**.

# Student Activity
## Plane Facts

The graph of a linear equation in three variables, such as $2x + 3y - 5z = 8$, is a flat surface called a **plane**. It is very difficult to imagine graphing in 3-D on 2-dimensional paper, but we'll give it a try for this activity. When we solve a system of three linear equations with three variables, there are many possible ways the three planes could be positioned in 3-dimensional space (each equation represents a plane).

Recall that the solution of a system of three equations is an ordered triple of numbers that satisfies each equation. Graphically, a point is a solution of a system of three equations in three variables if it lies on **each** of the corresponding planes. Below we show five ways that three planes could be positioned in space.

**Directions:** For each case, decide on the **number** of possible solutions (one, none, or infinitely many) and then use a colored pencil to indicate where the solution point (or points) lies.

**1.**

**2.**

**3.**

**4.**

**5.**

**6.** Which of the graphs represent a system that is consistent?

**7.** Which of the graphs represent a set of equations that are dependent?

# Student Activity
## Strange Systems

**Directions:** For each system of equations, **decide** what makes the system "strange" and then decide on a strategy for solving the system. **Carry out enough steps** to transform the system from a "strange" one to one that looks standard.

**1.** Solve: $\begin{cases} 3x + 4y - 2z = 2 \\ 6x - 4y + z = 0 \\ \qquad\qquad\quad 5z = 5 \end{cases}$

**2.** Solve: $\begin{cases} 0.02x + 0.1y + 0.02z = 0 \\ x + 10y + 3z = 1 \\ \dfrac{x}{2} + \dfrac{z}{8} = -\dfrac{1}{2} \end{cases}$

**3.** Solve: $\begin{cases} x + z = -y \\ y - 3 = z \\ 10 + 3z = 2y \end{cases}$

**4.** Solve: $\begin{cases} x + y = 0 \\ \qquad y + 2z = 3 \\ x \qquad - z = -1 \end{cases}$

**Still have time?** Go ahead and solve the systems using another sheet of paper.

# Student Activity
## Retirement Allocation

Last year, at her annual appointment with her financial advisor, Jackie agreed to reallocate a portion of the $500,000 that she had saved for retirement so that less money was allocated to high-risk investments. The high-risk account was projected to make 8% annual interest, a medium-risk account was projected to earn 6%, and the "safe" low-risk account was projected to earn 4.5%. Jackie and her advisor split the $500,000 into these three accounts. Looking at the performance statements at the end of the year, Jackie sees that her retirement funds now total $530,250. Jackie also notices that the interest earned by the money in the medium-risk account was twice the interest earned by the money in the low-risk account. How much of Jackie's retirement fund was allocated into each account at the beginning of last year?

**Define three variables:**

Let $H$ = _____.

Let $M$ = _____.

Let $L$ = _____.

**Write three equations:**

Total invested:

Interest earned:

Mid-risk / low-risk relationship:

**Solve the system:**

**Write a sentence to describe the solution:**

# Student Activity
## Blog Readers

Patrick is looking through the analytics on the number of readers for his popular math blog "From the Origin." In the blog statistics, Patrick can keep track of his readers from blog hits from inside the U.S., blog hits from outside the U.S., and blog hits from an RSS reader. For the month of January, he had a total of 4400 readers. The number of readers with blog hits from the United States was equal to the sum of the blog hits from readers outside the U.S. and the readers who used an RSS feed to read the blog. The number of RSS readers is 200 less than twice the number of blog readers from outside the U.S. Find the number of readers with blog hits from the U.S., the number of readers with blog hits from outside the U.S., and the number of readers who used an RSS feed.

Define three variables:

Let $x =$ _____ .

Let $y =$ _____ .

Let $z =$ _____ .

Write three equations:

Solve:

Write a sentence to describe the solution:

187

# Guided Learning Activity
## Fitting a Quadratic Curve

The equation of a parabola is of the form $y = ax^2 + bx + c$.

If we are given a graph of the parabola, we can use points from the graph to determine the equation of the parabola by setting up a system of three equations and three unknowns.

Begin by writing down the coordinates of **three** of the points that are indicated on the graph to the right.

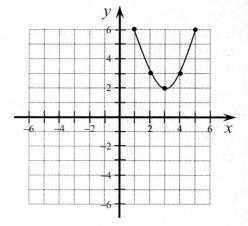

(__,__)   (__,__)   (__,__)

Now we write a "skeleton" of parentheses for the equation of a parabola:

$$y = ax^2 + bx + c$$
$$(\quad) = a(\quad)^2 + b(\quad) + c$$

Use the *x*- and *y*-coordinates of each point to complete an equation

(__,__)      $(\quad) = a(\quad)^2 + b(\quad) + c$      ___ $a$ + ___ $b$ + ___ $c$ = ___

(__,__)      $(\quad) = a(\quad)^2 + b(\quad) + c$      ___ $a$ + ___ $b$ + ___ $c$ = ___

(__,__)      $(\quad) = a(\quad)^2 + b(\quad) + c$      ___ $a$ + ___ $b$ + ___ $c$ = ___

Solve the system of three equations and three variables to find *a*, *b*, and *c*:

Write the equation of the parabola: $y = $ ___ $x^2 + $ ___ $x + $ ___

# Student Activity
## Tic-Tac-Toe on Row Echelon Form

A matrix is in **row echelon form** if it has 1's down its main diagonal and a zero below its main diagonal.

**Directions:** If the matrix in the square is in row echelon form, then circle it (put an **O** on the square).  If the matrix is **not** in row echelon form, then mark it with an **X**.

| | | |
|---|---|---|
| $\begin{bmatrix} 1 & 3 & \vert & 4 \\ 0 & 2 & \vert & 6 \end{bmatrix}$ | $\begin{bmatrix} 1 & 3 & \vert & 5 \\ 0 & 1 & \vert & 4 \end{bmatrix}$ | $\begin{bmatrix} 1 & 2 & 5 & \vert & -7 \\ 0 & 1 & 2 & \vert & 15 \\ 0 & 0 & 1 & \vert & \frac{1}{2} \end{bmatrix}$ |
| $\begin{bmatrix} 1 & 0 & 5 & \vert & 2 \\ 0 & 1 & 0 & \vert & 3 \\ 0 & 0 & 1 & \vert & 6 \end{bmatrix}$ | $\begin{bmatrix} 1 & 0 & 4 & \vert & 5 \\ 0 & -1 & 2 & \vert & 6 \\ 0 & 0 & 1 & \vert & 0 \end{bmatrix}$ | $\begin{bmatrix} 1 & 0 & 4 & \vert & 5 \\ 0 & 1 & 2 & \vert & 0 \\ 0 & 0 & 1 & \vert & 0 \end{bmatrix}$ |
| $\begin{bmatrix} 1 & 0 & \vert & \frac{3}{4} \\ 0 & 1 & \vert & \frac{1}{2} \end{bmatrix}$ | $\begin{bmatrix} 1 & 2 & \vert & 3 \\ 1 & 0 & \vert & 4 \end{bmatrix}$ | $\begin{bmatrix} 4 & 1 & \vert & 2 \\ 1 & 0 & \vert & -3 \end{bmatrix}$ |

# Student Activity

## Rowing to Freedom

**Directions:** Begin with the matrix in the box marked "Start". Follow each of the given row operations and shade the box that contains your result for each one (each row operation is carried out on the previous answer). At the "pit stop" you'll get a new matrix to use.

You can escape this matrix by "rowing" your way to freedom!

1. $-2R_2 \rightarrow R_2$

2. $R_2 + R_3 \rightarrow R_3$

3. $2R_1 + R_2 \rightarrow R_2$

4. $-\frac{3}{4}R_2 \rightarrow R_2$

5. $R_2 + R_3 \rightarrow R_3$

6. $-\frac{1}{3}R_2 \rightarrow R_2$ **(P.S.)**

7. $R_1 \leftrightarrow R_3$

8. $R_1 + (-2)R_2 \rightarrow R_2$

9. $R_2 + 2R_3 \rightarrow R_3$

10. $-1R_3 \rightarrow R_3$

11. $-\frac{1}{2}R_2 \rightarrow R_2$

12. $\frac{1}{4}R_1 \rightarrow R_1$

| | | | |
|---|---|---|---|
| **START** $\begin{bmatrix} 1 & 1 & 1 & 0 \\ 1 & -1 & 1 & -6 \\ 2 & 1 & 3 & -2 \end{bmatrix}$ | $\begin{bmatrix} 1 & 1 & 1 & 0 \\ -2 & 2 & -2 & 12 \\ 2 & 1 & 3 & -2 \end{bmatrix}$ | $\begin{bmatrix} 1 & 1 & 1 & 0 \\ -2 & 2 & -2 & 12 \\ 0 & 3 & 1 & 10 \end{bmatrix}$ | $\begin{bmatrix} 1 & 1 & 1 & 0 \\ 0 & 4 & 0 & 12 \\ 0 & 3 & 1 & 10 \end{bmatrix}$ |
| $\begin{bmatrix} 1 & 1 & 1 & 0 \\ 1 & -1 & 1 & -6 \\ -4 & -2 & -6 & 4 \end{bmatrix}$ | $\begin{bmatrix} -1 & 3 & -1 & 12 \\ -2 & 2 & -2 & 12 \\ 2 & 1 & 3 & -2 \end{bmatrix}$ | $\begin{bmatrix} 1 & 1 & 1 & 0 \\ 0 & -1 & 0 & -3 \\ 0 & 0 & 1 & 1 \end{bmatrix}$ | $\begin{bmatrix} 1 & 1 & 1 & 0 \\ 0 & -3 & 0 & -9 \\ 0 & 3 & 1 & 10 \end{bmatrix}$ |
| $\begin{bmatrix} 4 & 2 & 1 & 5 \\ 2 & 2 & 2 & 2 \\ 0 & 1 & 1 & 1 \end{bmatrix}$ | **PIT STOP** $\begin{bmatrix} 0 & 1 & 1 & 1 \\ 2 & 2 & 2 & 2 \\ 4 & 2 & 1 & 5 \end{bmatrix}$ | $\begin{bmatrix} 1 & 1 & 1 & 0 \\ 0 & 1 & 0 & 3 \\ 0 & 0 & 1 & 1 \end{bmatrix}$ | $\begin{bmatrix} 1 & 1 & 1 & 0 \\ 0 & -3 & 0 & -9 \\ 0 & 0 & 1 & 1 \end{bmatrix}$ |
| $\begin{bmatrix} 4 & 2 & 1 & 5 \\ 0 & -2 & -3 & 1 \\ 0 & 1 & 1 & 1 \end{bmatrix}$ | $\begin{bmatrix} 0 & 1 & 1 & 1 \\ 2 & 2 & 2 & 2 \\ 4 & 3 & 2 & 6 \end{bmatrix}$ | $\begin{bmatrix} 1 & 1 & 1 & 0 \\ 0 & 1 & 0 & 0 \\ 0 & 0 & 1 & 1 \end{bmatrix}$ | $\begin{bmatrix} 1 & 1 & 1 & 0 \\ 0 & 3 & 0 & 9 \\ 0 & 0 & 1 & 1 \end{bmatrix}$ |
| $\begin{bmatrix} 4 & 2 & 1 & 5 \\ 0 & -2 & -3 & 1 \\ 0 & 0 & -1 & 3 \end{bmatrix}$ | $\begin{bmatrix} 4 & 2 & 1 & 5 \\ 0 & 1 & \frac{3}{2} & -\frac{1}{2} \\ 0 & 0 & 1 & 0 \end{bmatrix}$ | $\begin{bmatrix} 1 & \frac{1}{2} & \frac{1}{4} & \frac{5}{4} \\ 0 & 1 & \frac{3}{2} & 0 \\ 0 & 0 & 1 & 0 \end{bmatrix}$ | $\begin{bmatrix} 1 & \frac{1}{2} & \frac{1}{4} & 0 \\ 0 & 1 & \frac{3}{2} & 0 \\ 0 & 0 & 1 & 0 \end{bmatrix}$ |
| $\begin{bmatrix} 4 & 2 & 1 & 5 \\ 0 & -2 & -3 & 1 \\ 0 & 0 & 1 & -3 \end{bmatrix}$ | $\begin{bmatrix} 4 & 2 & 1 & 5 \\ 0 & 1 & \frac{3}{2} & -\frac{1}{2} \\ 0 & 0 & 1 & -3 \end{bmatrix}$ | $\begin{bmatrix} 1 & \frac{1}{2} & \frac{1}{4} & \frac{5}{4} \\ 0 & 1 & \frac{3}{2} & -\frac{1}{2} \\ 0 & 0 & 1 & -3 \end{bmatrix}$ | **You've successfully "rowed" to freedom!** |

# Guided Learning Activity
**Parallel Operations**

**Directions:** Follow the steps to solve the system of equations, first by elimination, then using matrix operations. Note that the shaded directions are for the matrix operations column (on the right).

| | | |
|---|---|---|
| $x - y + 4z = 5$   (1) <br> $2x + y + z = 4$   (2) <br> $-x + y - z = -5$   (3) | **1.** Solve this system of equations. <br> **1.** This system can be represented by this coefficient matrix. | $\begin{bmatrix} 1 & -1 & 4 & \vline & 5 \\ 2 & 1 & 1 & \vline & 4 \\ -1 & 1 & -1 & \vline & -5 \end{bmatrix}$ |
| | **2.** Add equation #1 to equation #3. <br> **2.** Add row 1 to row 3 (replacing row 3). | |
| | **3.** Solve the resulting equation for $z$ by dividing both sides by 3. <br> **3.** Multiply row 3 by $\frac{1}{3}$. | |
| | **4.** Multiply equation #1 by $-2$ and add to equation #2. <br> **4.** Multiply row 1 by $-2$ and add to row 2 (replacing row 2). | |
| | **5.** Divide both sides of the resulting equation by 3. <br> **5.** Multiply row 2 by $\frac{1}{3}$. | |
| | **6.** Gather together equations found in steps 3 and 5, as well as equation 1. <br> **6.** Rewrite matrix as a system of equations. | |
| | **7.** Solve using back-substitution. <br> **7.** Solve using back-substitution. | |

# Student Activity
## Strategy Comparison

In this activity we look at two different strategies for row-reducing the same matrix.

**Strategy 1** will attempt to get the zeros in the matrix in this order:

$$2^{nd} \longrightarrow \quad 1^{st} \longrightarrow \begin{bmatrix} 4 & 3 & 2 \\ \boxed{5} & 1 & 3 \\ \boxed{1} & \boxed{2} & \boxed{0} \end{bmatrix} \quad 3^{rd}$$

**Strategy 2** will attempt to get the zeros in the matrix in this order:

$$2^{nd} \longrightarrow \quad 3^{rd} \longrightarrow \begin{bmatrix} 4 & 3 & 2 \\ \boxed{5} & 1 & 3 \\ \boxed{1} & \boxed{2} & \boxed{0} \end{bmatrix} \quad 1^{st}$$

| | |
|---|---|
| **Step 1:** $R_2 + (-5)R_3 \rightarrow R_3$ | **Step 1:** $-2R_2 + R_3 \rightarrow R_3$ |
| **Step 2:** $-5R_1 + 4R_2 \rightarrow R_2$ | **Step 2:** $-5R_1 + 4R_2 \rightarrow R_2$ |
| **Step 3:** $9R_2 + (-11)R_3 \rightarrow R_3$ | **Step 3:** $9R_1 + 4R_3 \rightarrow R_3$  What happens here? |
| **Step 4:** $\frac{1}{4}R_1 \rightarrow R_1$ $-\frac{1}{11}R_2 \rightarrow R_2$ $-\frac{1}{15}R_3 \rightarrow R_3$ | **Step 4:** $27R_2 + 11R_3 \rightarrow R_3$ |
| | **Step 5:** $\frac{1}{4}R_1 \rightarrow R_1$ $-\frac{1}{11}R_2 \rightarrow R_2$ $-\frac{1}{12}R_3 \rightarrow R_3$ |

Was one strategy easier than the other?  If so, which one?  If not, why?

# Student Activity
## Match Up on Determinants

**Match-up:** In each box of the grid below, you will find **mostly** determinants to evaluate. However, there are a few expressions that are absolute values! Match the result with the appropriate choice. If the result is not found among the choices A through D, then choose E (none of these).

**A** 4          **B** −7     **C** 5      **D** 0      **E** None of these

| $\begin{vmatrix} 3 & 5 \\ 5 & 6 \end{vmatrix}$ | $\begin{vmatrix} -2 & 5 \\ 1 & -5 \end{vmatrix}$ | $\begin{vmatrix} 2 & 1 & 3 \\ 0 & -1 & 2 \end{vmatrix}$ | $\begin{vmatrix} \frac{3}{4} & \frac{3}{2} \\ \frac{1}{3} & \frac{2}{3} \end{vmatrix}$ |
| $\begin{vmatrix} -2 & 5 \\ 3 & 1 \\ 1 & -5 \end{vmatrix}$ | $\begin{vmatrix} 3 & 5 \\ 1 & 3 \end{vmatrix}$ | $2 \cdot \begin{vmatrix} -2 & 6 \\ 1 & -4 \end{vmatrix}$ | $\begin{vmatrix} 4 & 2 \\ 6 & 3 \end{vmatrix}$ |
| $\begin{vmatrix} 5 & 0 \\ -1 & 0 \end{vmatrix}$ | $\left| 6-11 \right|$ | $\begin{vmatrix} 1 & 2 \\ -2 & 0 \end{vmatrix}$ | Find $k$ if $\begin{vmatrix} 3 & 1 \\ k & 3 \end{vmatrix} = 5$. |
| Evaluate $\begin{vmatrix} x & 8 \\ x & x \end{vmatrix}$ if $x=1$. | $\begin{vmatrix} 5 & -3 \\ -10 & 6 \end{vmatrix}$ | $\left| \frac{1}{4} + \frac{2}{3} \right|$ | Find $k$ if $\begin{vmatrix} 4 & -8 \\ 3 & k \end{vmatrix} = -4$. |

# Guided Learning Activity
## The Minor League

It is easy to find the determinant of a $2\times 2$ matrix: $\begin{vmatrix} a & b \\ c & d \end{vmatrix} = ad - bc$ .

It is much harder to find the determinant of a $3\times 3$ matrix. We use a technique called **expanding by minors**.

What is a **minor**?

In $\begin{vmatrix} a_1 & b_1 & c_1 \\ a_2 & b_2 & c_2 \\ a_3 & b_3 & c_3 \end{vmatrix}$, the minor of $a_1$ is $\begin{vmatrix} b_2 & c_2 \\ b_3 & c_3 \end{vmatrix}$.  Notice: $\begin{vmatrix} \cancel{a_1} & \cancel{b_1} & \cancel{c_1} \\ \cancel{a_2} & b_2 & c_2 \\ \cancel{a_3} & b_3 & c_3 \end{vmatrix}$

If you cross out the row and column of $a_1$, the minor of $a_1$ will be easier to see.

**1.** In $\begin{vmatrix} a_1 & b_1 & c_1 \\ a_2 & b_2 & c_2 \\ a_3 & b_3 & c_3 \end{vmatrix}$, what is the minor of $b_1$?

**2.** In $\begin{vmatrix} a_1 & b_1 & c_1 \\ a_2 & b_2 & c_2 \\ a_3 & b_3 & c_3 \end{vmatrix}$, what is the minor of $c_1$?

We use these minors to find the determinant of a $3\times 3$ matrix:

$$\begin{vmatrix} a_1 & b_1 & c_1 \\ a_2 & b_2 & c_2 \\ a_3 & b_3 & c_3 \end{vmatrix} = a_1\left(\text{Minor of } a_1\right) - b_1\left(\text{Minor of } b_1\right) + c_1\left(\text{Minor of } c_1\right).$$

**3.** $\begin{vmatrix} 1 & 0 & 2 \\ 0 & 3 & 2 \\ -2 & 1 & 1 \end{vmatrix} = \square \cdot \begin{vmatrix} \square & \square \\ \square & \square \end{vmatrix} - \square \cdot \begin{vmatrix} \square & \square \\ \square & \square \end{vmatrix} + \square \cdot \begin{vmatrix} \square & \square \\ \square & \square \end{vmatrix}$ .

$= \square \cdot \square - \square \cdot \square + \square \cdot \square = \square - \square + \square = \square$

**4.** $\begin{vmatrix} 3 & -2 & 5 \\ -9 & 1 & 4 \\ 6 & 7 & 0 \end{vmatrix} = \square \cdot \begin{vmatrix} \square & \square \\ \square & \square \end{vmatrix} - \square \cdot \begin{vmatrix} \square & \square \\ \square & \square \end{vmatrix} + \square \cdot \begin{vmatrix} \square & \square \\ \square & \square \end{vmatrix}$

$= \square \cdot \square - \square \cdot \square + \square \cdot \square = \square - \square + \square = \square$

By learning a pattern of signs (sometimes called the checkerboard pattern), we can use this method of expanding the minors for any row or column of the $3 \times 3$ determinant.

$$\begin{array}{ccc} + & - & + \\ - & + & - \\ + & - & + \end{array}$$

Here is the checkerboard pattern:

Look at the first expansion we learned:

$$\begin{vmatrix} \overset{+}{a_1} & \overset{-}{b_1} & \overset{+}{c_1} \\ a_2 & b_2 & c_2 \\ a_3 & b_3 & c_3 \end{vmatrix} = +a_1 \left( \text{Minor of } a_1 \right) - b_1 \left( \text{Minor of } b_1 \right) + c_1 \left( \text{Minor of } c_1 \right) \cdot$$

We can evaluate the same determinant using a minor expansion for the second row:

$$\begin{vmatrix} a_1 & b_1 & c_1 \\ \overset{-}{a_2} & \overset{+}{b_2} & \overset{-}{c_2} \\ a_3 & b_3 & c_3 \end{vmatrix} = -a_2 \left( \text{Minor of } a_2 \right) + b_2 \left( \text{Minor of } b_2 \right) - c_2 \left( \text{Minor of } c_2 \right).$$

**5.** Now try writing a minor expansion on the third <u>column</u>.

$$\begin{vmatrix} a_1 & b_1 & c_1 \\ a_2 & b_2 & c_2 \\ a_3 & b_3 & c_3 \end{vmatrix} = \underline{\quad} \left( \text{Minor of } \underline{\quad} \right) \underline{\quad} \left( \text{Minor of } \underline{\quad} \right) \underline{\quad} \left( \text{Minor of } \underline{\quad} \right)$$

Do all of these different minor expansions really result in the same determinant?

**6.** Evaluate $\begin{vmatrix} 2 & 8 & 4 \\ 5 & 3 & 0 \\ 6 & 1 & 7 \end{vmatrix}$ by expanding on the first row.

**7.** Evaluate $\begin{vmatrix} 2 & 8 & 4 \\ 5 & 3 & 0 \\ 6 & 1 & 7 \end{vmatrix}$ by expanding on the second row.

**8.** Evaluate $\begin{vmatrix} 2 & 8 & 4 \\ 5 & 3 & 0 \\ 6 & 1 & 7 \end{vmatrix}$ by expanding on the third column.

# Student Activity

## Cramer's Ruling

**Directions:** Follow the directions for each step and fill in the boxes with the missing pieces as you go. Careful – as you move from line to line, the boxes might contain different expressions! At the end, you should get a formula that you might recognize!

For the system: $\begin{cases} ax + by = e \\ cx + dy = f \end{cases}$, $D = \begin{vmatrix} a & b \\ c & d \end{vmatrix}$, $D_x = \begin{vmatrix} e & b \\ f & d \end{vmatrix}$ and $D_y = \begin{vmatrix} a & e \\ c & f \end{vmatrix}$.

**1.** Simplify each determinant:

$$D = ad - bc \qquad D_x = \underline{\hspace{1.5cm}} \qquad D_y = \underline{\hspace{1.5cm}}$$

Now we work through the process of solving the system:

$$\begin{cases} ax + by = e \\ cx + dy = f \end{cases} \text{ where } ad - bc \neq 0.$$

**2.** First, we eliminate the variable $y$ from the system:

Multiply both sides of the first equation by $d$: $\qquad adx + \boxed{\phantom{xx}} y = \boxed{\phantom{xx}}$

Multiply both sides of the second equation by $(-b)$: $\qquad \boxed{\phantom{xx}} x - \boxed{\phantom{xx}} = -bf$

**3.** Add the equations from step **1**: $\qquad \boxed{\phantom{xx}} = \boxed{\phantom{xx}}$

**4.** Factor $x$ from the left-hand side of the equation: $\qquad x\left(\boxed{\phantom{xxxx}}\right) = ed - bf$

**5.** Solve for $x$:: $\quad x = \dfrac{ed - bf}{\boxed{\phantom{xxx}}}$

**6.** Thus $x = \dfrac{D_x}{\boxed{\phantom{xx}}}$

Since the Commision switched over to Cramer's Rule, the outcomes have been a lot easier to determine.

# Student Activity
## D-lightful Determinants

**Directions:** For each system of equations, find the desired determinants. You do **not** need to solve the system of equations. As you find the missing determinants, shade them in to paint by Determinants.

**1.** $\begin{cases} 2x + 3y = -1 \\ y - x = 8 \end{cases}$

Find $D_x$ and $D_y$.

**2.** $\begin{cases} x + 2y + 4z = 4 \\ 3x + 2y + 6z = 1 \\ x + 5y + 4z = 1 \end{cases}$

Find $D_y$ and $D_z$.

**3.** $\begin{cases} 10x + 12y + 6z = 2 \\ 5x + 6y + 3z = 1 \\ x + 3z = 1 \end{cases}$

Find $D$, $D_x$, $D_y$ and $D_z$.

**4.** $\begin{cases} x + z = 4 \\ 3y + 3z = 12 \\ x + 2y = 6 \end{cases}$

Find $D$, $D_x$, $D_y$ and $D_z$.

| | | | | |
|---|---|---|---|---|
| $\begin{vmatrix} 1 & 0 & 1 \\ 0 & 3 & 3 \\ 1 & 2 & 0 \end{vmatrix}$ | $\begin{vmatrix} -1 & 3 \\ 8 & 1 \end{vmatrix}$ | $\begin{vmatrix} 4 & 0 & 1 \\ 12 & 3 & 3 \\ 6 & 2 & 0 \end{vmatrix}$ | $\begin{vmatrix} 2 & 3 \\ -1 & 1 \end{vmatrix}$ | $\begin{vmatrix} 4 & 1 & 0 \\ 12 & 3 & 3 \\ 6 & 2 & 0 \end{vmatrix}$ |
| $\begin{vmatrix} 2 & -1 \\ -1 & 8 \end{vmatrix}$ | $\begin{vmatrix} 1 & 1 & 0 \\ 0 & 3 & 3 \\ 1 & 2 & 0 \end{vmatrix}$ | $\begin{vmatrix} 1 & 2 & 4 \\ 3 & 2 & 6 \\ 1 & 5 & 4 \end{vmatrix}$ | $\begin{vmatrix} 10 & 12 & 6 \\ 5 & 6 & 3 \\ 1 & 0 & 3 \end{vmatrix}$ | $\begin{vmatrix} 10 & 12 & 6 \\ 5 & 6 & 3 \\ 1 & 3 & 0 \end{vmatrix}$ |
| $\begin{vmatrix} 10 & 12 & 2 \\ 5 & 6 & 1 \\ 1 & 0 & 1 \end{vmatrix}$ | $\begin{vmatrix} 10 & 12 & 2 \\ 5 & 6 & 1 \\ 1 & 3 & 1 \end{vmatrix}$ | $\begin{vmatrix} 10 & 2 & 6 \\ 5 & 1 & 3 \\ 1 & 1 & 0 \end{vmatrix}$ | $\begin{vmatrix} 10 & 2 & 6 \\ 5 & 1 & 3 \\ 1 & 1 & 1 \end{vmatrix}$ | $\begin{vmatrix} 4 & 2 & 4 \\ 1 & 2 & 6 \\ 1 & 5 & 4 \end{vmatrix}$ |
| $\begin{vmatrix} 1 & 0 & 4 \\ 0 & 3 & 12 \\ 1 & 2 & 6 \end{vmatrix}$ | $\begin{vmatrix} 1 & 1 & 4 \\ 0 & 3 & 12 \\ 1 & 2 & 6 \end{vmatrix}$ | $\begin{vmatrix} 1 & 3 \\ 8 & -1 \end{vmatrix}$ | $\begin{vmatrix} 1 & 2 & 4 \\ 3 & 2 & 1 \\ 1 & 5 & 1 \end{vmatrix}$ | $\begin{vmatrix} 2 & 3 \\ 1 & -1 \end{vmatrix}$ |
| $\begin{vmatrix} 1 & 4 & 4 \\ 3 & 1 & 6 \\ 1 & 1 & 4 \end{vmatrix}$ | $\begin{vmatrix} 2 & 12 & 6 \\ 1 & 6 & 3 \\ 1 & 0 & 3 \end{vmatrix}$ | $\begin{vmatrix} 1 & 4 & 1 \\ 0 & 12 & 3 \\ 1 & 6 & 0 \end{vmatrix}$ | $\begin{vmatrix} 2 & 12 & 6 \\ 1 & 6 & 3 \\ 1 & 3 & 0 \end{vmatrix}$ | $\begin{vmatrix} 1 & 4 & 0 \\ 0 & 12 & 3 \\ 1 & 6 & 0 \end{vmatrix}$ |

# Student Activity

**Strange Determinants**

**Directions:** Fill in the empty spaces in the table using Cramer's Rule. In the last column, write the solution (if one exists), or label the system as inconsistent or the equations as dependent. The first one has been done for you.

| | $D =$ | $D_x =$ | $D_y =$ | $D_z =$ | Solution? |
|---|---|---|---|---|---|
| **1.** | 2 | 3 | $-1$ | $-4$ | $\left(\frac{3}{2}, -\frac{1}{2}, -2\right)$ |
| **2.** | 0 | 0 | 0 | 0 | |
| **3.** | $\frac{1}{2}$ | 5 | 7 | 9 | |
| **4.** | 10 | 0 | 0 | 0 | |
| **5.** | 0 | 0 | 0 | 1 | |
| **6.** | 0 | $-1$ | $-3$ | $-5$ | |
| **7.** | | 15 | 25 | 30 | Inconsistent |
| **8.** | 0 | | | | Dependent |
| **9.** | 3 | | | | $\left(6, -1, \frac{1}{2}\right)$ |
| **10.** | | | 12 | | $(-2, -3, 1)$ |

# Assess Your Understanding

## More on Solving Systems

For each of the following, describe the strategies or key steps that will help you **start** the problem.  You do **not** have to complete the problems.

|     |                                                                                                                           | **What will help you to start this problem?** |
| --- | ------------------------------------------------------------------------------------------------------------------------- | --------------------------------------------- |
| **1.** | Solve the system: $$\begin{cases} x + y - z = 8 \\ 2x - y + 3z = -4 \\ x \quad + 2z = 0 \end{cases}$$                   |                                               |
| **2.** | Is the system consistent or inconsistent? $$\begin{cases} x - 4y = 6 \\ 2x = 8y + 10 \end{cases}$$                       |                                               |
| **3.** | Is the pair of equations dependent or independent? $$\begin{cases} 3x + y = 8 \\ 6x + 2y = 8 \end{cases}$$               |                                               |
| **4.** | Is $(1, -2, 0)$ a solution to the system? $$\begin{cases} 3x + y + 4z = 1 \\ x - y + z = -1 \\ 2x + y + 3z = 0 \end{cases}$$ |                                               |
| **5.** | Solve the system: $$\begin{cases} 5x + 4y + 9 = z \\ x = -(y + z) \\ 2x + y = 2z + 5 \end{cases}$$                        |                                               |
| **6.** | Find $D_x$ and $D_y$ for the system: $$\begin{cases} 3x - 4y = 7 \\ 5x - 8y = 8 \end{cases}$$                             |                                               |
| **7.** | Solve the system using matrices: $$\begin{cases} x + y + 2z = 1 \\ x + z = -1 \\ 3x - y - z = 13 \end{cases}$$            |                                               |
| **8.** | If $D = 0$, $D_x = 2$, and $D_y = 5$, describe the solution to the $2 \times 2$ system of equations.                     |                                               |
| **9.** | Write the matrix in row-reduced form: $\begin{bmatrix} 1 & 3 & \vdots & 3 \\ 1 & -4 & \vdots & 10 \end{bmatrix}$         |                                               |

# Metacognitive Skills
## More on Solving Systems

Metacognitive skills refer to the ability to judge how well you have learned something and to effectively direct your own learning and studying. This is a self-evaluation tool designed to help you focus your studying and to improve your metacognitive skills with regards to this math class.

Fill the 1st column out **before** you begin studying. Fill the 2nd column out after you study for your test.

Go back to this assessment after your test and circle any of the ratings that you would change – this identifies the "disconnects" between what you **thought** you knew well and what you **actually** knew well.

**Use the scale below to assign a number to each topic.**
5  *I am confident I can do any problems in this category correctly.*
4  *I am confident I can do most of the problems in this category correctly.*
3  *I understand how to do the problems in this category, but I still make a lot of mistakes.*
2  *I feel unsure about how to do these problems.*
1  *I know I don't understand how to do these problems.*

| Topic or Skill | Before Studying | After Studying |
|---|---|---|
| Checking whether an ordered pair or ordered triple is a solution of a system of equations. | | |
| Classifying a system of two or three equations as consistent or inconsistent. | | |
| Classifying a system of two or three equations as dependent or independent. | | |
| Understanding what the "solution" to a system of equations is graphically (both systems of two equations and systems of three equations). | | |
| Solving a system of two equations by substitution or elimination. | | |
| Solving a system of three equations with three variables. | | |
| Declaring the variables for an application involving a system of three equations. | | |
| Writing the system of equations for a problem involving $d = rt$. | | |
| Writing the system of equations for a value-mixture problem or a percent-solution problem. | | |
| Writing the system of equations for an investment problem. | | |
| Solving the system of equations from an application problem. | | |
| Writing the conclusion to an application problem. | | |
| Writing a system of equations as an augmented matrix. | | |
| Applying row operations to a matrix. | | |
| Using row operations to transform a matrix to row-reduced form. | | |
| Solving a system of equations by using matrices. | | |
| Knowing what types of row-reduced matrices indicate inconsistent systems or dependent equations. | | |
| Finding the determinant of a $2 \times 2$ matrix. | | |
| Finding the determinant of a $3 \times 3$ matrix by expanding the minors. | | |
| Knowing how to construct the determinants for using Cramer's Rule. | | |
| Knowing Cramer's Rule. | | |
| Solving a system of equations by using Cramer's Rule. | | |
| Knowing what types of results from using Cramer's Rule indicate inconsistent systems or dependent equations. | | |

# EXP: Exponent Rules

# Student Activity

**Exponent Duos**

**Directions:** In each of the "duos" below, place two equivalent exponential expressions of the following format:

| Expanded Expression using multiplication |
| --- |
| Compact exponential expression |

The first two duos have been done for you.

| $(3x)(3x)(3x)$ |
| --- |
| $(3x)^3$ |

| $3 \cdot x \cdot x \cdot x$ |
| --- |
| $3x^3$ |

| $(-5x)(-5x)$ |
| --- |
| |

| $-(5x)(5x)$ |
| --- |
| |

| |
| --- |
| $(2a^2)^3$ |

| |
| --- |
| $2a^5$ |

| $(x+1)(x+1)$ |
| --- |
| |

| |
| --- |
| $(a+b)^2$ |

| $\left(\dfrac{1}{2}\right)\left(\dfrac{1}{2}\right)\left(\dfrac{1}{2}\right)\left(\dfrac{1}{2}\right)$ |
| --- |
| |

| |
| --- |
| $\left(-\dfrac{2}{3}\right)^2$ |

| $(10ab)(10ab)$ |
| --- |
| |

| $6 \cdot x \cdot x \cdot y \cdot y$ |
| --- |
| |

| $7 \cdot m \cdot m \cdot m \cdot p$ |
| --- |
| |

| |
| --- |
| $\left(\dfrac{a}{b}\right)^3$ |

| $\left(\dfrac{1}{z}\right)\left(\dfrac{1}{z}\right)$ |
| --- |
| |

| $(a-2)(a-2)$ |
| --- |
| |

| $-(4x)(4x)(4x)$ |
| --- |
| |

| |
| --- |
| $(-2.1)^3$ |

| $\pi \cdot r \cdot r$ |
| --- |
| |

| |
| --- |
| $\dfrac{4}{3}\pi r^3$ |

# Student Activity
## Match Up on Basic Exponent Rules

**Match-up:** Match each of the expressions in the squares of the grid below with an equivalent simplified expression from the top. If an equivalent expression is not found among the choices A through D, then choose E (none of these).

**A** $x^5$

**B** $x^6$

**C** $9x^4$

**D** $12x^6$

**E** None of these

I haven't made it home yet, but I've made it to third base to like the 10th power.

| | | | | |
|---|---|---|---|---|
| $\left(6x^2\right)\left(2x^4\right)$ | $4x^4 - \left(-5x^4\right)$ | $x^3 \cdot x^2$ | $\left(-3x^3\right)\left(-3x\right)$ | $x^5 \cdot x$ |
| $6x^3 + 6x^3$ | $-4x^5 + 5x^5$ | $12\left(x^2\right)^3$ | $\left(-3x^3\right)\left(-4x^3\right)$ | $x^5 + x$ |
| $\left(x^2\right)\left(x^2\right)\left(x\right)$ | $\left(27x^3\right)\left(\dfrac{1}{3}x\right)$ | $\left(x^3\right)^2$ | $\left(12x^3\right)^2$ | $\left(-3x^2\right)^2$ |
| $\left(3x^2\right)^2$ | $3\left(-2x^3\right)^2$ | $3x^3 + 6x$ | $\left(5x^2\right)\left(\dfrac{1}{5}x^3\right)$ | $8x^6 + 4x^6$ |

# Guided Learning Activity
## Zero and Negative Exponents

To develop the concepts of zero and negative exponents, we will first look at several patterns of values.

**1.** Fill in the blanks to complete each sequence below:

$81, 27, 9,$ ___, ___, ___, ___    How do you get the next term? _____

$16, 8, 4,$ ___, ___, ___, ___    How do you get the next term? _____

$x^4, x^3, x^2,$ ___, ___, ___    How do you get the next term? _____

$4^4, 4^3, 4^2,$ ___, ___, ___    How do you get the next term? _____

$\dfrac{1}{125}, \dfrac{1}{25}, \dfrac{1}{5},$ ___, ___, ___    How do you get the next term? _____

**2.** Now complete each table below by finishing the patterns:

| 81 | $3^4$ |
|----|----|
| 27 | $3^3$ |
| 9 | $3^2$ |
| | |
| | |
| | |
| | |

| 16 | $2^4$ |
|----|----|
| 8 | $2^3$ |
| 4 | $2^2$ |
| | |
| | |
| | |
| | |

| $\frac{1}{125}$ | |
|----|----|
| $\frac{1}{25}$ | |
| $\frac{1}{5}$ | |
| | |
| | |
| | $5^2$ |
| | $5^3$ |

| 64 | $4^3$ |
|----|----|
| | $4^2$ |
| | $4^1$ |
| | |
| | |
| | |
| | |

**3.** Based on your observations in the patterns above, finish the exponent rules below:

It may seem like nothing, but it makes me ONE with the Universe.

### Rule for Zero Exponents:

$x^0 =$ _____ (for $x \neq 0$)

### Rule for Negative Exponents:

$x^{-1} =$ _____ (for $x \neq 0$)

$x^{-2} =$ _____

$x^{-3} =$ _____

$x^{-n} =$ _____

**4.** Apply the rules you have just developed to write each of these exponential expressions without zero or negative exponents.

$6^{-2} =$        $10^0 =$        $y^{-4} =$        $2^{-5} =$        $z^{-1} =$        $(4a)^0$

# Student Activity
## Match Up on Trickier Exponent Rules

**Match-up:** Match each of the expressions in the squares of the grid below with an equivalent simplified expression from the top. If an equivalent expression is not found among the choices A through D, then choose E (none of these).

**A** 1

**B** $\dfrac{4}{x^2}$

**C** $9x^2y^3$

**D** $\dfrac{-9x^4}{y^3}$

**E** None of these

Mark my words! You harness that negative power of yours, and you can make it to the top just like me!

| | | | | |
|---|---|---|---|---|
| $\left(4x^{-2}\right)^0$ | $\left(4x\right)^{-2}$ | $4x^{-2}$ | $4x^0$ | $\left(2x\right)^{-2}$ |
| $\left(\dfrac{x}{2}\right)^{-2}$ | $\dfrac{\left(-3xy\right)^2}{y^{-1}}$ | $\dfrac{3^{-2}y^{-3}}{x^{-4}}$ | $\dfrac{-\left(3x^2y\right)^2}{y^5}$ | $\left(\dfrac{100x^{27}y^{35}}{a^4b^5}\right)^0$ |
| $\left(\dfrac{x}{2}\right)^2$ | $\left(9x^2y^3\right)^0$ | $\left(-\dfrac{y}{a^4b^4}\right)\left(\dfrac{3xab}{y}\right)^4$ | $y^7\left(\dfrac{y^2}{3x}\right)^{-2}$ | $\left(2yz\right)^2\left(xyz\right)^{-2}$ |
| $4\left(\dfrac{1}{x^2}\right)^0$ | $8x^2\left(\dfrac{x^{-2}}{8}\right)$ | $3\left(x^2y^2\right)\left(3x^2y^2\right)^{-1}$ | $\dfrac{\left(2x^{-1}z^2\right)^2}{z^4}$ | $\dfrac{-12x^4}{5}\left(\dfrac{5}{-12x^4}\right)$ |

# Student Activity
## Double the Fun on Exponent Rules

**Directions:** For many problems involving the simplification of exponents, there are at least two ways to tackle the problem. For each of the problems below, try the problem both ways – your answers should be the same for both methods. If they are not, you'll have to go back and look for a mistake. The first one has been done for you.

| Simplify this expression | Apply exponent rules directly | First move factors to avoid working with negative exponents |
|---|---|---|
| $\dfrac{x^{-2}y^5}{x^{-6}y^{-1}}$ | $\dfrac{x^{-2}y^5}{x^{-6}y^{-1}} = x^{-2-(-6)}y^{5-(-1)}$ <br><br> $= x^{-2+6}y^{5+1} = x^4y^6$ | $\dfrac{x^{-2}y^5}{x^{-6}y^{-1}} = \dfrac{x^6y^5y^1}{x^2} = x^{6-2}y^{5+1} = x^4y^6$ |
| $\dfrac{16a^3b^{-2}}{8a^{-1}b^5}$ | | |
| $3x^{-2}x^3x^{-1}$ | | |
| $(2x)^{-2}\left(-6x^3\right)$ | | |
| $\left(\dfrac{x^{-2}y^5}{x^{-1}y^6}\right)^3$ | | |
| $\dfrac{\left(xy^3\right)^{-1}}{2x^4y^{-1}}$ | | |

# Student Activity

## Mathematical Heteronyms

**Directions:** In writing, there are words that are spelled the same but have different pronunciations and different definitions; these are called heteronyms. Many mathematical expressions look similar but are really very different (almost like mathematical heteronyms). In each set of expressions below, pay close attention to the use of parentheses and the mathematical operations and notation.

**1.**

| $3-1$ | $3(-1)$ | $3^{-1}$ | $3-(-1)$ | $-3^{-1}$ | $3 \div (-1)$ |
|---|---|---|---|---|---|
|  |  |  |  |  |  |

**2.**

| $2-5$ | $2(-5)$ | $2^{-5}$ | $2^{-1}-5$ | $2^{-1}-5^{-1}$ | $2-5^{-1}$ |
|---|---|---|---|---|---|
|  |  |  |  |  |  |

**3.**

| $x^2 x^{-3}$ | $x^2\left(-x^3\right)$ | $x^2 x^3$ | $x^{-2} x^3$ | $x^2 - x^3$ | $\left(x^2\right)^{-3}$ |
|---|---|---|---|---|---|
|  |  |  |  |  |  |

**4.**

| $2a^{-3}$ | $(-2a)^3$ | $2^3(-a)$ | $2-a^3$ | $(2a)^{-3}$ | $(-2a)\cdot 3$ |
|---|---|---|---|---|---|
|  |  |  |  |  |  |

**5.**

| $\left(\dfrac{3}{4}\right)^{-2}$ | $\dfrac{3^{-2}}{4}$ | $\left(\dfrac{3}{4}\right)(-2)$ | $\dfrac{3}{4}-2$ | $\dfrac{3}{4^{-2}}$ | $\left(-\dfrac{3}{4}\right)^2$ |
|---|---|---|---|---|---|
|  |  |  |  |  |  |

# Student Activity
## Exponents Using a Calculator

When you input exponents into a calculator, you must be careful to tell the calculator which part is the base and which part is the exponent. Each calculator requires a specific set of keystrokes to evaluate exponential expressions and this can become quite complicated when the expression also involves negatives. We will fill out the table below, first performing each calculation by hand, then finding the proper keystrokes to do the evaluation on your calculator.

First you need to locate your exponent button. It may look like $\boxed{\wedge}$ or $\boxed{x^y}$.

Reminders: There is a difference between the minus key and the negative key on your calculator. Fractions need to be placed inside parentheses for proper evaluation.

| | Expression | Evaluate by hand | Calculator Keystrokes to get equivalent result |
|---|---|---|---|
| **a.** | $2^3$ | | |
| **b.** | $2^{-1}$ | | |
| **c.** | $5^{-2}$ | | |
| **d.** | $\left(\dfrac{1}{2}\right)^3$ | | |
| **e.** | $\left(\dfrac{2}{3}\right)^{-2}$ | | |
| **f.** | $-2^4$ | | |
| **g.** | $(-2)^4$ | | |
| **h.** | $-4^{-2}$ | | |
| **i.** | $(-4)^{-2}$ | | |
| **j.** | $2^{-1} + 3^{-1}$ | | |
| **k.** | $(2+3)^{-1}$ | | |
| **l.** | $5 - 2^{-1}$ | | |

# Guided Learning Activity

**Charting Scientific Notation**

Scientific notation is based on using powers of 10. Fill in the Powers of 10 with their decimal notation in the table below. $10^3$ has been done for you.

| Power of 10 | $10^0$ | $10^1$ | $10^2$ | $10^3$ | $10^4$ | $10^5$ | $10^6$ |
|---|---|---|---|---|---|---|---|
| Value | | | | 1,000 | | | |

| Power of 10 | $10^{-6}$ | $10^{-5}$ | $10^{-4}$ | $10^{-3}$ | $10^{-2}$ | $10^{-1}$ | $10^0$ |
|---|---|---|---|---|---|---|---|
| Value | | | | | | | |

**Scientific notation:** $N \times 10^n$ where $1 \le N < 10$, and $n$ is an integer.
With your class, fill out the table below. The first one has been done for you.

| Number in standard notation (decimal notation) | N | 2nd factor in decimal notation | 2nd factor as a power of 10 | $N \times 10^n$ |
|---|---|---|---|---|
| 0.00495 | 4.95 | 0.001 | $10^{-3}$ | $4.95 \times 10^{-3}$ |
| 8,000,000,000 | | | | |
| 0.00000034 | | | | |
| | | | | $9.2 \times 10^5$ |
| | 6.843 | 0.0000001 | | |
| | | | | $8.04 \times 10^{-7}$ |
| | 2 | 10,000,000 | | |

Rather than remembering whether the decimal place moves left or right, it is easier to remember that 10 raised to negative powers means you are looking at a number that is small (between zero and 1) and 10 raised to positive powers means you are looking at a number that is large (>1). Simply move the decimal point the correct number of digits in the direction that creates the appropriate bigger or smaller number!

| Decimal Value | Is it a *small* or *large* number? | Is the power on the 10 *positive* or *negative?* | Scientific Notation |
|---|---|---|---|
| 0.000 072 5 | | | |
| | | | $1.2 \times 10^7$ |
| | | | $4.534 \times 10^{-6}$ |
| 84,000,000,000 | | | |
| 48 | | | |
| | | | $9 \times 10^{-4}$ |
| 674,000 | | | |
| | | | $2.99 \times 10^8$ |
| | | | $6.67 \times 10^{-11}$ |
| 0.000 000 056 704 | | | |

# Student Activity

## Escape the Matrix with Ten Power

**Directions:** Begin at the box marked START. By shading in pairs of adjacent squares that represent equal numbers, you will eventually find the path to "escape" this matrix of boxes. The first "step" in the path and two of the middle steps in the path have been shaded for you.

| START 0.00043 | $4.3 \times 10^{-4}$ | $5 \times 10^{6}$ | $5{,}000{,}000$ | $7.6 \times 10^{3}$ |
|---|---|---|---|---|
| | $4.3 \times 10^{-3}$ | $500{,}000$ | $5.0 \times 10^{7}$ | $50 \times 10^{6}$ |
| | $789 \times 10^{-2}$ | $7.89$ | $0.023$ | $2.3 \times 10^{-2}$ |
| $39 \times 10^{8}$ | $390{,}000{,}000$ | $3.90 \times 10^{7}$ | $2.3 \times 10^{2}$ | $23{,}000$ |
| $16 \times 10^{4}$ | $3.90 \times 10^{8}$ | $1{,}000{,}000$ | $10 \times 10^{5}$ | $5.555 \times 10^{-3}$ |
| $1.6 \times 10^{2}$ | $0.00016$ | $1 \times 10^{5}$ | $10 \times 10^{3}$ | $0.00555$ |
| $32{,}300$ | $32.30 \times 10^{5}$ | $3.230 \times 10^{6}$ | $0.003230 \times 10^{6}$ | $3230$ |
| $3.230 \times 10^{4}$ | $6.89 \times 10^{4}$ | $689$ | $689 \times 10^{-1}$ | $323 \times 10^{0}$ |
| $323.0 \times 10^{-2}$ | $3.230$ | $32.30$ | $3.230 \times 10^{1}$ | **ESCAPE** the Matrix |

I've got TEN power!

Scientific Notation

$9.7 \times 10^{3}$ Kg

$9.7 \times 10^{3}$ Kg

# Student Activity
## Scientific Notation Using a Calculator

There is a special button on most calculators that is used for scientific notation. Sometimes it is denoted with $\boxed{\text{EE}}$ or $\boxed{\text{EXP}}$. This button stands for "$\times 10$ to the," so, for example we could enter $4 \times 10^3$ as $\boxed{4}$ $\boxed{\text{EE}}$ $\boxed{3}$. When using the scientific notation button, you **do not need to** multiply by 10 too.

**1.** First we need to find the scientific notation button on your calculator. Try entering the expression: $4 \times 10^3$ then press $\boxed{\text{ENTER}}$ or $\boxed{=}$. You should see either 4000 or 4e3 or $4^{\ 03}$ or something like this. Ask your instructor if you are unsure about whether you have correctly located the scientific notation button.

Draw the series of keystrokes you used to enter $4 \times 10^3$ here:

**2.** Try another one: $2.5 \times 10^{12}$ and press $\boxed{\text{ENTER}}$ or $\boxed{=}$. This number is too big to display on most calculator screens, so you should see the calculator give the number to you in it's version of scientific notation, like 2.5e12, 2.5ᴇ12, or $2.5^{\ 12}$.

Draw how you entered $2.5 \times 10^{12}$ **and** the way the number appears on the screen:

**3.** Now we try a number with a negative exponent. Key in the number $3.4 \times 10^{-3}$ followed by $\boxed{\text{ENTER}}$ or $\boxed{=}$. This should display as 0.0034, 3.4e-3, or $3.4^{\ -03}$. Remember to use the negative key and not the minus key to enter the negative exponent.

Draw the series of keystrokes you used to enter $3.4 \times 10^{-3}$ here:

**4.** Try another one: $4.62 \times 10^{-14}$ and press $\boxed{\text{ENTER}}$ or $\boxed{=}$. This number has too many zeros to display on most calculator screens, so you should see the calculator give the number to you in it's version of scientific notation, like    4.62e-14, 4.62ᴇ-14, or $4.62^{\ -14}$.

Draw how you entered $4.62 \times 10^{-14}$ **and** the way it appears on the screen:

The real usefulness of the scientific notation button becomes obvious when we begin to perform calculations that involve scientific notation.

For example, if I wanted to calculate $\dfrac{2.4 \times 10^{-3}}{8 \times 10^{-12}}$ **without** the scientific notation button, here's what I'd have to enter on *my* calculator:

$$\boxed{(} \; 2.4 \; \boxed{\times} \; 10 \; \boxed{\wedge} \; \boxed{(-)} 3 \boxed{)} \; \boxed{\div} \; \boxed{(} \; 8 \; \boxed{\times} \; 10 \; \boxed{\wedge} \; \boxed{(-)} 12 \boxed{)}$$

Using the scientific notation button, this is much easier:

$$2.4 \; \boxed{\text{EE}} \; \boxed{(-)} \; 3 \; \boxed{\div} \; 8 \; \boxed{\text{EE}} \; \boxed{(-)} 12$$

**5.** Try the calculation $\dfrac{2.4 \times 10^{-3}}{8 \times 10^{-12}}$ on your calculator. (the answer is $3 \times 10^{8}$)

Draw how you entered $\dfrac{2.4 \times 10^{-3}}{8 \times 10^{-12}}$ here:

For practice, try these calculations and write down the answers using the proper scientific notation. In other words, don't write an answer like $8e3$; write the answer as $8 \times 10^{3}$ instead.

**6.** $\left(8.2 \times 10^{9}\right)\left(2 \times 10^{4}\right)$

**7.** $\dfrac{1.5 \times 10^{-4}}{2.5 \times 10^{3}}$

**8.** $\left(1.1 \times 10^{-5}\right)^{2}$

**9.** $\dfrac{\left(5.4 \times 10^{-6}\right)\left(2 \times 10^{3}\right)}{9 \times 10^{8}}$

**10.** $\dfrac{3.6 \times 10^{-4}}{\left(1.2 \times 10^{3}\right)\left(1.5 \times 10^{-10}\right)}$

# Assess Your Understanding

## Exponent Rules

For each of the following, describe the strategies or key steps that will help you **start** the problem. You do **not** have to complete the problems.

|     |                                                              | **What will help you to start this problem?** |
| --- | ------------------------------------------------------------ | --------------------------------------------- |
| **1.**  | Evaluate: $3^{-2}$                                        |                                               |
| **2.**  | Simplify: $\dfrac{3a^2b^5}{a^3b}$                         |                                               |
| **3.**  | Simplify: $\left(-4x^3\right)^2$                          |                                               |
| **4.**  | Write $0.000027$ in scientific notation.                 |                                               |
| **5.**  | Simplify: $5x^0$                                          |                                               |
| **6.**  | Multiply: $\left(3.1\times10^5\right)\left(2\times10^{-3}\right)$ |                                       |
| **7.**  | Simplify $\dfrac{6x^{-3}}{-2x^2}$ and write without negative exponents. |                          |
| **8.**  | Write $3.14\times10^5$ in standard notation.             |                                               |
| **9.**  | Simplify: $\left(-4x^2y\right)\left(2x^3y^4\right)$       |                                               |
| **10.** | Simplify: $\left(3x^{-2}\right)^3\left(2x^4\right)$       |                                               |

# Metacognitive Skills
## Exponent Rules

Metacognitive skills refer to the ability to judge how well you have learned something and to effectively direct your own learning and studying. This is a self-evaluation tool designed to help you focus your studying and to improve your metacognitive skills with regards to this math class.

Fill the 1$^{st}$ column out **before** you begin studying. Fill the 2$^{nd}$ column out after you study for your test.

Go back to this assessment after your test and circle any of the ratings that you would change – this identifies the "disconnects" between what you **thought** you knew well and what you **actually** knew well.

**Use the scale below to assign a number to each topic.**
5 *I am confident I can do any problems in this category correctly.*
4 *I am confident I can do most of the problems in this category correctly.*
3 *I understand how to do the problems in this category, but I still make a lot of mistakes.*
2 *I feel unsure about how to do these problems.*
1 *I know I don't understand how to do these problems.*

| Topic or Skill | Before Studying | After Studying |
|---|---|---|
| Understanding what an exponent represents (what does $2^5$ mean?). | | |
| Understanding how expressions like $(-3)^2$ and $-3^2$ OR $2x^2$ and $(2x)^2$ are different. | | |
| Knowing the product, quotient and power rules for exponents. | | |
| Correctly applying the product, quotient, and power rules for exponents. | | |
| Knowing the exponent rules for zero and negative exponents. | | |
| Rewriting an expression to eliminate the negative exponents. | | |
| Simplifying exponential expressions involving negative exponents. | | |
| Simplifying exponential expressions that have zero exponents. | | |
| Simplifying expressions involving many exponent rules all mixed up (terms in parentheses, fractions, negative exponents, zero exponents, etc.). | | |
| Converting back and forth between standard notation and scientific notation. | | |
| Performing operations (like multiplication and division) using scientific notation | | |
| Using the scientific notation button on a calculator to perform calculations involving scientific notation. | | |

# POLY: Simplifying Polynomials

# Guided Learning Activity
## Language of Polynomials

A **polynomial** is a single term or a sum of terms in which all variables have whole-number exponents and no variable appears in a denominator. Recall that the **terms** of an algebraic expression are separated by addition (remember that subtraction can be rewritten as addition of a negative term). A polynomial with exactly one term is called a **monomial**; exactly two terms, a **binomial**; and exactly three terms, a **trinomial**. A polynomial can be written with one or more variables.

With your class, fill out the table below. The first one has been done for you.

| Expression (rewrite with addition symbols if necessary) | How many variables? | Classification | | | |
|---|---|---|---|---|---|
| | | Monomial | Binomial | Trinomial | Polynomial |
| $5x^2 - 2x + 7$<br>$5x^2 + (-2x) + 7$ | 1 | | | X | X |
| $y - 2.5$ | | | | | |
| $-5x^2 y$ | | | | | |
| $4x^2 - \dfrac{2}{x^3}$ | | | | | |
| $b^3 + \dfrac{1}{3}b^2 - \dfrac{1}{2}b + 12$ | | | | | |
| $x^2 - 3xy - 10y^2$ | | | | | |

Polynomials are often written in **descending powers** of the variable (the variable exponents decrease from left to right). When a polynomial is written in descending powers, the first term is called the **lead term**. The coefficient of the lead term is called the **lead coefficient**. Recall that a term that consists of a single number is called the **constant term**.

| Expression | Rewrite the expression in descending powers of $x$ | Lead term | Lead coefficient | Constant term (if there is one) |
|---|---|---|---|---|
| $4x - x^2$ | $-x^2 + 4x$ | $-x^2$ | $-1$ | None |
| $3x + 5 - 2x^2$ | | | | |
| $x^3 + 200x + 300x^2$ | | | | |
| $7x^3 - 2 - 3x^4$ | | | | |

The **degree of a term** of a polynomial in one variable is the value of the exponent on the variable. Thus, the degree of $7x^4$ is equal to 4. If the polynomial has more than one variable, the degree of a term is the *sum* of the exponents on the variables in that term. Thus, the degree of $7x^2 y^3$ is 5. For a constant term, we can imagine an unwritten variable with a zero power, consider that 7 could be written as $7x^0$. Thus, the **degree of a nonzero constant** is zero. We can also discuss the **degree of a polynomial** which is the same as the degree of the highest degree term of the polynomial.

| Expression | Degree of... | | | | Degree of polynomial |
|---|---|---|---|---|---|
| | $1^{st}$ term | $2^{nd}$ term | $3^{rd}$ term | $4^{th}$ term | |
| $5x^2 - 3x + 7$ | 2 | 1 | 0 | None | 2 |
| $y - 2.5$ | | | | | |
| $-5x^4 y + 2x^2 y^2$ | | | | | |
| $b^3 + \dfrac{1}{3}b^2 - \dfrac{1}{2}b + 12$ | | | | | |
| $x^2 - 3xy - 10 y^2$ | | | | | |
| $x^3 + 300x^2 + 200x - 1000$ | | | | | |

# Student Activity
## Match Up on Polynomial Evaluation

Remember that it may be helpful to first create a parentheses skeleton for each polynomial before you substitute the designated values.

For example, $x^2 - 3x$ would be first rewritten as $(\quad)^2 - 3(\quad)$.

**Match-up:** Match each of the answers in the squares of the grid below with its result in choices A through D. If you do not see the result in choices A through D, then choose choice E (none of these).

**A** 9      **B** 1      **C** $-2$      **D** 0      **E** None of these

| | | |
|---|---|---|
| Evaluate $5x^2 + 10x$ for $x = -2$. | Evaluate $5x^2 - 5x + \dfrac{9}{4}$ for $x = \dfrac{1}{2}$. | Simplify: $3x^2 + 5 - 2x^2 - 7 - x^2$ |
| Evaluate $x^2 y$ for $x = -1$ and $y = -2$. | Evaluate $x^2 + 9x + \dfrac{23}{9}$ for $x = \dfrac{2}{3}$. | Evaluate $x^3 + x^2 + x - 2$ for $x = -2$ |
| Simplify: $\dfrac{x^5}{2} - 2x^5 + \dfrac{3x^5}{2}$ | Let $y = 3x^2 - 2x - 7$; find $y$ when $x = 2$. | Let $y = 2.25x^3 + 7.75x - 1$; find $y$ when $x = 1$. |

# Student Activity
## Caught in the Net

**Directions:** Which fish get "trapped" between the two nets?

Graph $y = \dfrac{1}{9}x^3 + 1$ and $y = -\dfrac{3}{4}x^2 + 10$ (the "nets") to find out.

# Guided Learning Activity
## Vertical Form of Polynomial Addition and Subtraction

The expressions $342 + 605$ and $(3x^2 + 4x + 2) + (6x^2 + 5)$ are simplified in almost the same way: For both expressions, we line up the like terms vertically; then add.

**Example 1:**

$$
\begin{array}{rrr}
 & 3 & 4 & 2 \\
+ & 6 & 0 & 5 \\
\hline
 & 9 & 4 & 7
\end{array}
$$

| | 3 hundreds | 4 tens | 2 ones |
|---|---|---|---|
| + | 6 hundreds | 0 tens | 5 ones |
| | 9 hundreds | 4 tens | 7 ones |

$$
\begin{array}{rrrr}
 & 3x^2 & +4x & +2 \\
+ & 6x^2 & +0x & +5 \\
\hline
 & 9x^2 & +4x & +7
\end{array}
$$

One of the differences between the vertical addition of the expressions is that we cannot "carry" coefficients greater than 9 to the next column like we carry numbers greater than 9. Another difference is that we can have negative coefficients if that is part of the original expression.

**Example 2:** $(4x^2 + 5x + 9) + (-2x^2 + 5x - 6)$ becomes

$$
\begin{array}{rrrr}
 & 4x^2 & +5x & +9 \\
+ & -2x^2 & +5x & -6 \\
\hline
 & 2x^2 & +10x & +3
\end{array}
$$

Now try these. Make sure to **line up the like terms** in the same columns and keep the signed coefficients with the terms in the columns.

**a.** $(y^3 + 3y^2 - 4y + 2) + (9y^2 + 4y - 7)$     **b.** $(a^2 + 4ab - 2b^2) + (a^2 + 8ab - 3b^2)$

In subtraction, we have to remember that the second row of each column is being subtracted from the first row **for every entry**. Below we look at the similar expressions $749 - 615$ and $(7x^2 + 4x + 9) - (6x^2 + x + 5)$.

**Example 3:**

$$
\begin{array}{rrr}
 & 7 & 4 & 9 \\
- & 6 & 1 & 5 \\
\hline
 & 1 & 3 & 4
\end{array}
$$

| | 7 hundreds | 4 tens | 9 ones |
|---|---|---|---|
| − | 6 hundreds | 1 ten | 5 ones |
| | 1 hundred | 3 tens | 4 ones |

$$
\begin{array}{rrrr}
 & 7x^2 & +4x & +9 \\
- & 6x^2 & +1x & +5 \\
\hline
 & 1x^2 & +3x & +4
\end{array}
$$

We've learned on many occasions that subtraction of grouped terms can be tricky as the $-1$ must be distributed to the grouped terms. Because some of the terms in the second row have a + in front of the coefficients or constants, it is confusing to think of this as a subtraction problem. To be clear, we should write the vertical format of subtraction like this:

$$
\begin{array}{rrrr}
 & 7x^2 & +4x & +9 \\
- & (6x^2 & +x & +5)
\end{array}
$$

And since polynomial subtraction can be rewritten without the grouping as polynomial addition (after the distribution), we can further alter the problem:

$$
\begin{array}{l}
\phantom{-}\,7x^2 \quad +4x \quad +9 \quad \text{becomes} \\
\underline{- \left(6x^2 \quad\; +x \quad +5\right)}
\end{array}
\qquad
\begin{array}{l}
\phantom{+}\,7x^2 \quad +4x \quad +9 \\
\underline{+ \;\; -6x^2 \quad\; -x \quad -5} \\
\phantom{+}\,1x^2 \quad +3x \quad +4
\end{array}
$$

**Example 4:** Let's go through this process with $\left(4x^2 + 3x - 6\right) - \left(3x^2 - 2\right)$

First we line up the like terms, still using the grouping symbols for the subtraction:

$$
\begin{array}{l}
\phantom{-}\,4x^2 \quad +3x \quad -6 \quad \text{becomes} \\
\underline{- \left(3x^2 \quad +0x \quad -2\right)}
\end{array}
\qquad
\begin{array}{l}
\phantom{+}\,4x^2 \quad +3x \quad -6 \\
\underline{+ \;\; -3x^2 \quad -0x \quad +2} \\
\phantom{+}\,1x^2 \quad +3x \quad -4
\end{array}
$$

**Now try these.** Write the polynomial subtraction in the vertical format, first with grouping symbols as subtraction, then performing the distribution by $-1$ and rewriting as addition.

**c.** $\left(y^3 + 9y^2 - 4y + 2\right) - \left(6y^2 + 2y - 7\right)$

**d.** $\left(a^2 + 4ab - 2b^2\right) - \left(a^2 - 8ab + 3b^2\right)$

**e.** $\left(6x^3 - 5x - 4\right) - \left(2x^2 - 8x + 3\right)$

# Student Activity
## When Does Order Matter?

**Directions:** Add or subtract in each of the expressions below and then combine the like terms. Write the simplified expression in descending powers of $x$.

**1.** Add: $\left(x^2 + 3x - 7\right) + \left(4x^2 + 5x - 2\right)$

**2.** Add: $\left(4x^2 + 5x - 2\right) + \left(x^2 + 3x - 7\right)$

**3.** What lesson can we learn from problems **1** and **2**?

**4.** Subtract: $\left(x^2 + 3x - 7\right) - \left(4x^2 + 5x - 2\right)$

**5.** Subtract: $\left(4x^2 + 5x - 2\right) - \left(x^2 + 3x - 7\right)$

**6.** What lesson can we learn from problems **4** and **5**?

**7.** Subtract: $\left(x^2 + 3x - 7\right) - \left(4x^2 + 5x - 2\right)$

**8.** Simplify: $x^2 + 3x - 7 - 4x^2 + 5x - 2$

**9.** What lesson can we learn from problems **7** and **8**?

**10.** Add: $\left(x^2 + 3x - 7\right) + \left(4x^2 + 5x - 2\right)$

**11.** Simplify: $x^2 + 3x - 7 + 4x^2 + 5x - 2$

**12.** What lesson can we learn from problems **10** and **11**?

# Student Activity
## Monomial Addition and Multiplication Tables

Here are simple addition and multiplication tables with monomial inputs.

**Addition:**

| + | $x$ | $2x$ | $x^2$ | $2x^2$ |
|---|---|---|---|---|
| $x$ | $2x$ | $3x$ | | |
| $2x$ | $3x$ | $4x$ | | |
| $x^2$ | | | $2x^2$ | $3x^2$ |
| $2x^2$ | | | $3x^2$ | $4x^2$ |

**Multiplication:**

| $\bullet$ | $x$ | $2x$ | $x^2$ | $2x^2$ |
|---|---|---|---|---|
| $x$ | $x^2$ | $2x^2$ | $x^3$ | $2x^3$ |
| $2x$ | $2x^2$ | $4x^2$ | $2x^3$ | $4x^3$ |
| $x^2$ | $x^3$ | $2x^3$ | $x^4$ | $2x^4$ |
| $2x^2$ | $2x^3$ | $4x^3$ | $2x^4$ | $4x^4$ |

**Notice that there are shaded spaces left in the addition table where unlike terms can not be combined.**

**Directions:** The following tables are addition or multiplication tables involving monomial inputs. Fill in the missing squares with the appropriate monomials. Write NL (for "not like") or shade the grid spaces where the terms *cannot* be combined.

| + | $-3x^2$ | $-x^2$ | $-3x$ | $-x$ | $0$ | $x$ | $3x$ | $x^2$ | $3x^2$ |
|---|---|---|---|---|---|---|---|---|---|
| $-3x^2$ | | | | | | | | | |
| $-x^2$ | | | | | | | | | |
| $-3x$ | | | | | | | | | |
| $-x$ | | | | | | | | | |
| $0$ | | | | | | | | | |
| $x$ | | | | | | | | | |
| $3x$ | | | | | | | | | |
| $x^2$ | | | | | | | | | |
| $3x^2$ | | | | | | | | | |

| • | $-3x^2$ | $-x^2$ | $-3x$ | $-x$ | $0$ | $x$ | $3x$ | $x^2$ | $3x^2$ |
|---|---|---|---|---|---|---|---|---|---|
| $-3x^2$ | | | | | | | | | |
| $-x^2$ | | | | | | | | | |
| $-3x$ | | | | | | | | | |
| $-x$ | | | | | | | | | |
| $0$ | | | | | | | | | |
| $x$ | | | | | | | | | |
| $3x$ | | | | | | | | | |
| $x^2$ | | | | | | | | | |
| $3x^2$ | | | | | | | | | |

**Directions:** The following tables are addition or multiplication tables with missing information. Fill in the missing squares with the appropriate monomials. Write NL for "not like" in the squares where unlike terms cannot be added.

| + | $7x$ | | | | $x^2$ |
|---|---|---|---|---|---|
| $5x$ | | | | $-2x$ | |
| | $7x$ | $-2x^2$ | $0$ | | $x^2$ |
| | | $-6x^2$ | $-4x^2$ | | $-3x^2$ |
| $-4x$ | | | | $-11x$ | |
| | | | $6x^2$ | | $7x^2$ |

| • | | $x^6$ | | |
|---|---|---|---|---|
| $x^4$ | $-5x^6$ | $x^{10}$ | | |
| | $-15x^7$ | $3x^{11}$ | | $24x^8$ |
| $-7x$ | $35x^3$ | | | |
| | | $x^6$ | | $8x^3$ |
| | | $x^7$ | $2x^2$ | |

# Guided Learning Activity
## Vertical Form of Polynomial Multiplication

The expressions $304 \cdot 21$ and $(3x^2 + 4) + (2x + 1)$ are simplified in almost the same way: For both expressions, we line up the like terms vertically; then multiply.

**Example 1:**

$$
\begin{array}{r}
3\ \ 0\ \ 4 \\
\times \quad\ 2\ \ 1 \\
\hline
3\ \ 0\ \ 4 \\
6\ \ 0\ \ 8 \\
\hline
6\ \ 3\ \ 8\ \ 4
\end{array}
$$

| | | | |
|---|---|---|---|
| 3 hundreds | 0 tens | 4 ones | |
| × | | 2 tens | 1 one |
| 3 hundreds | 0 tens | 4 ones | |
| 6 thousands | 0 hundreds | 8 tens | |
| 6 thousands | 3 hundreds | 8 tens | 4 ones |

$$
\begin{array}{rrrr}
& 3x^2 & +0x & +4 \\
\times & & +2x & +1 \\
\hline
& +3x^2 & +0x & +4 \\
6x^3 & +0x^2 & +8x & \\
\hline
6x^3 & +3x^2 & +8x & +4
\end{array}
$$

One of the differences between the vertical multiplications of the expressions is that we cannot "carry" coefficients greater than 9 to the next column like we carry numbers greater than 9. Another difference is that we can have negative coefficients if that is part of the original expression.

**Example 2:** $(4x + 5)(2x - 6)$ becomes

$$
\begin{array}{rrr}
& 4x & +5 \\
\times & 2x & -6 \\
\hline
& -24x & -30 \\
8x^2 & +10x & \\
\hline
8x^2 & -14x & -30
\end{array}
$$

Now try these. Make sure to **line up the like terms** in the same columns and keep the signed coefficients with the terms in the columns.

**a.** $(7y - 6)(y + 5)$

**b.** $(x^2 - 3x + 6)(x - 9)$

**c.** $(a^2 + 8a + 6)(a^2 + 5)$

**d.** $(2x^2 + 5x - 4)(x^2 - 6x + 7)$

# Student Activity

## Not All Multiples are the Same

**Directions:** All the expressions below are written in sets of triples. First multiply and simplify each expression. Then circle the pairs from each problem that are really equivalent (if there are any).

**1.** $(x+3)(x-5)$        $(x-5)(x+3)$        $(x-3)(x+5)$

**2.** $(x-2)(x+6)$        $(x+2)(x-6)$        $(x-6)(x+2)$

**3.** $(x-2)(x-2)$        $(2-x)(2-x)$        $(x-2)(2-x)$

**4.** $(x+3)(x+4)$        $(3+x)(4+x)$        $(x-3)(x-4)$

**5.** $(x+2)(x^2-2x+4)$        $(x-2)(x^2+2x-4)$        $(x-2)(x-2)(x-2)$

**6.** $(x+5)(x-5)$        $(x-5)^2$        $(5+x)(5-x)$

# Student Activity
## Multiplying Binomials with FOIL

When we multiply two **binomials**, we can use a shortcut method called the FOIL method. FOIL stands for **First Outer Inner Last**. The first one has been done for you.

Trent began looking for new ways to use his algebra vocabulary.

| Expression | Find these terms of the multiplication… | | | | Combine like terms and simplify |
| --- | --- | --- | --- | --- | --- |
| | First | Outer | Inner | Last | |
| $(x+3)(x-5)$ | $x^2$ | $-5x$ | $3x$ | $-15$ | $x^2-2x-15$ |
| $(u-4)(u-8)$ | | | | | |
| $(y+3)(y+3)$ | | | | | |
| $(x+7)(x-7)$ | | | | | |
| $\left(a+\dfrac{1}{3}\right)\left(a+\dfrac{2}{3}\right)$ | | | | | |
| $(2x+5)(3x-4)$ | | | | | |
| $(4x+3)(6x+7)$ | | | | | |
| $(x^2+4)(x^2+6)$ | | | | | |
| $\left(2w+\dfrac{1}{2}\right)\left(w-\dfrac{3}{2}\right)$ | | | | | |
| $(s^3-2)(s^3-8)$ | | | | | |
| $(3.6x+1)(2x-4.5)$ | | | | | |

# Student Activity
## Charting Squared Binomials

When we **square a binomial**, we are really just multiplying two binomials and can use the FOIL method to do it. For example, $(x-6)^2$ is really the same as $(x-6)(x-6)$. Fill in the empty boxes in the table below. The first one has been done for you.

| Expression | Find these terms of the multiplication... | | | | Combine like terms and simplify |
|---|---|---|---|---|---|
| | First | Outer | Inner | Last | |
| $(x-6)^2$ <br> $(x-6)(x-6)$ | $x^2$ | $-6x$ | $-6x$ | $36$ | $x^2-12x+36$ |
| $(u+4)^2$ <br> $(u+4)(u+4)$ | | | | | |
| $(a-5)^2$ <br> $(a-5)(a-5)$ | | | | | |
| $\left(x+\frac{1}{2}\right)^2$ <br> $\left(x+\frac{1}{2}\right)\left(x+\frac{1}{2}\right)$ | | | | | |
| $(A+B)^2$ <br> $(A+B)(A+B)$ | | | | | |
| $(A-B)^2$ <br> $(A-B)(A-B)$ | | | | | |

Notice that in the first row, $-6x-6x=-12x$ is the same as $2(-6x)=-12x$. If you look back through the rows in the table above, you should see a similar pattern in every row.

We can thus use the formula below to determine the product of a binomial squared.
$$(A+B)^2 = A^2 + 2\cdot A\cdot B + B^2$$

The middle term is found by taking twice the product of $A$ and $B$. Make sure that you use parentheses when you are squaring the $A$-term and the $B$-term! For example, the term like $2x^2$ is not the same as $(2x)^2$.

**Directions:** Using the binomial squared formulas, fill in the empty boxes in the table below. The first one has been done for you.

| Expression | Find these terms of binomial squared. | | | Final expression $A^2 + 2AB + B^2$ |
| --- | --- | --- | --- | --- |
| | The square of the first term: $A^2$ | Twice the product of the first and second terms: $2 \cdot A \cdot B$ | The square of the last term: $B^2$ | |
| $(x-6)^2$ | $(x)^2$ | $2(x)(-6)$ | $(-6)^2$ | $x^2 - 12x + 36$ |
| $(u+4)^2$ | | | | |
| $(a-5)^2$ | | | | |
| $\left(x+\dfrac{1}{2}\right)^2$ | | | | |
| $(2y+5)^2$ | | | | |
| $(3a-1)^2$ | | | | |
| $(5x+4y)^2$ | | | | |
| $(9-u)^2$ | | | | |
| $(v^3+2)^2$ | | | | |
| $\left(6z-\dfrac{4}{3}\right)^2$ | | | | |
| $(x^4-y^3)^2$ | | | | |

# Student Activity

## Charting a Sum-Difference

There's one more special product to look at, but again, we'll start by going back to the FOIL method to figure it out. Fill in the empty boxes in the table below. The first one has been done for you.

| Expression | Find these terms of the multiplication... | | | | Combine like terms and simplify |
|---|---|---|---|---|---|
| | First | Outer | Inner | Last | |
| $(x+3)(x-3)$ | $x^2$ | $-3x$ | $3x$ | $-9$ | $x^2-9$ |
| $(x-3)(x+3)$ | | | | | |
| $(x-3)(x-3)$ | | | | | |
| $(x+3)(x+3)$ | | | | | |
| $(a-5)(a+5)$ | | | | | |
| $(a+5)(a-5)$ | | | | | |
| $(a-5)(a-5)$ | | | | | |
| $(a+5)(a+5)$ | | | | | |
| $\left(x+\frac{1}{2}\right)\left(x-\frac{1}{2}\right)$ | | | | | |
| $(A+B)(A-B)$ | | | | | |

Circle all the rows where the inner and outer terms were additive inverses (where their sum was zero). Did it happen in every row? _____

What is special about the beginning expressions where the middle terms **were** additive inverses?

# Student Activity
## Visual Representation for Multiplying Binomials

Multiplication can be represented using a diagram called an area model. Find an expression for the area of each of the squares and rectangles below. Remember that the formula for finding the area of a rectangle is Area = (length)(width). Write the expression for the area inside each square or rectangle, and then write it inside the same square or rectangle in the composite figure below. **Thus each area will be written in two places.**

**Problem 1:**

To find the area of the composite rectangle, we could either add the areas of all the smaller pieces, which would give this area:

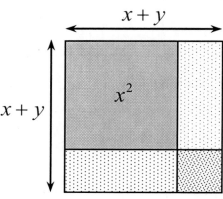

_____

Or we could find the area by using the area formula, which would give Area $= (x+y)(x+y) = (x+y)^2$

This tells us that $(x+y)^2 =$ _____ .

**Problem 2:** Do this one the same way.

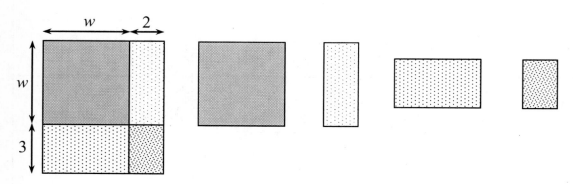

What mathematical equation does this set of figures tell us?

_____  =  _____

# Student Activity
## Exponential and Polynomial Heteronyms

**Directions:** In writing, there are words that are spelled the same but have different pronunciations and different definitions; these are called heteronyms. Many polynomial and exponential expressions look similar but are really very different (almost like mathematical heteronyms). In each set of expressions below, pay close attention to the use of parentheses and the mathematical operations and notation.

**1.**

| $(3x)^2$ | $2(3x)$ | $2(x+3)$ | $(x+3)^2$ | $(2x)^3$ |
|---|---|---|---|---|
|  |  |  |  |  |

**2.**

| $(x-4)^2$ | $2(x-4)$ | $2(-4x)$ | $(-2x)^4$ | $(4x)^2$ |
|---|---|---|---|---|
|  |  |  |  |  |

**3.**

| $(x+3)(x-2)$ | $(3x)(-2x)$ | $3x(x-2)$ | $-2x(x+3)$ | $x^3(-2x)$ |
|---|---|---|---|---|
|  |  |  |  |  |

**4.**

| $(x+5)(x-5)$ | $(5x)(x-5)$ | $5x(-5x)$ | $x^5(x-5)$ | $x^5 \cdot x^5$ |
|---|---|---|---|---|
|  |  |  |  |  |

**5.**

| $5x^2(4x^3)$ | $(5+2x)(4x+3)$ | $5x^2(4x+3)$ | $(5x)^2(4x^3)$ | $(5x)^2(4x+3)$ |
|---|---|---|---|---|
|  |  |  |  |  |

# Guided Learning Activity
## Polynomial Long Division

The expressions $168 \div 14$ and $(1x^2 + 6x + 8) \div (1x + 4)$ are simplified in almost the same way using long division. For both expressions, we line up the like terms vertically; then divide.

**Example 1:**

$$
\begin{array}{r}
1 \quad 2 \\
1\;4\overline{)1\;6\;8} \\
-1\;4 \\
\hline
2\;8 \\
-2\;8 \\
\hline
0
\end{array}
\qquad
\begin{array}{r}
1x \quad +2 \\
1x\;+4\overline{)1x^2\;+6x\;+8} \\
-(1x^2\;+4x) \\
\hline
2x\;+8 \\
-(2x\;+8) \\
\hline
0
\end{array}
$$

Notice that in both long division problems, the like terms line up in the same columns.

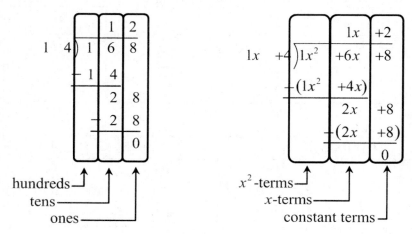

As you perform the operations in polynomial long division, you will want to make sure that you line up the like terms as well.

**If you are missing a column of terms, you will need to insert a placeholder.** For example, $12\overline{)1008}$ would not give the same quotient as $12\overline{)18}$, the two zeros in $1008$ act as placeholders for the tens place and the hundred place. Also, we would not change the order of the digits in the columns, for example $12\overline{)1008}$ is not the same as $12\overline{)8010}$.

In a similar fashion, we would not write $x+2\overline{)x^3 + 8}$ because the $x$-term and the $x^2$-term columns are missing. Instead we write $x+2\overline{)x^3 + 0x^2 + 0x + 8}$ to hold the places of the necessary columns.

**In polynomial long division, the terms of the dividend and the divisor must be written in descending order with no missing terms.**

**1.** Rewrite these long division problems so that all the columns are accounted for and the terms are in the proper order.  Write in any unwritten ones.  For example, write $x^2$ as $1x^2$.

**a.** $x-2\overline{\smash{)}x^3-4x+5}$     **b.** $x-1\overline{\smash{)}4-5x+x^2}$     **c.** $2+a\overline{\smash{)}a^2-4}$

Here is a polynomial long division example shown step by step.  Your instructor will walk you through the steps.  You should add your own notes so that you will remember what has happened on each step.  At the very least, highlight the new part of each step in color so that it is easier to see later on when you go back through your notes.

**Example 2:**

$$\begin{array}{r} x \\ x+8\overline{\smash{)}x^2+5x-24} \end{array}$$

$$\begin{array}{r} x \\ x+8\overline{\smash{)}x^2+5x-24} \\ \underline{x^2+8x} \end{array}$$

$$\begin{array}{r} x \\ x+8\overline{\smash{)}\,x^2+5x-24} \\ -\left(x^2+8x\right) \\ \hline \end{array}$$

$$\begin{array}{r} x \\ x+8\overline{\smash{)}\,x^2+5x-24} \\ \underline{-x^2-8x} \end{array}$$

$$\begin{array}{r} x \\ x+8\overline{\smash{)}\,x^2+5x-24} \\ \underline{-x^2-8x} \\ -3x \end{array}$$

$$\begin{array}{r} x\ -3 \\ x+8\overline{\smash{)}\,x^2+5x-24} \\ \underline{-x^2-8x} \\ -3x-24 \end{array}$$

$$\begin{array}{r} x\ -3 \\ x+8\overline{\smash{)}\,x^2+5x-24} \\ \underline{-x^2-8x} \\ -3x-24 \\ \underline{-3x-24} \end{array}$$

$$\begin{array}{r} x\ -3 \\ x+8\overline{\smash{)}\,x^2+5x-24} \\ \underline{-x^2-8x} \\ -3x-24 \\ -(-3x-24) \\ \hline \end{array}$$

$$\begin{array}{r} x\ -3 \\ x+8\overline{\smash{)}\,x^2+5x-24} \\ \underline{-x^2-8x} \\ -3x-24 \\ +3x+24 \end{array}$$

$$\begin{array}{r} x\ -3 \\ x+8\overline{\smash{)}\,x^2+5x-24} \\ \underline{-x^2-8x} \\ -3x-24 \\ \underline{+3x+24} \\ 0 \end{array}$$

Usually, in practice, we do not write this step: $x+8 \overline{)\phantom{x}} \begin{array}{c} x \\ x^2 + 5x - 24 \\ -\left(x^2 + 8x\right) \end{array}$

We skip directly from $x+8 \overline{)\phantom{x}} \begin{array}{c} x \\ x^2 + 5x - 24 \\ x^2 + 8x \end{array}$ to $x+8 \overline{)\phantom{x}} \begin{array}{c} x \\ x^2 + 5x - 24 \\ -x^2 - 8x \\ \hline -3x \end{array}$ by changing the signs.

**Change the signs using a colored pencil**, so that you can see that you have done this step *after* writing the terms for the original row.

**Directions:** Try each of the problems below. Stop after each problem and wait for your instructor to go over the answer. If you make mistakes, correct them using a colored pencil so that you will see the corrections (and original mistakes) when you go back to study later.

**2.** $\left(2x^2 - 5x - 12\right) \div \left(x - 4\right)$

If you think that's amazing, you should see the polonomial long division I just did!

**3.** $\left(7x^2 + x^3 - 28 + 5x\right) \div \left(x + 4\right)$

If there is a remainder after the long division is done, we write this as a ratio with the divisor. Consider the following example to see why it is done this way.

**Example 3:**

$\frac{25}{4} = 6\frac{1}{4}$ or we could write it like this $4\overline{)25}$ , which gives the result $6\text{ R }1$.

$$\begin{array}{r} 6 \\ 4\overline{)25} \\ \underline{-24} \\ 1 \end{array}$$

If we want to write $6\,\text{R}\,1$ mathematically, it should be $6\frac{1}{4}$, which is the same as $6+\frac{1}{4}$.

**Example 4:**

Likewise, when we do the following division, we get a remainder:

$$\begin{array}{r} x +3 \\ x-5\overline{)x^2 -2x -16} \\ \underline{-x^2 +5x} \\ 3x -16 \\ \underline{-3x +15} \\ -1 \end{array}$$

So we write the answer: $x+3+\dfrac{-1}{x-5}$  $\leftarrow$ remainder
$\leftarrow$ divisor

**Directions:** Try each of the problems below. Stop after each problem and wait for your instructor to go over the answer. If you make mistakes, correct them using a colored pencil so that you will see the corrections (and original mistakes) when you go back to study later.

**4.** $\left(x^2 +4x -8\right) \div \left(x+6\right)$

**5.** $\left(x^3 +4x^2 -5x -22\right) \div \left(x+4\right)$

# Student Activity

## Match Up on Polynomial Division

**Directions:** Match each of the expressions in the squares of the grid below with an equivalent simplified expression from the top. If an equivalent expression is not found among the choices A through D, then choose E (none of these). Be careful – you may have to simplify rational expression terms.

**A** $x+4$      **B** $2x-3$      **C** $x+4+\dfrac{3}{x-1}$      **D** $2x-3-\dfrac{4}{x+3}$      **E** None of these

| | | |
|---|---|---|
| $x-1\overline{)x^2+3x-1}$ | $x+7\overline{)x^2+11x+28}$ | $x+3\overline{)x^2+7x+8}$ |
| $x+3\overline{)2x^2+3x-9}$ | $\left(4x^2+6x-26\right)\div\left(2x+6\right)$ | $\left(x^2+7x+16\right)\div\left(x+3\right)$ |
| $\left(x^2-16\right)\div\left(x-4\right)$ | $\dfrac{6x^2y+2x^3y}{2x^2y}$ | $\dfrac{36a^2b^2cx-54a^2b^2c}{18a^2b^2c}$ |

# Student Activity
## Pay Attention!

**Directions:** When you're simplifying polynomials, many problems look similar. You **must** pay attention to the operation and directions for the problems. Try the problems below, but please, **pay attention** to what you're doing!

**1.** Multiply: $\left(x^2 + 5x - 14\right)\left(x + 7\right)$

**2.** Add: $\left(x^2 + 5x - 14\right) + \left(x + 7\right)$

**3.** Divide using long division: $\left(x^2 + 5x - 14\right) \div \left(x + 7\right)$

**4.** Subtract: $\left(x^2 + 5x - 14\right) - \left(x + 7\right)$

**5.** Add: $\left(27x^3 + 12x^2\right) + \left(3x^2\right)$

**6.** Divide: $\left(27x^3 + 12x^2\right) \div \left(3x^2\right)$

**7.** Multiply: $\left(27x^3 + 12x^2\right)\left(3x^2\right)$

**8.** Subtract: $\left(27x^3 + 12x^2\right) - \left(3x^2\right)$

# Student Activity
## Language of Polynomial Operations

**Directions:** Now that we've looked at adding, subtracting, multiplying, and dividing polynomials, let's practice with the language of polynomials. You **do not** have to carry out the mathematics in each problem; simply translate the sentence into mathematical notation. The first one has been done for you.

| | Directions in words | Write the expression with mathematical notation, using the proper grouping terms, fraction bars, or long division notation where necessary. |
|---|---|---|
| 1. | Find the sum of $x+4$ and $x-1$. | $(x+4)+(x-1)$ |
| 2. | Divide $x^2-3x+2$ by $x-1$. | |
| 3. | Find the difference of $x^2-2x-7$ and $x^2-4$. | |
| 4. | Find the quotient of $-6x^2y^3$ and $2x^2y$. | |
| 5. | Find the product of $x^2+4x+3$ and $x+5$. | |
| 6. | Find the product of $x+6$ and $x-9$. | |
| 7. | Find the total of $x+3$, $x+4$, and $x-2$. | |
| 8. | Find twice the sum of $x^2$ and $x^2-4$. | |
| 9. | Find the quotient of $10a^3b-25a^4$ and $5a^2$. | |
| 10. | What is $x^2-10x+16$ less $2x+4$? | |

# Assess Your Understanding
## Simplifying Polynomials

For each of the following, describe the strategies or key steps that will help you **start** the problem. You do **not** have to complete the problems.

| | | What will help you to start this problem? |
|---|---|---|
| **1.** | Write $6 - 5x^2 + 3x$ in descending powers of $x$. | |
| **2.** | What is the leading coefficient of $x^2 + 5x$ ? | |
| **3.** | Expand the binomial: $(3x - 5)^2$ | |
| **4.** | Divide: $(4x^2 - 9) \div (2x + 3)$ | |
| **5.** | Multiply: $(x - 5)(x - 6)$ | |
| **6.** | Subtract: $(2x^2 - 4x) - (3x - 7)$ | |
| **7.** | Multiply: $(12 + a)(12 - a)$ | |
| **8.** | Divide: $(14x^3 - 21x^2 + 28x) \div (7x)$ | |
| **9.** | Evaluate $y = x^3 - 4x^2 + 12$ for $x = -3$. | |
| **10.** | Multiply: $(x + 3)(x^2 - 6x + 8)$ | |

# Metacognitive Skills
## Simplifying Polynomials

Metacognitive skills refer to the ability to judge how well you have learned something and to effectively direct your own learning and studying. This is a self-evaluation tool designed to help you focus your studying and to improve your metacognitive skills with regards to this math class.

Fill the 1$^{st}$ column out **before** you begin studying. Fill the 2$^{nd}$ column out after you study for your test.

Go back to this assessment after your test and circle any of the ratings that you would change – this identifies the "disconnects" between what you **thought** you knew well and what you **actually** knew well.

**Use the scale below to assign a number to each topic.**
5  *I am confident I can do any problems in this category correctly.*
4  *I am confident I can do most of the problems in this category correctly.*
3  *I understand how to do the problems in this category, but I still make a lot of mistakes.*
2  *I feel unsure about how to do these problems.*
1  *I know I don't understand how to do these problems.*

| Topic or Skill | Before Studying | After Studying |
|---|---|---|
| Identifying the number of terms in a polynomial. | | |
| Writing a polynomial in descending order. | | |
| Identifying the degree or coefficient of a term or of a polynomial. | | |
| Evaluating a polynomial for a given value (especially negative values). | | |
| Graphing an equation that involves $x^2$ or $x^3$ by making a table of values and plotting points. | | |
| Adding or subtracting polynomials. | | |
| Multiplying a monomial by a polynomial. Example: $3x^2(x^2 - 2x + 4)$. | | |
| Multiplying two binomials using FOIL. Example: $(x+3)(x+7)$ | | |
| Multiplying any two polynomials. Example: $(x+3)(x^2 - 4x + 7)$ | | |
| Multiplying a special sum-difference product. Example: $(a+7)(a-7)$ | | |
| Squaring a binomial. Example: $(x-4)^2$ (be careful, the answer's **not** $x^2 - 16$) | | |
| Distinguishing between addition and multiplication, because the rules are not the same! For example, $(x+5)^2$ vs. $(5x)^2$ OR $(x+2)(x+3)$ vs. $(x+2)+(x+3)$. | | |
| Dividing a polynomial by a monomial. Like $(6x^3 + 3x^2 - 9x) \div 3x$ | | |
| Using polynomial long division, including what to do with a remainder! Example: $(x^2 - 6x + 8) \div (x-2)$ | | |
| Knowing how to make adjustments in polynomial long division when there's a term missing. | | |

# FACT: Factoring and More

# Guided Learning Activity
## Charting Factors and GCFs

**Directions Part I:** For each expression, decide if each of the terms in the top row is a factor of the given expression. Place an **X** in any column for which the term **would** be a factor, then use those decisions to help you find the GCF.

| | | 2 | 3 | 4 | 6 | 8 | 12 | $x$ | $x^2$ | $x^3$ | $x^4$ | $y$ | $y^2$ | $y^3$ | $y^4$ | GCF |
|---|---|---|---|---|---|---|---|---|---|---|---|---|---|---|---|---|
| **a.** | $6x^3 - 3x^2$ | | X | | | | | X | X | | | | | | | $3x^2$ |
| **b.** | $18y^2 + 24y$ | | | | | | | | | | | | | | | |
| **c.** | $19x^4y - 3xy^2$ | | | | | | | | | | | | | | | |
| **d.** | $16 + 40x^3$ | | | | | | | | | | | | | | | |
| **e.** | $24x^4y^4 + 36x^2y^2$ | | | | | | | | | | | | | | | |
| **f.** | $x^4 + y^4$ | | | | | | | | | | | | | | | |
| **g.** | $36x + 2x^2$ | | | | | | | | | | | | | | | |
| **h.** | $9y^4 + 3xy^4$ | | | | | | | | | | | | | | | |
| **i.** | $x^3y^2w + x^2y^3z$ | | | | | | | | | | | | | | | |

**Directions Part II:** Given each expression, find the numerical GCF, the $x$ GCF, and the $y$ GCF. Then use the product to find the overall GCF for the expression. The first one has been done for you.

| | Expression | Constant GCF | $x$ GCF | $y$ GCF | Overall GCF | Factored Expression |
|---|---|---|---|---|---|---|
| **a.** | $15x^4 - 25x^3$ | 5 | $x^3$ | | $5x^3$ | $5x^3(3x - 5)$ |
| **b.** | $7y^3 + 49y^2$ | | | | | |
| **c.** | $16xy^2 + 32y^4$ | | | | | |
| **d.** | $24x^8 - 46x^7$ | | | | | |
| **e.** | $4x^6 + 4y^6$ | | | | | |
| **f.** | $12x^2 + 144$ | | | | | |
| **g.** | $33xy - 44x^2y^2$ | | | | | |
| **h.** | $9x^{10}y^5 + 3x^7y^4$ | | | | | |

# Student Activity
## Hatch the Missing Factors

**Directions:** Fill in the blank in each equation to make each statement true. All of the solutions are provided inside the "egg". When you have correctly placed all of the factors, you have "hatched" the egg.

$$8x^2 \cdot \underline{\hspace{1.5cm}} = 24x^3$$

$$-4x \cdot \underline{\hspace{1cm}} = 36x^4 \qquad\qquad 9y^2 \cdot \underline{\hspace{1cm}} = 54y^6$$

$$\underline{\hspace{1cm}} \cdot 6x^3 = -36x^3 \qquad\qquad 2x^2y \cdot \underline{\hspace{1cm}} = -8x^2y^3$$

$$5x^2 \cdot \underline{\hspace{1cm}} = 5x^2 \qquad\qquad 5x^2 \cdot \underline{\hspace{1cm}} = -5x^2$$

$$\underline{\hspace{0.5cm}} \cdot (-3z) = 3z^2 \qquad\qquad 7w^3 \cdot \underline{\hspace{1cm}} = -49w^5$$

$$5uv \cdot \underline{\hspace{0.5cm}} = 5u^4v \qquad\qquad 16ab^2 \cdot \underline{\hspace{0.5cm}} = 32a^2b^2$$

$$9r^2 \cdot \underline{\hspace{0.5cm}} = 36r^3s^2 \qquad\qquad t^7u \cdot \underline{\hspace{0.5cm}} = 16t^8u^2v$$

$$\underline{\hspace{0.5cm}} \cdot (-7x) = 14x \qquad\qquad c \cdot \underline{\hspace{1cm}} = mc^2$$

$$12x^4 \cdot \underline{\hspace{0.5cm}} = -144x^{12} \qquad\qquad 18x^4 \cdot \underline{\hspace{0.5cm}} = -36x^5y^3$$

$$-10x^3y^2z \cdot \underline{\hspace{1cm}} = -120x^5y^3z^4$$

Inside the egg:

$6y^4$    $-z$    $-2xy^3$    $-1$    $16tuv$    $-4y^2$    $-6$    $-7w^2$    $12x^2yz^3$    $-12x^8$    $mc$    $u^3$    $-9x^3$    $2a$    $-2$    $1$    $3x$    $4rs^2$

# Student Activity
## The First Factoring Matchup

**Directions (READ them):** In each box of the grid, you will find an expression that needs to be factored. Once you have factored each expression, look to see whether the factor appears in the list at the top. If it does, list that letter, if none of the factors are listed, then choose F (none of these are factors). The first one has been done for you. Some boxes may have more than one answer.

**A** $x+3$     **B** $x-2$    **C** $x+2y$    **D** $y-5$    **E** $x+4$    **F** None of these are factors

| | | | |
|---|---|---|---|
| $x^2-3x+2xy-6y$<br>$x(x-3)+2y(x-3)$<br>Ans: $(x+2y)(x-3)$<br><br>C | $3x^2y^2+9xy^2$ | $4xy^4+8y^4$ | $xy+3y-5x-15$ |
| $x^2+4x+2xy+8y$ | $3x^2y^2z^3-6xy^2z^3$ | $xy+4y-5x-20$ | $7xy-14y-3x+6$ |
| $6xy^2+10y^2+3xy+5y$ | $x^2+3x+2xy+6y$ | $3a^2b^3x+12a^2b^3$ | $10x^2y-50x^2+9xy-45x$ |
| $xy+3y-2x-6$ | $x^2+2xy+2xy+4y^2$ | $xy+2y^2-5x-10y$ | $xy-2y-5x+10$ |

# Student Activity
## Revenge of Factor Pairings

**Directions:** In each diagram, there is a number in the top box and exactly enough spaces beneath it to write all the possible factor-pairs involving **integers**. The number −12 has been done for you. See if you can find all the missing factor-pairs.

| −12 | |
|---|---|
| 1(−12) | −1(12) |
| 2(−6) | −2(6) |
| 3(−4) | −3(4) |

| −32 | |
|---|---|
| | |
| | |

| −45 | |
|---|---|
| | |
| | |

| 50 | |
|---|---|
| | |
| | |

| 63 | |
|---|---|
| | |
| | |

| −18 | |
|---|---|
| | |
| | |

| 28 | |
|---|---|
| | |
| | |

| −30 | |
|---|---|
| | |
| | |

| 100 | |
|---|---|
| | |
| | |
| | |

| −36 | |
|---|---|
| | |
| | |
| X | |

| 56 | |
|---|---|
| | |
| | |
| | |

| −42 | |
|---|---|
| | |
| | |
| | |

**Directions:** In the tables below, you are given a factor pair. If you **ADD** the two factors instead of multiplying them, what is the result? Complete the tables to find the sum of each factor pair. The first one has been started for you.

| −12 | Sum |
|---|---|
| −1(12) | 11 |
| 1(−12) | −11 |
| −2(6) | |
| 2(−6) | |
| −3(4) | |
| 3(−4) | |

| 32 | Sum |
|---|---|
| 1(32) | |
| −1(−32) | |
| 2(16) | |
| −2(−16) | |
| 4(8) | |
| −4(−8) | |

| −64 | Sum |
|---|---|
| −1(64) | |
| 1(−64) | |
| −2(32) | |
| 2(−32) | |
| −4(16) | |
| 4(−16) | |
| −8(8) | |

| 30 | Sum |
|---|---|
| 1(30) | |
| −1(−30) | |
| 2(15) | |
| −2(−15) | |
| 3(10) | |
| −3(−10) | |
| 5(6) | |
| −5(−6) | |

# Student Activity
## Find the Prime Trinomial

**Directions:** In each row there are three trinomials; two are factorable and one is not. If the trinomial is factorable, factor it. If it is not factorable, write "PRIME."

| | | | |
|---|---|---|---|
| **1.** | $x^2 - 5x - 6$ | $x^2 - x + 6$ | $x^2 - 5x + 6$ |
| **2.** | $x^2 - 5x + 24$ | $x^2 - 5x - 24$ | $x^2 + 5x - 24$ |
| **3.** | $x^2 - 8x + 12$ | $x^2 - x - 12$ | $x^2 + 4x + 12$ |
| **4.** | $x^2 - 11x - 12$ | $x^2 + 11 + 12$ | $x^2 + 13x + 12$ |
| **5.** | $x^2 + 3x + 2$ | $x^2 + 3x + 10$ | $x^2 + 3x - 28$ |
| **6.** | $x^2 - 19x + 60$ | $x^2 - 19x - 60$ | $x^2 + 19x + 60$ |
| **7.** | $x^2 - 4x - 60$ | $x^2 + 4x - 60$ | $x^2 - 4x + 60$ |
| **8.** | $x^2 - 16x + 60$ | $x^2 + 16x + 60$ | $x^2 - 16x - 60$ |

# Student Activity
## Match Up on Factoring Trinomials

**Directions (READ them):** In each box of the grid, you will find an expression that needs to be factored. Once you have factored each expression, look to see whether the factor appears in the list at the top. If it does, list that letter, if none of the factors are listed, then choose F (none of these are factors). The first one has been done for you. Some boxes may have more than one answer.

**A** $x+4$      **B** $x-2$   **C** $x-1$   **D** $x+5$   **E** $x-7$      **F** None of these are factors

| | | | |
|---|---|---|---|
| $x^2-3x+2$ $(x-2)(x-1)$ Ans: B, C | $x^2+12x+35$ | $x^2-2x-35$ | $x^2+8x+16$ |
| $x^2-3x-4$ | $x^2-4x+4$ | $x^2+x-2$ | $x^2-x-42$ |
| $x^2+13x+42$ | $x^2-6x-7$ | $x^2+2x-8$ | $x^2+x-20$ |
| $3x^2y-18xy+24y$ | $5x^2+10x-15$ | $5x^2+20x+15$ | $10x^2y^2-30xy^2-280y^2$ |

# Student Activity

## Life After the GCF, Part I

**Directions:** For each expression below, find and factor out the GCF. Then categorize the resulting expression in the table as one of the following:

- GCF only: if no other factoring can be done within the parentheses
- GCF and form $x^2 + bx + c$: expression within parentheses is of the form $x^2 + bx + c$
- GCF and form $ax^2 + bx + c$ expression within parentheses is of the form $ax^2 + bx + c$

The first one has been done for you.

| | Expression (and factor out the GCF) | GCF only | GCF and form $x^2 + bx + c$ | GCF and form $ax^2 + bx + c$ |
|---|---|---|---|---|
| **1.** | $3x^2 - 12x - 15$ <br> $3(x^2 - 4x - 5)$ | | **X** | |
| **2.** | $8x^3 + 16x^2 - 24x$ | | | |
| **3.** | $20x^2 yz + 15x^2$ | | | |
| **4.** | $9x^3 y + 15x^2 y + 6xy$ | | | |
| **5.** | $60t^2 x^2 + 78t^2 x + 24t^2$ | | | |
| **6.** | $27x^2 + 27x + 6$ | | | |
| **7.** | $150a^2 b^2 c^2 - 125ab^2 c$ | | | |
| **8.** | $25x^5 + 75x^4 + 50x^3$ | | | |
| **9.** | $26y^2 + 13y$ | | | |
| **10.** | $36x^2 + 99x + 54$ | | | |

# Student Activity
## Is it Completely Factored?

Is $(8x-2)(x+6)$ the completed factoring for the expression $8x^2+46x-12$? Let's see.

1. Multiply $(8x-2)(x+6)$: _____

2. Is the result of the multiplication equivalent to $8x^2+46x-12$? _____

3. From the expression $8x^2+46x-12$, first factor out the 2: _____

4. Then continue to factor by examining the trinomial within the parentheses:



5. Was the original expression, $(8x-2)(x+6)$, completely factored? _____

6. How can you make $(8x-2)(x+6)$ a completely factored expression?


7. **Tic-tac-toe Directions:** If the expression in the square **IS** completely factored, then circle it (thus putting an **O** on the square). If the expression **IS NOT** completely factored, then put an **X** on the square and finish the factoring.

| | | |
|---|---|---|
| $(x-2)(3x-12)$ | $(7y-35)(2y+3)$ | $(9x-28)(x+2)$ |
| $(4x-9)(2x-5)$ | $(3x+4)(4x-9)$ | $(2a+8)(3a+12)$ |
| $(2x+6)(2x-6)$ | $(3w-16)(6w+21)$ | $(5x-9)(6x+25)$ |

# Student Activity
## Spotting Perfect Squares and Cubes

**Directions:** For each expression below, determine whether it might be a perfect square or a perfect cube.

Then write the term in its squared $(\phantom{x})^2$ or cubed $(\phantom{x})^3$ form.

Be careful, a couple of these terms are both squares and cubes.

The first two have been done for you.

Look! A perfect Cube!

No, I think it's a perfect square.

It's an $(x^2)^3$.

No, It's an $(x^3)^2$.

| | | | | |
|---|---|---|---|---|
| $-27$ $(-3)^3$ | $125$ | $9$ | $512$ | $216x^3y^3$ |
| $16b^2$ $(4b)^2$ or $(-4b)^2$ | $-64$ | $1$ | $100$ | $1000x^3$ |
| $729$ | $49a^4b^4$ | $-1$ | $8x^3$ | $81h^6$ |
| $144$ | $9x^2$ | $25z^2$ | $-125z^3$ | $64$ |
| $27a^3$ | $0$ | $169$ | $216x^3y^9$ | $1000a^3$ |

# Student Activity
## Skeleton of a Perfect Square Trinomial

Previously, you may have practiced squaring a binomial to get a perfect square trinomial. In this section, we take this process backwards.

**Example:** Factor: $x^2 - 12x + 36$

Notice that both the first and last terms are squared terms: $(x)^2$ and $(6)^2$. This is the first clue that the trinomial might be a perfect square trinomial.

In a perfect square trinomial, the middle term is supposed to be twice the product of the bases (in this case, $x$ and $6$).

$2(x)(6)$ is $12x$, so the trinomial is a perfect square trinomial.

Remember the formula for a perfect square trinomial is the sum (or difference) of the bases, the quantity squared:

$$A^2 + 2AB + B^2 = (A + B)^2 \qquad A^2 - 2AB + B^2 = (A - B)^2$$

Write the correct factored form: $(x)^2 - 2(x)(6) + (6)^2 = (x - 6)^2$.

**Now you try these!** For each problem, fill in the skeleton for the sum or difference of squares, then factor appropriately. If the trinomial is NOT a perfect square trinomial, say so!

**1.** Factor: $x^2 - 10x + 25$ $\quad = (\quad)^2 - 2(\quad)(\quad) + (\quad)^2 = (\qquad)^2$

**2.** Factor: $y^2 + 2y + 1$ $\quad = (\quad)^2 + 2(\quad)(\quad) + (\quad)^2 = (\qquad)^2$

**3.** Factor: $25x^2 - 60x + 36$ $\quad = (\quad)^2 + 2(\quad)(\quad) + (\quad)^2 = (\qquad)^2$

**4.** Factor: $x^2 + 7x + 49$ $\quad = (\quad)^2 + 2(\quad)(\quad) + (\quad)^2 = (\qquad)^2$

**5.** Factor: $v^4 + 6v^2 + 9$ $\quad = (\quad)^2 + 2(\quad)(\quad) + (\quad)^2 = (\qquad)^2$

**6.** Factor: $16 - 8x + x^2$ $\quad = (\quad)^2 + 2(\quad)(\quad) + (\quad)^2 = (\qquad)^2$

**7.** Factor: $w^4 - 4w + 4$ $\quad = (\quad)^2 + 2(\quad)(\quad) + (\quad)^2 = (\qquad)^2$

**8.** Factor: $a^2b^2 + 14abc + 49c^2$ $\quad = (\quad)^2 + 2(\quad)(\quad) + (\quad)^2 = (\qquad)^2$

# Student Activity
## Skeleton of the Difference or Sum of Two Squares

The best way to learn the sum and
difference of squares formulas is to learn a
procedure for constructing the formulas
rather than memorizing them.  Here are
three examples, with the description of each
step:

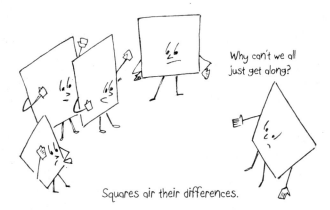

Why can't we all just get along?

Squares air their differences.

**Example 1:** Factor: $x^2 - 25$

Construct a skeleton of the difference of squares: $(\quad)^2 - (\quad)^2$

Fill in the skeleton with the appropriate bases: $(x)^2 - (5)^2$

Construct the skeleton of the difference of squares formula:  $(\quad + \quad)(\quad - \quad)$

Insert the bases into both factors: $(x+5)(x-5)$

**Example 2:** Factor: $64y^2 - 81z^2$

Construct a skeleton of the difference of squares: $(\quad)^2 - (\quad)^2$

Fill in the skeleton with the appropriate bases: $(8y)^2 - (9z)^2$

Construct the skeleton of the difference of squares formula:  $(\quad + \quad)(\quad - \quad)$

Insert the bases into both factors: $(8y+9z)(8y-9z)$

**Example 3:** Factor: $4x^2 + 9$

Construct a skeleton of the sum of squares: $(\quad)^2 + (\quad)^2$

Fill in the skeleton with the appropriate bases: $(2x)^2 + (3)^2$

A sum of squares is prime.  This binomial does not factor.

**Now you try these!**  For each problem, fill in the skeleton for the sum or difference of
squares, then factor appropriately.

**1.** Factor: $36 - a^2$  $= (\quad)^2 + (\quad)^2 =$ _____

**2.** Factor: $49y^2 + x^2 = (\quad)^2 + (\quad)^2 =$ _____

**3.** Factor: $a^2b^2 - 4$  $= (\quad)^2 - (\quad)^2 =$ _____

**4.** Factor: $w^4 - 49v^2 = (\quad)^2 - (\quad)^2 =$ _____

**5.** Factor: $81 + u^2$  $= (\quad)^2 + (\quad)^2 =$ _____

**6.** Factor: $9w^2 - z^6$  $= (\quad)^2 - (\quad)^2 =$ _____

# Guided Learning Activity
## Learning the Cubes Procedure

The best way to learn the sum and difference of cubes formulas is to learn a **procedure** for constructing the formulas rather than memorizing the formulas.

Here are two examples, with a description of each step.

After you learn the procedure, try your own hand at using the cubes procedures by doing the four problems at the end.

The operation may seem a bit complicated, but believe me, you're the perfect candidate for the Difference of Cubes Procedure.

**Example 1:** Factor: $x^3 - 125$

Step 1:  Construct a skeleton of the difference of cubes: $(\quad)^3 - (\quad)^3$

Step 2:  Fill in the skeleton with the appropriate bases: $(x)^3 - (5)^3$

Step 3:  Construct the skeleton of the cubes formula: $(\quad - \quad)(\quad + \quad + \quad)$

same sign as original expression ⟶

opposite of first sign ⟶

always positive ⟶

Step 4:  Insert the bases: $(x - 5)(\quad + \quad + \quad)$

Step 5:  Insert the squares of those bases: $(x - 5)(x^2 + \quad + 25)$

$(x)^2 \qquad (5)^2$

Step 6:  Insert the product of the bases: $(x - 5)(x^2 + 5x + 25)$

$(x)(5)$

**Example 2:** Factor: $8w^3 + 27x^6$

Step 1: Construct a skeleton of the sum of cubes: $(\quad)^3 + (\quad)^3$

Step 2: Fill in the skeleton with the appropriate bases: $(2w)^3 + (3x^2)^3$

Step 3: Construct the skeleton of the cubes formula: $(\quad + \quad)(\quad - \quad + \quad)$

same sign as original expression ⟶

opposite of first sign ⟶

always positive ⟶

Step 4: Insert the bases: $(2w + 3x^2)(\quad - \quad + \quad)$

Step 5: Insert the squares of those bases: $(2w - 3x^2)(4w^2 + \quad + 9x^4)$

$(2w)^2 \qquad (3x^2)^2$

Step 6: Insert the product of the bases: $(2w - 3x^2)(4w^2 + 6x^2w + 9x^4)$

$(2w)(3x^2)$

---

**Now you try these!**

**1.** Factor: $x^3 - 8 = (\quad)^3 - (\quad)^3$

$= (\qquad)(\qquad\qquad)$

$\uparrow \qquad \uparrow \qquad \uparrow$

$(\quad)^2 \ (\quad)(\quad) \ (\quad)^2$

**2.** Factor: $64y^3 + 125 = (\quad)^3 + (\quad)^3$

$= (\qquad)(\qquad\qquad)$

$\uparrow \qquad \uparrow \qquad \uparrow$

$(\quad)^2 \ (\quad)(\quad) \ (\quad)^2$

**3.** Factor: $a^9 + 8b^6 = (\quad)^3 + (\quad)^3$

$= (\qquad)(\qquad\qquad)$

$\uparrow \qquad \uparrow \qquad \uparrow$

$(\quad)^2 \ (\quad)(\quad) \ (\quad)^2$

# Student Activity

## Match Up on Factoring Binomials

**Match-up:** In each box of the grid, you will find an expression. Decide how to classify the expression and look to see whether the classification appears in the list at the top. If it does, list that letter, if the classification is not listed, then choose E (none of these). The first one has been done for you. **Some boxes may have more than one answer.**

Also, draw the appropriate skeleton and insert the appropriate bases.

**A** Sum of Squares: $(\quad)^2 + (\quad)^2$

**B** Difference of Squares: $(\quad)^2 - (\quad)^2$

**C** Sum of Cubes: $(\quad)^3 + (\quad)^3$

**D** Difference of Cubes: $(\quad)^3 - (\quad)^3$

**E** None of these

| | | | |
|---|---|---|---|
| $x^2 - 16$ | $4z^4 - 9$ | $9x^2 + 27$ | $y^2 + 8$ |
| $w^4 + 4z^2$ | $8x^3 - 27$ | $x^6 - 64$ | $8x^9 + 125$ |
| $729 - x^6$ | $4a^2 + 144b^6$ | $169x^4 - 121y^2$ | $343x^3 + 64$ |
| $x^2 - 3$ | $x^3 - 1000$ | $w^6 - 27x^3$ | $w^6 + 1$ |

# Student Activity
## Life After the GCF, Part II

**Directions:** For each expression below, find and factor out the GCF. Then categorize the resulting expression in the table by examining the expression that is within the parentheses. Some expressions may have more than one categorization. The first one has been done for you.

| | Expression (factor out the GCF) | GCF only | GCF and form $x^2 + bx + c$ | GCF and form $ax^2 + bx + c$ | GCF and perfect-square trinomial | GCF and sum or difference of squares | GCF and sum or difference of cubes |
|---|---|---|---|---|---|---|---|
| **1.** | $12x^3 + 12x^2 - 72x$  $12x(x^2 + x - 6)$ | | X | | | | |
| **2.** | $125x^5 - 625x^3$ | | | | | | |
| **3.** | $12x^3y^3 + 42x^2y^3 + 18xy^3$ | | | | | | |
| **4.** | $3t^2x^3 + 24t^2$ | | | | | | |
| **5.** | $40x^2 - 250$ | | | | | | |
| **6.** | $2x^2y^2 + 4xy^2 + 2y^2$ | | | | | | |
| **7.** | $24y^3 + 27$ | | | | | | |
| **8.** | $27x^4y^6 - 8x^4$ | | | | | | |
| **9.** | $4x^4 - 16x^3 - 16x^2$ | | | | | | |
| **10.** | $32t^2x^3 - 2t^2x$ | | | | | | |

# Student Activity
## Leftovers

**Directions:** Factor each of the binomials below and then mark the factors that they contain. Make sure to look for GCFs first! The first one has been done for you.

| Factor these binomials. | $x+2$ | $x-3$ | $x+5$ | $x-4$ | $2x-3$ | $x^2+4$ | Leftovers |
|---|---|---|---|---|---|---|---|
| **1.** $x^3+125$ <br> $(x+5)(x^2-5x+25)$ | | | X | | | | $(x^2-5x+25)$ |
| **2.** $3x^2-27$ | | | | | | | |
| **3.** $4x^3-9x$ | | | | | | | |
| **4.** $x^3-64$ | | | | | | | |
| **5.** $2x^4-32$ | | | | | | | |
| **6.** $4x^3+32$ | | | | | | | |
| **7.** $2x^3-54$ | | | | | | | |
| **8.** $8x^3-27$ | | | | | | | |
| **9.** $x^6+64$ | | | | | | | |
| **10.** $128x-8x^5$ | | | | | | | |

# Student Activity
## Skeletons of Tricky Binomials

**Directions:** Categorize each of the expressions using the choices below, then write the appropriate skeleton for the problem, and factor the expression. If the expression cannot be factored, say so. The first one has been done for you.

**A** Sum of Squares: $(\ )^2 + (\ )^2$

**B** Difference of Squares: $(\ )^2 - (\ )^2$

**C** Sum of Cubes: $(\ )^3 + (\ )^3$

**D** Difference of Cubes: $(\ )^3 - (\ )^3$

**E** None of these

| Polynomial | Category | Skeleton | Factored Form |
|---|---|---|---|
| 1. $x^4 - (a+b)^2$ | B | $(x^2)^2 - (a+b)^2$ | $(x^2 + (a+b))(x^2 - (a+b))$ or $(x^2 + a + b)(x^2 - a - b)$ |
| 1. $(a-b)^2 - x^4$ | | | |
| 2. $1 + (x+y)^3$ | | | |
| 3. $(a-b)^2 + x^2$ | | | |
| 4. $\dfrac{8}{27} - x^6$ | | | |
| 5. $x^3 - 36$ | | | |
| 6. $x^2 + 2xy + y^2 - 100$ | | | |
| 7. $(x+y)^2 + 27$ | | | |
| 8. $(a+b)^3 - (c-d)^3$ | | | |

# Guided Learning Activity

## GCF to the Rescue

It's tempting to start a factoring problem by immediately setting up a factor table or two sets of parentheses. However, you really should look for a GCF first. Yes, it's true that you can still factor a GCF out at the end of the problem, but it's so much easier if you find the GCF first!

**Directions (READ them):** In each box of the grid, you will find an expression that needs to be factored. Once you have completely factored the expression, look to see whether any of the factors appear in the list at the top. If none of the factors are listed, then choose E (none of these are factors). Some boxes may have more than one answer. The first one has been done for you.

**A** $x+1$  **B** $2x+3$  **C** $4x-1$  **D** $x-5$  **E** None of these are factors

| $\begin{array}{c} 40x^2 + 100x + 60 \\ 20\left(2x^2 + 5x + 3\right) \\ 20\left(2x^2 + 2x + 3x + 3\right) \\ 20\left[2x(x+1) + 3(x+1)\right] \\ 20(2x+3)(x+1) \\ \mathbf{A,\ B} \end{array}$ | $120x^2 + 90x - 30$ | $70 - 70x^2$ |
|---|---|---|
| $-60x^2 + 315x - 75$ | $40x^2 + 80x + 40$ | $8x^2 + 10x - 3$ |
| $60x^2 - 260x - 200$ | $20x^2 - 40x + 20$ | $40x^2 - 140x - 300$ |

# Student Activity

## Finding Factors

**Directions:** Factor each polynomial and mark which factors it contained. The first one has been done for you.

| Factor these polynomials. | Mark the factors that appear in the polynomials. | | | | | | | |
|---|---|---|---|---|---|---|---|---|
| | $x$ | $x+2$ | $x-3$ | $x+5$ | $2x+1$ | $3x-2$ | $x^2+4$ | $x^2-2$ |
| **1.** $x^2+7x+10$ <br> $(x+2)(x+5)$ | | X | | X | | | | |
| **2.** $x^3-2x$ | | | | | | | | |
| **3.** $x^2+2x-15$ | | | | | | | | |
| **4.** $2x^2-5x-3$ | | | | | | | | |
| **5.** $x^3-3x^2-2x+6$ | | | | | | | | |
| **6.** $6x^2-x-2$ | | | | | | | | |
| **7.** $3x^3-2x^2+12x-8$ | | | | | | | | |
| **8.** $2x^2+5x+2$ | | | | | | | | |
| **9.** $x^4+2x^2-8$ | | | | | | | | |
| **10.** $3x^3-11x^2+6x$ | | | | | | | | |

# Student Activity
## Factoring Strategizing

For each type of factoring problem, describe the classification of the factoring problem and the strategies and key steps (like the parentheses skeleton) to remember while doing the problem. You do not have to complete the problems.

Factor HQ

GCF

$3x^2 + 16x + 5$

Ok, Let's see who gets this one...

**Choose from these classifications** (more than one may apply):

| | | |
|---|---|---|
| Factor out a GCF | Trinomial $ax^2 + bx + c$ | Sum of Cubes |
| Factor by Grouping | Sum of Squares | Difference of Cubes |
| Trinomial $x^2 + bx + c$ | Difference of Squares | Perfect Square Trinomial |

| Expression to be factored | Classification | Strategies and Key Steps |
|---|---|---|
| **1.** $9x^2 - 4$ | | |
| **2.** $4u^4 + 32u$ | | |
| **3.** $t^2 - 12t + 35$ | | |
| **4.** $x^2 + 49$ | | |
| **5.** $x^3 + 5x^2 - 3x - 15$ | | |
| **6.** $15y^2 - 36y + 12$ | | |
| **7.** $x^2 + 14x + 49$ | | |
| **8.** $64z^3 - 27$ | | |
| **9.** $a^2 + 15ab + 54b^2$ | | |
| **10.** $4x^4 + 4x^3 - 80x^2$ | | |

# Student Activity

## Factoring Variations on a Number

**Directions:** In each set there are six expressions to be factored, all ending with the same constant (positive or negative). If the expression is factorable, factor it. If it is not factorable, write "PRIME."

| Variations on 1 | $x^2-1$ | $x^3-1$ | $x^2-2x+1$ |
| --- | --- | --- | --- |
| | $x^2+1$ | $x^3+1$ | $x^2+2x+1$ |

| Variations on 64 | $x^2-64$ | $x^2+64$ | $x^2-16x+64$ |
| --- | --- | --- | --- |
| | $x^3-64$ | $x^3+64$ | $x^2+16x+64$ |

| Variations on 36 | $x^2-36$ | $9x^2+36$ | $x^2+12x+36$ |
| --- | --- | --- | --- |
| | $4x^2-36$ | $9x^2-36$ | $x^2-12x+36$ |

# Student Activity
## Factors in Hiding

**Directions:** Factor each polynomial and then find each factor in the grid at the bottom of the page. You should find every factor in the grid.

1. Factor: $(a+b)^2 + 5(a+b) + 6$    HINT: Treat it like a trinomial.

2. Factor: $(a-b)^2 + 9(a-b) - 22$

3. Factor: $(a+b)^2 - 4(a+b) + 4$

4. Factor: $4(a-b) + 11(a-b) - 3$

5. Factor: $(a-b)(x+y) + 2(x+y)$    HINT: Is there a GCF?

6. Factor: $(a+b)^2 + x(a+b)$

7. Factor: $a^2b^2 + 5a^2b + 3a^2 + 10b^2 + 50b + 30$    HINT: Factor by grouping.

8. Factor: $a^2b + 4ab + 2b - 3a^2 - 12a - 6$

| $a+b-2$ | $a+b-2$ | $a-b-2$ | $a+b$ |
|---------|---------|---------|-------|
| $a+b+2$ | $b^2+5b+3$ | $b-3$ | $x+y$ |
| $a^2+10$ | $a-b+11$ | $a-b+2$ | $a-b+3$ |
| $a+b+3$ | $4a-4b-1$ | $a^2+4a+2$ | $a+b+x$ |

# Student Activity
## Checking the Factoring with a Calculator

**Directions:** One way you can check your factoring is to evaluate both the original expression and the factored expressions for the same value of the variable. It is not wise to use the values of 0, 1, or 2 for these types of checks since these three numbers have some quirky properties:

$$0+0=0 \text{ and } 0\cdot 0=0 \qquad 1\cdot 1=1 \qquad 2+2=4 \text{ and } 2\cdot 2=4$$

If a check by evaluation results in different answers, then you should reexamine the factoring. Note that getting the same result from evaluation does not **ensure** that your factors are correct, but it is likely to find a mistake in your work if there is one.

In each table below, check the student's factoring by evaluating for the given values with a calculator. It will help to write out the skeleton with the value substituted first. Then input the numerical expression into your calculator to get the result. If you find a factoring mistake, correct it! The first one has been started for you.

1.

| | Evaluate for $x=3$. | Evaluate for $x=2.4$. |
|---|---|---|
| $3x^2+7x+2$ | $3(3)^2+7(3)+2=50$ | |
| $(x+2)(3x+1)$ | $(3+2)(3(3)+1)=50$ | |

2.

| | Evaluate for $x=4$. | Evaluate for $x=1/3$. |
|---|---|---|
| $x^2+4x-21$ | | |
| $(x-7)(x+3)$ | | |

3.

| | Evaluate for $x=1/2$. | Evaluate for $x=-3$. |
|---|---|---|
| $4x^2-25$ | | |
| $(2x+5)(2x-5)$ | | |

4.

| | Evaluate for $x=3.5$. | Evaluate for $x=-5/4$. |
|---|---|---|
| $4x^2+20x+25$ | | |
| $(4x+5)^2$ | | |

5.

| | Evaluate for $x=5$. | Evaluate for $x=3$. |
|---|---|---|
| $x^3-8$ | | |
| $(x-2)(x^2+4x+4)$ | | |

# Student Activity
## Escape the Matrix by Solving Quadratic Equations

**Directions:** In each box of the grid, you will find a quadratic equation that needs to be solved. Once you have solved the equations, use pairs of matching solution numbers to navigate your way out of the matrix. The first one has been done for you. For example, the repeated solution of $-3$ leads you to the box on the right, with solutions of $-3$ and 4. To move the next step on the escape route, you need to find an adjacent box with a solution of 4 and something else.

| START HERE<br>Solve: $(x+3)^2 = 0$<br><br>Solutions: $-3$ and $-3$ | Solve: $x^2 - x - 12 = 0$<br><br>$(x-4)(x+3) = 0$<br>Solutions 4 and $-3$ | Solve: $x^2 - 8x - 48 = 0$ | Solve: $(x+12)^2 = 0$ |
|---|---|---|---|
| Solve: $x^2 + 2x - 8 = 0$ | Solve:<br>$(x-4)\left(x-\frac{1}{2}\right) = 0$ | Solve:<br>$x^2 - \dfrac{5}{6}x + \dfrac{1}{6} = 0$ | Solve:<br>$3x^2 - 19x + 6 = 0$ |
| Solve:<br>$x^2 - 8x + 15 = 0$ | Solve: $x^2 - x - 42 = 0$ | Solve: $x^2 + x = 2$ | Solve:<br>$(x-6)(x+2) = 0$ |
| ESCAPE<br>the Matrix<br>Solve:<br>$x^2 + 8x + 16 = 0$ | Solve: $x^2 - 16 = 0$ | Solve: $x^2 - 5x + 4 = 0$ | Solve: $x^2 - 11x = -24$ |

# Student Activity
## Multiply, Factor, or Solve

**Directions:** The procedures for multiplying polynomials, factoring polynomials, and solving quadratic equations using factoring are most likely stored very near to each other in your memory because the procedures are entwined. Here is some practice to make sure that the procedures are clearly separated in your mind.

**1.** Factor: $y^2 - 49$

**2.** Solve: $21x^2 - 25x - 4 = 0$

**3.** Multiply: $(x-2)^2$

**4.** Solve: $x^2 - 16 = 0$

**5.** Factor: $2x^5 - 2y^2$

**6.** Solve: $x^2 = 8x - 16$

**7.** Multiply: $x(x-2)(x+3)$

**8.** Factor: $x^2 - 3xy - 4y^2$

**9.** Multiply: $(x-5)(x^2 + 5x + 25)$

**10.** Factor: $2x^4 + 16x^3 - 40x^2$

**Describe what each of the directions mean in your own words.**

Factor:

Multiply:

Solve:

# Student Activity
## Parentheses, Brackets, and Braces, Oh My!

Some mathematical situations require very specific notation with parentheses, brackets, and braces. In these cases, the notation is **not** interchangeable.

- Ordered pairs are written with parentheses. For example, $(-2, 5)$ is an ordered pair with $x$-coordinate $-2$ and $y$-coordinate 5.

- Interval solutions use parentheses and brackets. For example $(-2, 5]$ is an interval in which 5 is included in the solution.

- Solution sets use braces. For example $\{-2, 5\}$ shows two solution values, $-2$ and 5.

1. Match the solution with the appropriate graph:

___ **a.** $(-2, 5)$       ___ **b.** $\{-2, 5\}$       ___ **c.** $(-2, 5]$

**I**

**II**

**III**

2. Solve $3 - 2x \le 9$ and write the solution in interval notation.

3. Find the ordered pair solution for $3x - y = 4$ if $x = 3$.

**4.** Solve $x^2 - 5x - 36 = 0$ and write the answer as a solution set.

**5.** Solve $-2 \le x + 4 \le 7$ and write the solution in interval notation.

**6.** Solve $5(x - 2) + 3x = 4x - 8$ and write the solution as a solution set.

**7.** Is $(-1, 3)$ a solution of $4x + y = 1$?

**8.** Write the solution that is represented by the graph below.

**9.** Write the ordered pair that is represented by the graph below.

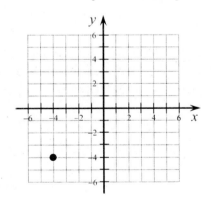

**10.** Write the solution that is represented by the graph below.

# Guided Learning Activity

## Using the Problem Solving Strategy with Factoring

1. **Analyze.** Draw diagrams, tables, and charts to help make sense of the problem.
2. **Declare.** Declare the variables and the quantitative givens in the problem.
3. **Formulate.** Write equations to represent the problem.
4. **Solve.** Solve the equation or system of equations.
5. **Validate.** Check the result. Does it make sense? Is it reasonable?
6. **Conclude.** State the conclusion. Don't forget units.

**Problem 1:** The area of a vegetable garden is $32 \text{ m}^2$. If the length is 4 meters longer than the width, what are the dimensions (length and width) of the garden?

**Problem 2:** The area of a right triangle is $10 \text{ ft}^2$. If the base of the triangle is 1 foot longer than the height, what are the dimensions of the triangle?

**Problem 3:** The base of a right triangle is one less than twice the height. If the length of the hypotenuse is 17 cm, what are the measurements of the other two sides?

# Student Activity
## Proof of the Pythagorean Theorem

**Materials required:** Paper for tracing, scissors, and tape.

This is a geometric proof of the Pythagorean Theorem (by dissection) given by the Arabian Mathematician Thâbit ibn Kurrah, who lived from 836 to 901 A.D. in the Middle East.

Sources: http://mathworld.wolfram.com/PythagoreanTheorem.html,
http://www.britannica.com/eb/article-9071897/Thabit-ibn-Qurra

**Step 1:** Trace the figure below onto a **new sheet of paper**.
Cut out the square with side *a* and the square with side *b*.

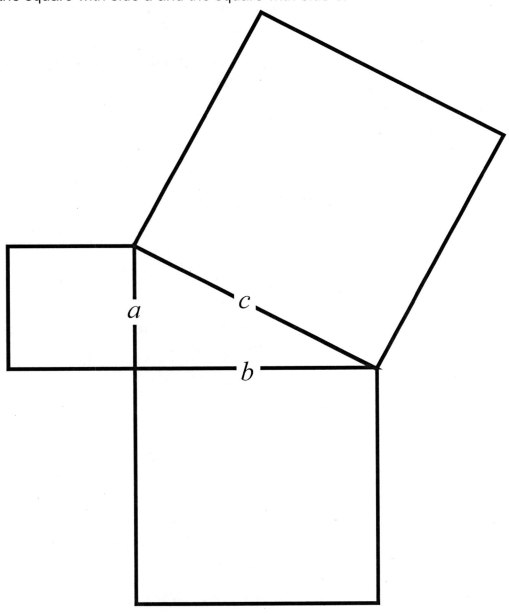

**Step 2:** Securely tape these two squares together like so.

This figure now represents $a^2 + b^2$.

**Step 3:** Mark the midpoint of the base of the *b*-square.

**Step 4:** Draw in these two dashed lines (shown below), and **cut** along these lines. You should now have three geometric figures (two triangles and one with an irregular shape

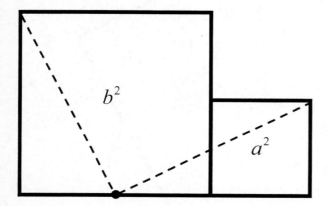

**Step 5:** Use these three pieces like pieces of a puzzle and reassemble the three pieces to fit in the square with side *c* **that was on the first page**.

**Question:** Why does this "prove" the Pythagorean Theorem?

---

**More Information:** There are hundreds of different proofs of the Pythagorean Theorem. If you are interested in learning more about proofs of the Pythagorean Theorem or the mathematician who first documented this particular proof, conduct a simple Internet search – there is plenty of information available.

# Student Activity
## Pumpkin Launch

Ballistics is the study of the flight characteristics of projectiles. For ballistic flights, Littlewood's Law can be used to *approximate* the peak height of a projectile by using the total flight time.

Littlewood's Law: $h = \frac{g}{8} \cdot t^2$ where $h$ is the peak height in meters, $t$ is the total flight time in seconds, and $g$ is the gravity constant, which is $g = 9.8$ m/s$^2$ near the surface of the earth.

### The "Pumpkin Rocket II"
(source: http://www.et.byu.edu/~wheeler/benchtop/pumpkin2.php)

The "Pumpkin Rocket II" was launched using a hydrogen-power method to pressurize the launch vehicle (a large water bottle).

**Question 1:** The pumpkin and bottle reached a peak height of 18 meters. Try substituting different values of $t$ into Littlewood's Law to find how long the flight must have been to reach this peak height.

| $t$ (seconds) | $h$ (meters) |
|---|---|
| 1 | |
| 2 | |
| 3 | |
| 4 | |
| 5 | |
| 6 | |

**Question 2:** Assuming we could build a pumpkin rocket that launch a pumpkin to a height of 78.4 meters, what would the total flight time be for this rocket? **This time, simplify the equation and write it in standard quadratic form, then use factoring to solve the problem.**

# Assess Your Understanding
**Factoring and More**

For each of the following, describe the strategies or key steps that will help you **start** the problem. You do **not** have to complete the problems.

| | | What will help you to start this problem? |
|---|---|---|
| **1.** | Factor: $y^2 + 7y - 18$ | |
| **2.** | Factor: $3x^2 - 12$ | |
| **3.** | Factor: $3x^2 - 14x - 24$ | |
| **4.** | Factor: $xy^2 - 9x + 4y^2 - 36$ | |
| **5.** | Is $-2$ a solution of $x^2 - 5x = 14$? | |
| **6.** | Factor: $4x^2 + 11x - 20$ | |
| **7.** | Factor: $5x^2 - 80$ | |
| **8.** | The area of a rectangle is $18\,\text{cm}^2$. If the length is twice the width, find the dimensions of the rectangle. | |
| **9.** | Solve: $x^2 - 8x + 12 = 3(x - 4)$ | |
| **10.** | Factor: $x^3 y^3 - 64z^6$ | |

# Metacognitive Skills

## Factoring and More

Metacognitive skills refer to the ability to judge how well you have learned something and to effectively direct your own learning and studying. This is a self-evaluation tool designed to help you focus your studying and to improve your metacognitive skills with regards to this math class.

Fill the 1$^{st}$ column out **before** you begin studying. Fill the 2$^{nd}$ column out after you study for your test.

Go back to this assessment after your test and circle any of the ratings that you would change – this identifies the "disconnects" between what you **thought** you knew well and what you **actually** knew well.

**Use the scale below to assign a number to each topic.**
**5** *I am confident I can do any problems in this category correctly.*
**4** *I am confident I can do most of the problems in this category correctly.*
**3** *I understand how to do the problems in this category, but I still make a lot of mistakes.*
**2** *I feel unsure about how to do these problems.*
**1** *I know I don't understand how to do these problems.*

| Topic or Skill | Before Studying | After Studying |
|---|---|---|
| Finding the GCF for several terms. | | |
| Factoring out the GCF. | | |
| Factoring out a $-1$. Why is it sometimes necessary to factor out a $-1$? | | |
| Factoring by grouping and knowing when to employ this method. | | |
| Distinguishing between trinomials of the form $x^2 + bx + c$ and the form $ax^2 + bx + c$. | | |
| Factoring a trinomial of the form $x^2 + bx + c$. | | |
| Factoring a trinomial of the form $ax^2 + bx + c$. | | |
| Knowing the numbers that are perfect squares and perfect cubes. | | |
| Recognizing expressions that are perfect squares and perfect cubes. | | |
| Recognizing binomials that are either a sum or difference of squares. | | |
| Factoring a difference of squares. | | |
| Knowing about the factorability of a sum of squares. | | |
| Recognizing binomials that are either a sum or difference of cubes. | | |
| Factoring a sum or difference of cubes. | | |
| Recognizing a perfect-square trinomial and properly factoring it. | | |
| Using the general factoring strategy. What should you always do first? | | |
| How can the number of terms in a polynomial help you to make decisions about how to factor it? | | |
| Understanding when an expression is not yet completely factored, and how to complete it. Ex: Is $x(2x+4)(x^2-9)$ completely factored? | | |

*Continued on next page.*

**Use the scale below to assign a number to each topic.**

**5**  *I am confident I can do any problems in this category correctly.*
**4**  *I am confident I can do most of the problems in this category correctly.*
**3**  *I understand how to do the problems in this category, but I still make a lot of mistakes.*
**2**  *I feel unsure about how to do these problems.*
**1**  *I know I don't understand how to do these problems.*

| Topic or Skill | Before Studying | After Studying |
|---|---|---|
| Being able to check the answer to a factoring problem by multiplying out the answer. | | |
| Understanding why you must first get "=0" on one side of a quadratic equation before solving by factoring. | | |
| Rewriting a quadratic equation in standard quadratic form. | | |
| Solving a quadratic equation by factoring. | | |
| Knowing how to properly write a solution set. | | |
| Checking the solution to a quadratic equation. | | |
| Distinguishing between the instructions: multiply, factor, or solve. | | |

# RAT: Rational Expressions and Equations

## Undefined Expressions

A **rational expression** is an expression of the form $\dfrac{A}{B}$

where $A$ and $B$ are polynomials and $B$ does not equal 0. To evaluate a rational expression, you may find it helpful to first create the parentheses skeleton.

**Example 1:**

Evaluate $\dfrac{x^2 + 4x - 5}{x^2 - 9}$ for $x = 2$ and $x = -3$.

Parentheses skeleton: $\dfrac{(\ \ )^2 + 4(\ \ ) - 5}{(\ \ )^2 - 9}$

Evaluate for $x = 2$: $\dfrac{(2)^2 + 4(2) - 5}{(2)^2 - 9} = \dfrac{4 + 8 - 5}{4 - 9} = \dfrac{7}{-5}$

Evaluate for $x = -3$: $\dfrac{(-3)^2 + 4(-3) - 5}{(-3)^2 - 9} = \dfrac{9 - 12 - 5}{9 - 9} = \dfrac{-8}{0} = $ undefined

*You just stand there with that blank expression. Sometimes I seriously wonder if you are even trying to be rational...*

| Expression | Parentheses Skeleton | Evaluate the expression for ... | | | | |
|---|---|---|---|---|---|---|
| | | $x = 1$ | $x = 4$ | $x = 0$ | $x = -2$ | $x = \frac{1}{3}$ |
| $\dfrac{3x - 12}{x^2 + 2x}$ | | | | | | |
| $\dfrac{3x^2 + 8x - 3}{x^2 - 5x + 4}$ | | | | | | |
| $\dfrac{x^2 - 16}{x^2 + 2x}$ | | | | | | |

It is often easier to perform the evaluation if the rational expression is factored first. This way, you can quickly see the values that create a factor of zero in the numerator or denominator.

**Example 2:** Evaluate $\dfrac{x^2 + 4x - 5}{x^2 - 9}$ for $x = 0$, $x = 1$, and $x = 3$.

Factored form: $\dfrac{(x + 5)(x - 1)}{(x + 3)(x - 3)}$.

Evaluate for $x = 0$: $\dfrac{(0 + 5)(0 - 1)}{(0 + 3)(0 - 3)} = \dfrac{(5)(-1)}{(3)(-3)} = \dfrac{-5}{-9} = \dfrac{5}{9}$

Evaluate for $x = 1$: $\dfrac{(1 + 5)(1 - 1)}{(1 + 3)(1 - 3)} = \dfrac{(6)(0)}{(4)(-2)} = \dfrac{0}{-8} = 0$

Evaluate for $x = 3$: $\dfrac{(3 + 5)(3 - 1)}{(3 + 3)(3 - 3)} = \dfrac{(8)(2)}{(6)(0)} = \dfrac{16}{0} = $ undefined

As soon as you see a **factor** of zero in the numerator or denominator, you can quickly find the value of the answer.

| Expression | Factored Form of the Expression | Evaluate the expression for ... | | | | |
|---|---|---|---|---|---|---|
| | | $x = 1$ | $x = 4$ | $x = 0$ | $x = -2$ | $x = \frac{1}{3}$ |
| $\dfrac{3x - 12}{x^2 + 2x}$ | | | | | | |
| $\dfrac{3x^2 + 8x - 3}{x^2 - 5x + 4}$ | | | | | | |
| $\dfrac{x^2 - 16}{x^2 + 2x}$ | | | | | | |

Recall that we can solve an equation like $(x+3)(x-5)=0$ using the zero property. Because of the Zero Factor Property, either $x+3=0$ or $x-5=0$. This leads us to solutions of $-3$ and $5$.

**Example 3:** Where is the expression $\dfrac{x^2 + 4x - 5}{x^2 - 9}$ undefined?

This expression is undefined when the denominator is equal to zero.

We can answer the question by solving the equation $x^2 - 9 = 0$.

Factor first: $(x+3)(x-3)=0$

Set each factor equal to zero and solve:

$$x + 3 = 0 \qquad \text{OR} \qquad x - 3 = 0$$
$$x = -3 \qquad\qquad\qquad x = 3$$

The expression $\dfrac{x^2 + 4x - 5}{x^2 - 9}$ is undefined for $-3$ and $3$.

| Expression | Factored Form of the Expression | Set the denominator = 0. Solve the resulting equation. | Where is the expression undefined? |
|---|---|---|---|
| $\dfrac{3x - 12}{x^2 + 2x}$ | | | |
| $\dfrac{3x^2 + 8x - 3}{x^2 - 5x + 4}$ | | | |
| $\dfrac{x^2 - 16}{x^2 + 2x}$ | | | |

# Student Activity

## Match Up on Simplifying Rational Expressions

**Directions:** Match each of the expressions in the squares in the table below with an equivalent simplified expression from the top. If an equivalent expression is not found among the choices A through E, then choose F (none of these).

**A** $1$          **B** $-1$

**C** $x+5$        **D** $\dfrac{x}{3}$

**E** $3x$          **F** None of these

| | | | |
|---|---|---|---|
| $\dfrac{9x^3+15x}{3x^2+5}$ | $\dfrac{x^2-25}{x-5}$ | | |
| $\dfrac{x-1}{1-x}$ | $\dfrac{x^2+x}{3x+3}$ | | |
| $\dfrac{(3x+2)(x+1)}{3x^2+5x+2}$ | $\dfrac{x-1}{1+x}$ | $\dfrac{3x^3-6x^2}{x-2}$ | $\dfrac{3x+1}{1+3x}$ |
| $\dfrac{3x^3-27x}{(x+3)(x-3)}$ | $\dfrac{x^3+2x^2+x}{3x^2+6x+3}$ | $\dfrac{x^2+10x+25}{x+5}$ | $\dfrac{3x-1}{1-3x}$ |
| $\dfrac{x^2+6x+5}{x+1}$ | $\dfrac{x^2+25}{x^2-25}$ | $\dfrac{x-8}{-x+8}$ | $\dfrac{18x^2-3x}{-1+6x}$ |

*I can't believe I'm saying this, but I really miss the good old days when fractions only involved numbers ...*

# Student Activity

## The Ones Recycling Center

**Directions:** In each pair of expressions, make a decision about whether the expression is equivalent to 1 or −1. Then sort each expression into the correct recycling bin below. If an expression does not belong in either recycling bin, just leave it out! The first one has been done for you.

**The 1 Bin:** The numerator and denominator are equivalent.

**The -1 Bin:** The numerator and denominator are opposites.

| | | | |
|---|---|---|---|
| $\dfrac{x-1}{1-x}$ | $\dfrac{x+2}{2+x}$ | $\dfrac{x-3}{x+3}$ | $\dfrac{x^2-1}{1-x^2}$ |
| $\dfrac{(x+3)^2}{x^2+6x+9}$ | $\dfrac{3(x+4)}{(x+4)(3)}$ | $\dfrac{v^2-1}{v^2+1}$ | $\dfrac{-x-3}{x+3}$ |
| $\dfrac{-x+3}{3-x}$ | $\dfrac{y^2+1}{1+y^2}$ | $\dfrac{x^2+2x-1}{x^2+2x+1}$ | $\dfrac{x^2-2x-3}{3+2x-x^2}$ |
| $\dfrac{a-b}{-a+b}$ | $\dfrac{x^2+3x+1}{3x+1+x^2}$ | $\dfrac{5-x}{5-x}$ | $\dfrac{(x+3)^2}{x^2+9}$ |

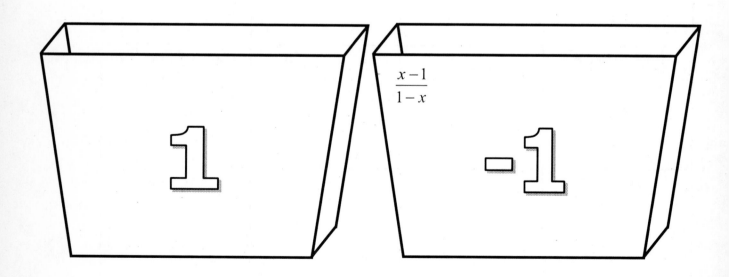

# Student Activity

## Which of These is Not Like the Others?

**Directions:** When you look at your answer to a problem and compare it to the answer in the back of the book or a friend's answer, you might find that they are not quite the same. This does not mean that one of them is *really* different though. In each row of the table, all the expressions are equivalent except for one of them. **Circle the expressions that are the same and place an X over the "oddball" expression in each row.** The first one has been done for you.

Here's two hints if you're really stuck:

- You could evaluate all the expressions for the same given value and see which expressions have the same result.
- You could try factoring out a –1 if the numerator and denominator look suspiciously like they might be opposites.

| | | | | |
|---|---|---|---|---|
| **1.** | $\dfrac{-4}{-x}$ (X) | $\left(\dfrac{-4}{x}\right)$ | $\left(\dfrac{4}{-x}\right)$ | $\left(-\dfrac{4}{x}\right)$ |
| **2.** | $-\dfrac{x+3}{x-4}$ | $\dfrac{-x-3}{x-4}$ | $\dfrac{-x-3}{-x+4}$ | $\dfrac{x+3}{4-x}$ |
| **3.** | $\dfrac{x-5}{5-x}$ | $\dfrac{x-5}{x-5}$ | $-\dfrac{x-5}{x-5}$ | $\dfrac{x-5}{-x+5}$ |
| **4.** | $-\dfrac{-x-4}{-x-4}$ | $\dfrac{x+4}{x+4}$ | $\dfrac{x+4}{4+x}$ | $-\dfrac{-x-4}{x+4}$ |
| **5.** | $-\dfrac{x+1}{1-x}$ | $\dfrac{1+x}{x-1}$ | $\dfrac{x+1}{x-1}$ | $\dfrac{1+x}{1-x}$ |
| **6.** | $\dfrac{x-4}{(x+1)(x+3)}$ | $\dfrac{x-4}{x^2+4x+3}$ | $\dfrac{4-x}{x^2+4x+3}$ | $-\dfrac{4-x}{x^2+4x+3}$ |

# Student Activity
## Does Position Matter?

**Directions:** Simplify each expression below.  Circle the multiplication example that matches the result of the shaded division problem.

| Expression | Simplified Fraction | Decimal Equivalent |
|---|---|---|
| $\dfrac{1}{5} \cdot \dfrac{3}{4}$ | $\dfrac{3}{20}$ | 0.15 |
| $\dfrac{1}{5} \cdot \dfrac{4}{3}$ | | |
| $\dfrac{5}{1} \cdot \dfrac{4}{3}$ | | |
| $\dfrac{5}{1} \cdot \dfrac{3}{4}$ | | |
| $\dfrac{1}{5} \div \dfrac{3}{4}$ | | |

| Expression | Simplify the Expression |
|---|---|
| $\dfrac{1}{x+3} \cdot \dfrac{x+3}{2}$ | |
| $\dfrac{1}{x+3} \cdot \dfrac{2}{x+3}$ | |
| $\dfrac{x+3}{1} \cdot \dfrac{x+3}{2}$ | |
| $\dfrac{x+3}{1} \cdot \dfrac{2}{x+3}$ | |
| $\dfrac{1}{x+3} \div \dfrac{2}{x+3}$ | |

| Expression | Simplify the Expression |
|---|---|
| $\dfrac{2x+3}{x} \cdot \dfrac{1}{x}$ | |
| $\dfrac{x}{2x+3} \cdot \dfrac{1}{x}$ | |
| $\dfrac{2x+3}{x} \cdot \dfrac{x}{1}$ | |
| $\dfrac{x}{2x+3} \cdot \dfrac{x}{1}$ | |
| $\dfrac{2x+3}{x} \div \dfrac{1}{x}$ | |

**Question:** When dividing fractions, does it matter whether you use the reciprocal of the first or second fraction?  In other words, do you get the same result?

# Student Activity
## Paint by Factors of 1

**Directions:** For each expression, perform the multiplication or division and simplify by factoring the numerator and denominator **completely** and removing factors equal to 1.

As you remove factors of 1, like $\dfrac{x+2}{x+2}$ or $\dfrac{x^2}{x^2}$, shade in the corresponding squares in the grid below. The first one has been done for you. There's a surprise when you're finished!

1. Multiply: $\dfrac{2x-4}{x+3} \cdot \dfrac{x+3}{2x+8} = \dfrac{2(x-2)(x+3)}{2(x+3)(x+4)} = \dfrac{\cancel{2}^1 (x-2)\cancel{(x+3)}^1}{\cancel{2}_1 \cancel{(x+3)}_1 (x+4)} = \dfrac{x-2}{x+4}$ (shade $\dfrac{x+3}{x+3}$ & $\dfrac{2}{2}$)

2. Divide: $\dfrac{3x^2-6x}{x^3+2x^2} \div \dfrac{-x^2-2x+8}{x^4+6x^3+8x^2}$

3. Multiply: $\dfrac{5x^4+2x^3}{2x^4+3x^3} \cdot \dfrac{2x^3+x^2-3x}{5x^2-3x-2}$

4. Multiply: $\dfrac{5x^2-125}{x^3+5x^2} \cdot \dfrac{6x^2+2x}{5x-25}$

5. Divide: $\dfrac{21x^2-49x-42}{4x^2-28x+48} \div \dfrac{21x+14}{8x^2-32x}$

| $\dfrac{12}{12}$ | $\dfrac{4x-3}{4x-3}$ | $\dfrac{x+3}{x+3}$ | $\dfrac{x^2}{x^2}$ | $\dfrac{2-5x}{2-5x}$ | $\dfrac{x^2+16}{x^2+16}$ |
|---|---|---|---|---|---|
| $\dfrac{7x+3}{7x+3}$ | $\dfrac{x+2}{x+2}$ | $\dfrac{x^3}{x^3}$ | $\dfrac{x-2}{x-2}$ | $\dfrac{-9}{-9}$ | $\dfrac{10x+1}{10x+1}$ |
| $\dfrac{3}{3}$ | $\dfrac{x-7}{x-7}$ | $\dfrac{5}{5}$ | $\dfrac{2x+3}{2x+3}$ | $\dfrac{4x+3}{4x+3}$ | $\dfrac{8x+3}{8x+3}$ |
| $\dfrac{4-5x}{4-5x}$ | $\dfrac{5x-7}{5x-7}$ | $\dfrac{x+5}{x+5}$ | $\dfrac{7}{7}$ | $\dfrac{13}{13}$ | $\dfrac{25}{25}$ |
| $\dfrac{7x+2}{7x+2}$ | $\dfrac{x+25}{x+25}$ | $\dfrac{3x+2}{3x+2}$ | $\dfrac{x+4}{x+4}$ | $\dfrac{3x-4}{3x-4}$ | $\dfrac{x+8}{x+8}$ |
| $\dfrac{2-9x}{2-9x}$ | $\dfrac{8x-3}{8x-3}$ | $\dfrac{4}{4}$ | $\dfrac{2}{2}$ | $\dfrac{x^2+36}{x^2+36}$ | $\dfrac{4x-17}{4x-17}$ |
| $\dfrac{4x-5}{4x-5}$ | $\dfrac{x+10}{x+10}$ | $\dfrac{x-3}{x-3}$ | $\dfrac{x-4}{x-4}$ | $\dfrac{x^2+4}{x^2+4}$ | $\dfrac{7x-3}{7x-3}$ |
| $\dfrac{3x}{3x}$ | $\dfrac{5x+2}{5x+2}$ | $\dfrac{x-5}{x-5}$ | $\dfrac{x}{x}$ | $\dfrac{x-1}{x-1}$ | $\dfrac{2x-1}{2x-1}$ |

# Student Activity
## Testing Equality

**Directions:** In each table there is a rational expression and a simplified expression. Use a test value to check the equality (or equivalency) of the two expressions. The first one has been started for you!

**Note:** Because of some unique properties of 0, 1 and 2 ($0+0=0\cdot 0$, $1\cdot 1=1$, $2+2=2\cdot 2$), it is usually best **not** to use these for test values.

1.
| Given expression: | | Possible simplification: |
|---|---|---|
| $\dfrac{x^2+3}{x}\cdot\dfrac{1}{x+2}$ | $\overset{?}{=}$ | $\dfrac{x^2+3}{x^2+2}$ |

What values make the denominator of the **given** expression zero?  $\underline{\ 0,-2\ }$

**Choose** a simple test value for $x$ that does **not** result in a zero denominator: $\underline{\text{Let } x = 3.}$

| Evaluate the rational expression: | | Evaluate the possible simplification: |
|---|---|---|
| $\dfrac{(3)^2+3}{(3)}\cdot\dfrac{1}{(3)+2}=\dfrac{12}{3}\cdot\dfrac{1}{5}=\dfrac{12}{15}=\dfrac{4}{5}$ | $\overset{?}{=}$ | |

Are the two expressions equal? _____

2.
| Given expression: | | Possible simplification: |
|---|---|---|
| $\dfrac{x^3+3x^2}{x}\div\dfrac{x}{x+7}$ | $\overset{?}{=}$ | $\dfrac{x^2}{\left(x^3+3x^2\right)(x+7)}$ |

What values make the denominator of the **given** expression zero? _____

**Choose** a simple test value for $x$ that does **not** result in a zero denominator: _____

| Evaluate the rational expression: | | Evaluate the possible simplification: |
|---|---|---|
| | $\overset{?}{=}$ | |

Are the two expressions equal? _____

**3.**

| Given expression: |
|---|
| $$\dfrac{x^2 - x - 12}{x - 2} \div \dfrac{x + 3}{x - 2}$$ |

$\overset{?}{=}$

| Possible simplification: |
|---|
| $x - 4$ |

What values make the denominator of the **given** expression zero? _____

**Choose** a simple test value for $x$ that does **not** result in a zero denominator: _____

| Evaluate the rational expression: |
|---|
| |

$\overset{?}{=}$

| Evaluate the possible simplification: |
|---|
| |

Are the two expressions equal? _____

**4.**

| Given expression: |
|---|
| $$\dfrac{2x^2 - 5x - 12}{x - 4} \cdot \dfrac{2x + 3}{x - 4}$$ |

$\overset{?}{=}$

| Possible simplification: |
|---|
| $$\dfrac{4x^2 + 12x + 9}{x - 4}$$ |

What values make the denominator of the **given** expression zero? _____

**Choose** a simple test value for $x$ that does **not** result in a zero denominator: _____

| Evaluate the rational expression: |
|---|
| |

$\overset{?}{=}$

| Evaluate the possible simplification: |
|---|
| |

Are the two expressions equal? _____

# Student Activity
## Thread of Like Terms

**Directions:** Simplify each of the expressions that follow.

1.

| 7 miles + 5 miles | $7x + 5x$ | $7x^2 + 5x^2$ | $7x^2y + 5x^2y$ |
|---|---|---|---|
| $\dfrac{7}{13} + \dfrac{5}{13}$ | $\dfrac{7}{x} + \dfrac{5}{x}$ | $\dfrac{7}{xy} + \dfrac{5}{xy}$ | $\dfrac{7}{x+2} + \dfrac{5}{x+2}$ |

2.

| $\dfrac{9}{11} - \dfrac{2}{11}$ | $9w^2 - 2w^2$ | $\dfrac{9}{ab^2} - \dfrac{2}{ab^2}$ | $\dfrac{9}{y} - \dfrac{2}{y}$ |
|---|---|---|---|
| $\dfrac{9}{x-4} - \dfrac{2}{x-4}$ | $9ab - 2ab$ | $9y - 2y$ | $9\ \text{ft}^2 - 2\ \text{ft}^2$ |

3.

| $\dfrac{8}{a^2b} - \dfrac{1}{a^2b}$ | $8xy - xy$ | $\dfrac{8}{z} - \dfrac{1}{z}$ | $8a^2 - a^2$ |
|---|---|---|---|
| $\dfrac{8}{x+8} - \dfrac{1}{x+8}$ | $8w - w$ | $8\ \text{cm} - 1\ \text{cm}$ | $\dfrac{8}{9} - \dfrac{1}{9}$ |

4.

| $\dfrac{4}{a} - \dfrac{3}{a}$ | $4\ \text{in}^3 - 3\ \text{in}^3$ | $\dfrac{4}{5} - \dfrac{3}{5}$ | $\dfrac{4}{5x^2} - \dfrac{3}{5x^2}$ |
|---|---|---|---|
| $\dfrac{4}{x+1} - \dfrac{3}{x+1}$ | $4x^2y^2 - 3x^2y^2$ | $4a - 3a$ | $4b^3 - 3b^3$ |

**Question:** What is the lesson to be learned here about addition and subtraction in algebra and mathematics in general?

# Student Activity
## Match Up with Like Denominators

**Match-up:** Complete each addition or subtraction problem and simplify the result.  Then choose the letter that corresponds to the **numerator** of this result.  If the numerator is not among the choices, then choose E (None of these).  The first one has been done for you.

| **A** $x+1$ | **B** $-1$ | **C** $1$ | **D** $x-1$ | **E** None of these |
|---|---|---|---|---|

| | | | |
|---|---|---|---|
| $\dfrac{x}{3x+3}+\dfrac{1}{3x+3}$ <br><br> $=\dfrac{x+1}{3x+3}=\dfrac{\cancel{(x+1)}^{1}}{3\cancel{(x+1)}_{1}}$ <br><br> $=\dfrac{1}{3}$ <br><br> **C** | $\dfrac{1}{x^3}-\dfrac{x^2+1}{x^3}$ | $\dfrac{x^2}{x^2+x}-\dfrac{1}{x^2+x}$ | $\dfrac{-3x^2+1}{(2x+1)^2}+\dfrac{3x^2+2x}{(2x+1)^2}$ |
| $\dfrac{2x-1}{x}-\dfrac{3x-2}{x}$ | $\dfrac{1-2x}{x}-\dfrac{2-3x}{x}$ | $\dfrac{x^2+x}{x^2+1}+\dfrac{1-x^2}{x^2+1}$ | $\dfrac{x^2+x}{x^2-1}+\dfrac{1-x^2}{x^2-1}$ |
| $\dfrac{x^2}{x^2-1}+\dfrac{2x+1}{x^2-1}$ | $\dfrac{x}{2x+1}-\dfrac{x+1}{2x+1}$ | $\dfrac{2x^2-2x}{(x-1)^2}-\dfrac{2x-2}{(x-1)^2}$ | $\dfrac{2x^2-2x}{2(x-1)^2}-\dfrac{2x-2}{2(x-1)^2}$ |

# Student Activity

## Determining the LCD

**Directions:** For each of the expressions below, start by factoring both denominators. If a denominator **doesn't** factor, write that denominator **in parentheses**. Then find the LCD for the expression.

For example, the denominators in $\dfrac{x}{x+1} + \dfrac{3}{4x+4}$ would be factored like this:

$\dfrac{x}{(x+1)} + \dfrac{3}{4(x+1)}$ and the LCD would be $4(x+1)$.

| | Expression | Expression with Factored Denominators | LCD |
|---|---|---|---|
| **1.** | $\dfrac{4}{x+3} + \dfrac{2}{x^2+3x}$ | $\dfrac{4}{\rule{2cm}{0.4pt}} + \dfrac{2}{\rule{2cm}{0.4pt}}$ | |
| **2.** | $\dfrac{5x}{x^2-25} - \dfrac{5}{x+5}$ | $\dfrac{5x}{\rule{2cm}{0.4pt}} - \dfrac{5}{\rule{2cm}{0.4pt}}$ | |
| **3.** | $\dfrac{3}{x^2+x-12} + \dfrac{2x}{x^2-x-6}$ | $\dfrac{3}{\rule{2cm}{0.4pt}} + \dfrac{2x}{\rule{2cm}{0.4pt}}$ | |
| **4.** | $\dfrac{x+3}{5x^3+25x^2} + \dfrac{2x}{3x+15}$ | $\dfrac{x+3}{\rule{2cm}{0.4pt}} + \dfrac{2x}{\rule{2cm}{0.4pt}}$ | |
| **5.** | $\dfrac{x}{x-2} - \dfrac{3}{2-x}$ | $\dfrac{x}{\rule{2cm}{0.4pt}} - \dfrac{3}{\rule{2cm}{0.4pt}}$ | |
| **6.** | $\dfrac{10}{8x^2-14x-15} - \dfrac{15}{8x^2-20x}$ | $\dfrac{10}{\rule{2cm}{0.4pt}} - \dfrac{15}{\rule{2cm}{0.4pt}}$ | |
| **7.** | $\dfrac{x^2+9}{3x^4-27x^2} + \dfrac{3x}{x^3-3x^2}$ | $\dfrac{x^2+9}{\rule{2cm}{0.4pt}} + \dfrac{3}{\rule{2cm}{0.4pt}}$ | |
| **8.** | $\dfrac{4}{15x^3+12x^2} - \dfrac{y}{3x^2}$ | $\dfrac{4}{\rule{2cm}{0.4pt}} - \dfrac{y}{\rule{2cm}{0.4pt}}$ | |

# Student Activity
## The Missing Form of 1

**Directions:** In each equation below, there is a missing multiple with a form of 1 and a missing numerator. Fill in the empty spaces to complete each expression.

Example: $\dfrac{x-2}{x+3} \cdot \boxed{\phantom{---}} = \dfrac{}{(x+3)(x-2)}$ becomes $\dfrac{x-2}{x+3} \cdot \boxed{\dfrac{x-2}{x-2}} = \dfrac{x^2-4x+4}{(x+3)(x-2)}$.

**Hint for problems 6-10: Try factoring the denominator on the right hand side first.**

1. $\dfrac{x+1}{x+2} \cdot \boxed{\phantom{---}} = \dfrac{}{3x(x+2)}$

2. $\dfrac{x-5}{x+5} \cdot \boxed{\phantom{---}} = \dfrac{}{(x+5)(x-5)}$

3. $\dfrac{x}{2-x} \cdot \boxed{\phantom{---}} = \dfrac{}{x-2}$

4. $\dfrac{x^2+3x}{3x+2} \cdot \boxed{\phantom{---}} = \dfrac{}{(3x+2)(x+3)}$

5. $\dfrac{4}{x+3} \cdot \boxed{\phantom{---}} = \dfrac{}{(x+3)(x+4)(x+5)}$

6. $\dfrac{3x}{x+1} \cdot \boxed{\phantom{---}} = \dfrac{}{x^2-1}$

7. $\dfrac{2x+3}{x+7} \cdot \boxed{\phantom{---}} = \dfrac{}{x^2+14x+49}$

8. $\dfrac{16x}{4-x} \cdot \boxed{\phantom{---}} = \dfrac{}{16-8x+x^2}$

9. $\dfrac{x^2}{5x+4} \cdot \boxed{\phantom{---}} = \dfrac{}{30x^3+24x^2}$

10. $\dfrac{8+x}{4x+3} \cdot \boxed{\phantom{---}} = \dfrac{}{20x^2+7x-6}$

# Student Activity

## The Reunion

**Directions:** Now that we've made it through addition, subtraction, multiplication, and division, it seems only fair to bring the whole gang back together for a reunion. Simplify each expression according to its operation!

| | |
|---|---|
| $\dfrac{x-1}{x+2} + \dfrac{x+3}{x+2}$ | |
| $\dfrac{x-1}{x+2} - \dfrac{x+3}{x+2}$ | |
| $\dfrac{x-1}{x+2} \cdot \dfrac{x+3}{x+2}$ | |
| $\dfrac{x-1}{x+2} \div \dfrac{x+3}{x+2}$ | |

| | |
|---|---|
| $\dfrac{x}{x-3} + \dfrac{x-1}{3-x}$ | |
| $\dfrac{x}{x-3} - \dfrac{x-1}{3-x}$ | |
| $\dfrac{x}{x-3} \cdot \dfrac{x-1}{3-x}$ | |
| $\dfrac{x}{x-3} \div \dfrac{x-1}{3-x}$ | |

| | |
|---|---|
| $\dfrac{12}{x+2} + \dfrac{4x+8}{x+1}$ | |
| $\dfrac{12}{x+2} - \dfrac{4x+8}{x+1}$ | |
| $\dfrac{12}{x+2} \cdot \dfrac{4x+8}{x+1}$ | |
| $\dfrac{12}{x+2} \div \dfrac{4x+8}{x+1}$ | |

# Student Activity
## Tempting Expressions

**Directions:** In all the expressions below, you will find yourself tempted to do incorrect mathematics by visually-pleasing expressions. Of course, just because it looks good, that doesn't mean it is the right thing to do. ☺ Think carefully about the steps involved in each expression.

**1.** Add: $\dfrac{5}{x} + \dfrac{x}{5}$

**6.** Multiply: $\dfrac{x^2 + 4x - 12}{x^2 + 4x - 32} \cdot \dfrac{x^2 + 10x + 16}{x^2 + 10x + 24}$

**2.** Subtract: $\dfrac{2}{x+2} - \dfrac{1}{x}$

**7.** Subtract: $\dfrac{3x}{x+2} - \dfrac{x+2}{3x+3}$

**3.** Add: $\dfrac{x}{x+4} + \dfrac{x+4}{x}$

**8.** Simplify: $\dfrac{5x^2 + 25x}{5x^2 - 125}$

**4.** Divide: $\dfrac{x-1}{x+6} \div \dfrac{x+6}{x^2 - 36}$

**9.** Subtract: $\dfrac{x^2}{x^2 - 9} - 9$

**5.** Simplify: $\dfrac{2x+8}{2x+4}$

**10.** Add: $\dfrac{2}{x+1} + \dfrac{2}{1-x}$

## Student Activity

### Tic-Tac-Toe on Complex Fraction Pieces

**Directions:** In every box in the tic-tac-toe grid, there is a rational expression and a simplified form. If the two are equivalent, then circle the simplified form (thus placing an **O** on the square). If the expression has **not** been simplified correctly, then put an **X** over the incorrect simplification.

**Tic-tac-toe Game#1:**

| $x^2\left(\dfrac{3}{x}\right)$ | $(2-x)\left(\dfrac{1}{x-2}\right)$ | $\left(\dfrac{2}{x}\right)\left(\dfrac{x^2}{4}\right)$ |
|---|---|---|
| $3x$ | $-1$ | $2x$ |
| $\left(24x^3\right)\left(\dfrac{3}{8x^2}\right)$ | $(15x)\left(\dfrac{4}{25x^2}\right)$ | $\left(\dfrac{x+3}{2(x-3)}\right)(x-3)$ |
| $9x$ | $\dfrac{4}{5x}$ | $\dfrac{x+3}{2}$ |
| $\left(\dfrac{3-2x}{x^3}\right)\left(x^3\right)$ | $\left(81x^4\right)\left(\dfrac{9}{9x^4}\right)$ | $\left(\dfrac{1}{x-1}\right)(1-x)$ |
| $3x^3-2x^4$ | $81$ | $1$ |

**Tic-tac-toe Game#2:**

| $\left(8x^2\right)\left(\dfrac{3}{4x}-2\right)$ | $x^2\left(\dfrac{5}{x^2}-\dfrac{2}{x}\right)$ | $x^3\left(\dfrac{5}{x^2}+4x\right)$ |
|---|---|---|
| $6x-16x^2$ | $5-\dfrac{2}{x}$ | $5x+4x^2$ |
| $x\left(4+\dfrac{x+1}{x}\right)$ | $\left(\dfrac{x}{5}\right)\left(\dfrac{25}{x}+5\right)$ | $\left(\dfrac{x^2}{3}\right)\left(6-\dfrac{9}{10x^2}\right)$ |
| $5x+1$ | $5+5x$ | $2x^2-\dfrac{3}{10}$ |
| $\left(32x^4\right)\left(\dfrac{1}{16x}-\dfrac{1}{32x^4}\right)$ | $\left(\dfrac{24x}{y}\right)\left(\dfrac{y}{8x}-y\right)$ | $\left(42x^2\right)\left(\dfrac{1}{6x}+\dfrac{1}{7x}\right)$ |
| $2x^3-1$ | $3-24x$ | $5$ |

# Student Activity

## Double the Fun on Complex Fractions

**Directions:** For each of the complex fractions below, simplify the expression using both methods. The result should be the same for both methods. If they are not, go back and look for a mistake. The first one has been **started** for you.

**Method I:** Using division.  **Method II:** Multiplying by the LCD.

| 1. | Method I | $\dfrac{\dfrac{1}{x}+4}{\dfrac{5}{x}-1} = \dfrac{\dfrac{1}{x}+4\left(\dfrac{x}{x}\right)}{\dfrac{5}{x}-1\left(\dfrac{x}{x}\right)} = \dfrac{\dfrac{1+4x}{x}}{\dfrac{5-x}{x}} =$ |
|----|----------|---|
|    | Method II | $\dfrac{\dfrac{1}{x}+4}{\dfrac{5}{x}-1} = \dfrac{\left(\dfrac{1}{x}+4\right)}{\left(\dfrac{5}{x}-1\right)} \cdot \dfrac{\dfrac{x}{1}}{\dfrac{x}{1}} =$ |

| 2. | Method I | $\dfrac{5-\dfrac{1}{x^2}}{\dfrac{2}{x}+x}$ |
|----|----------|---|
|    | Method II | $\dfrac{5-\dfrac{1}{x^2}}{\dfrac{2}{x}+x}$ |

| 3. | Method I | $\dfrac{\dfrac{1}{x+1}+1}{1+\dfrac{2}{x}}$ |
|----|----------|---|
|    | Method II | $\dfrac{\dfrac{1}{x+1}+1}{1+\dfrac{2}{x}}$ |

# Student Activity
## One of Us is Wrong!

**Directions:** Lou and Stu have both simplified the given complex fraction. Unfortunately, they have different answers. 1) Choose a test value for $x$ and use it to evaluate the complex fraction and the students' answers. Give the student that is correct a ☺ for their work. 2) Then determine where the unfortunate student made the error.

| 1. Given expression | Lou's Work | Stu's Work |
|---|---|---|
| $$\dfrac{\dfrac{2x+5}{x+2}}{\dfrac{x+2}{2x+5}}$$ <br><br> Test value: $x =$ ___ | $$\dfrac{\dfrac{2x+5}{x+2}}{\dfrac{x+2}{2x+5}} = \dfrac{2x+5}{x+2} \div \dfrac{x+2}{2x+5}$$ <br><br> $$= \dfrac{2x+5}{x+2} \cdot \dfrac{2x+5}{x+2} = \dfrac{(2x+5)^2}{(x+2)^2}$$ | $$\dfrac{\dfrac{2x+5}{x+2}}{\dfrac{x+2}{2x+5}} = \dfrac{\dfrac{2x+5}{\cancel{x+2}^{\,1}}}{\dfrac{\cancel{x+2}^{\,1}}{2x+5}}$$ <br><br> $$= \dfrac{\cancel{2x+5}^{\,1}}{\cancel{2x+5}^{\,1}} = 1$$ |
| Expression value: | Expression value: | Expression value: |

| 2. Given expression | Lou's Work | Stu's Work |
|---|---|---|
| $$\dfrac{2x - \dfrac{1}{2x}}{\dfrac{1}{2x} + 2x}$$ <br><br> Test value: $x =$ ___ | $$\dfrac{2x - \dfrac{1}{2x}}{\dfrac{1}{2x} + 2x} = \dfrac{\cancel{2x}^{\,1} - \dfrac{1}{\cancel{2x}^{\,1}}}{\dfrac{1}{\cancel{2x}^{\,1}} + \cancel{2x}^{\,1}}$$ <br><br> $$= \dfrac{1-1}{1+1}$$ <br><br> $$= \dfrac{0}{2} = 0$$ | $$\dfrac{2x - \dfrac{1}{2x}}{\dfrac{1}{2x} + 2x} = \dfrac{\left(2x - \dfrac{1}{2x}\right)\dfrac{2x}{1}}{\left(\dfrac{1}{2x} + 2x\right)\dfrac{2x}{1}}$$ <br><br> $$= \dfrac{2x\left(\dfrac{2x}{1}\right) - \dfrac{1}{\cancel{2x}_{\,1}}\left(\dfrac{\cancel{2x}^{\,1}}{1}\right)}{\dfrac{1}{\cancel{2x}_{\,1}}\left(\dfrac{\cancel{2x}^{\,1}}{1}\right) + 2x\left(\dfrac{2x}{1}\right)} = \dfrac{4x^2 - 1}{1 + 4x^2}$$ |
| Expression value: | Expression value: | Expression value: |

| 3. Given expression | Lou's Work | Stu's Work |
|---|---|---|
| $$\dfrac{\dfrac{x^2}{2}}{\dfrac{2}{x} + \dfrac{4}{x}}$$ <br><br> Test value: $x =$ ___ | $$\dfrac{\dfrac{x^2}{2}}{\dfrac{2}{x} + \dfrac{4}{x}} = \dfrac{x^2}{2} \div \left(\dfrac{2}{x} + \dfrac{4}{x}\right)$$ <br><br> $$= \dfrac{x^2}{2} \cdot \left(\dfrac{x}{2} + \dfrac{x}{4}\right) = \dfrac{x^2}{2} \cdot \left(\dfrac{2x}{4} + \dfrac{x}{4}\right)$$ <br><br> $$= \dfrac{x^2}{2} \cdot \left(\dfrac{3x}{4}\right) = \dfrac{3x^3}{8}$$ | $$\dfrac{\dfrac{x^2}{2}}{\dfrac{2}{x} + \dfrac{4}{x}} = \dfrac{\dfrac{x^2}{2}}{\dfrac{6}{x}} = \dfrac{x^2}{2} \div \dfrac{6}{x}$$ <br><br> $$= \dfrac{x^2}{2} \cdot \dfrac{x}{6} = \dfrac{x^3}{12}$$ |
| Expression value: | Expression value: | Expression value: |

# Student Activity
## Checking Solutions to Rational Equations with a Calculator

You can use a calculator to check if the solution that you find to a rational equation is correct. For example, let's see if $-10$ is a solution of the equation $\dfrac{4}{x+2} = \dfrac{5}{x}$.

**a)** Construct the parentheses skeleton for the equation: $\dfrac{4}{(\ )+2} = \dfrac{5}{(\ )}$

**b)** Use extra parentheses or brackets to group the numerator or denominator if they contain more than one term: $\dfrac{4}{\left[(\ )+2\right]} = \dfrac{5}{(\ )}$

**c)** Insert the possible solution: $\dfrac{4}{\left[(-10)+2\right]} = \dfrac{5}{(-10)}$

**d)** Enter each side into your calculator, using extra parentheses to group the numerator or denominator if there is more than one term:

Left side: **4/((-10)+2)**          Right side: **5/(-10)**

In this case, both sides result in $-0.5$, so $-10$ **is** a solution to the equation $\dfrac{4}{x+2} = \dfrac{5}{x}$.

**Directions:** Fill in the table below. If the one side of the equation is *undefined* for a value, say so (your calculator will likely give you an "error" message if this is the case). The first row has been done for you.

| Equation | Parentheses skeleton and extra parentheses needed for calculator | Evaluate for... | Left side of equation | Right side of equation | Is this value a solution? |
|---|---|---|---|---|---|
| $\dfrac{5}{3+2x} = \dfrac{2}{x+1}$ | $\dfrac{5}{\left[3+2(\ )\right]} = \dfrac{2}{\left[(\ )+1\right]}$ | $x=1$ | 1 | 1 | yes |
|  |  | $x=0$ |  |  |  |
|  |  | $x=-3$ |  |  |  |
| $\dfrac{9}{x} = \dfrac{12}{x-1}$ |  | $x=1$ |  |  |  |
|  |  | $x=0$ |  |  |  |
|  |  | $x=-3$ |  |  |  |
| $\dfrac{1-x}{x} = \dfrac{x-1}{3}$ |  | $x=1$ |  |  |  |
|  |  | $x=0$ |  |  |  |
|  |  | $x=-3$ |  |  |  |
| $\dfrac{x}{x+3} = 3x + \dfrac{x}{x+1}$ |  | $x=1$ |  |  |  |
|  |  | $x=0$ |  |  |  |
|  |  | $x=-3$ |  |  |  |

# Student Activity

## Can't Use That!

**Directions:** Examine each of the equations below and decide what values cannot be allowed because of a "division by zero" problem. For the "disallowed" values, shade in the corresponding value in the grid below. The first one has been done for you.

1. $\dfrac{x+5}{x-2} = \dfrac{3}{x}$; $x \neq 0, 2$

2. $\dfrac{x-4}{x+3} + \dfrac{2}{x-2} = 5$

3. $\dfrac{x}{3} + \dfrac{2}{x-6} = 1$

4. $\dfrac{3}{2x-1} = \dfrac{4}{x+5}$

5. $\dfrac{3}{4x} + \dfrac{1}{2} = \dfrac{2-x}{4-x}$

6. $\dfrac{5}{x+7} - \dfrac{3x}{x-10} = 6$

7. $\dfrac{x+3}{x-3} = \dfrac{4}{3x+2}$

8. $1 + \dfrac{x}{5} = \dfrac{5x}{x-8}$

9. $\dfrac{1}{4x-1} = \dfrac{4}{x}$

10. $\dfrac{9}{9+x} - 2 = \dfrac{1-x}{2x}$

11. $\dfrac{4}{x} = \dfrac{2x+1}{10}$

12. $\dfrac{x+6}{x-7} - \dfrac{9}{3x} = 4$

13. $\dfrac{x+1}{x+5} + \dfrac{2}{5x} = 3$

14. $\dfrac{x-3}{3x-4} + \dfrac{2x}{x+3} = 8$

15. $\dfrac{x}{12} + 5 = \dfrac{x+2}{x-2}$

| | | | | |
|---|---|---|---|---|
| $x \neq 0, 1, 4$ | $x \neq -3, 0, 2$ | $x \neq 0, 2, 4$ | $x \neq -5, 0, 2$ | $x \neq 0, 2$ |
| $x \neq 0, \dfrac{1}{4}, 1$ | $x \neq -3, 2, 4$ | $x \neq 0, 8$ | $x \neq 0, 4$ | $x \neq 2, 4$ |
| $x \neq 0, \dfrac{1}{4}$ | $x \neq -3, 2$ | $x \neq 8$ | $x \neq 6$ | $x \neq -5, 0$ |
| $x \neq -2, 3$ | $x \neq -3, -\dfrac{2}{3}, 3$ | $x \neq -7, 10$ | $x \neq 0, 6$ | $x \neq 3, 6$ |
| $x \neq 0$ | $x \neq -\dfrac{2}{3}, 3$ | $x \neq 0, 7$ | $x \neq \dfrac{4}{3}, 3$ | $x \neq 2$ |
| $x \neq 0, 1, 9$ | $x \neq -5, \dfrac{1}{2}$ | $x \neq -6, 0$ | $x \neq -7, 0, 10$ | $x \neq -2, 0, 2$ |
| $x \neq 0, 9$ | $x \neq -5, 1$ | $x \neq 0, -1$ | $x \neq -3, 0, 3, 4$ | $x \neq -1, 0, 5$ |

# Student Activity
## One Step at a Time for Rational Equations

When you are solving rational equations, the first steps can be the hardest ones. In this activity we practice taking those first few steps. Find the LCD, multiply by the LCD on both sides of the equation, distribute (where necessary), simplify each side of the equation, and then you finally have an equation that can be solved with prior methods.

**Example:**

| Equation: | $\dfrac{3}{x}+\dfrac{4}{x+2}=3-\dfrac{1}{x}$ | LCD: $x(x+2)$ |
|---|---|---|
| | $x(x+2)\left(\dfrac{3}{x}+\dfrac{4}{x+2}\right)=\left(3-\dfrac{1}{x}\right)\cdot x(x+2)$ | |

| | **Left side** | **Right side** |
|---|---|---|
| **Distribute** | $x(x+2)\left(\dfrac{3}{x}\right)+x(x+2)\left(\dfrac{4}{x+2}\right)$ | $(3)(x)(x+2)-\dfrac{1}{x}(x)(x+2)$ |
| **Simplify** | $3(x+2)+4x$ <br> $3x+6+4x$ <br> $7x+6$ | $3x(x+2)-(x+2)$ <br> $3x^2+6x-x-2$ <br> $3x^2+5x-2$ |
| **And reunite!** | $7x+6=3x^2+5x-2$ | |

**1.**

| Equation: $\dfrac{3x+1}{x-3}=\dfrac{6x+2}{x^2-4x+3}$ | LCD: _____ |
|---|---|

| | **Left side** | **Right side** |
|---|---|---|
| **Distribute** | | |
| **Simplify** | | |
| **And reunite!** | | |

**2.**

**Equation:** $1 - \dfrac{27}{x^2 + x - 12} = -\dfrac{1}{x-3}$   LCD: _____

| | Left side | Right side |
|---|---|---|
| **Distribute** | | |
| **Simplify** | | |
| **And reunite!** | | |

Working to keep your equations
clear of fractions

**3.**

**Equation:** $\dfrac{2x+5}{4x+5} = \dfrac{-2}{2x+1} - \dfrac{5}{4x+5}$   LCD: _____

| | Left side | Right side |
|---|---|---|
| **Distribute** | | |
| **Simplify** | | |
| **And reunite!** | | |

# Student Activity
## Match Up on Simple Rational Equations

**Directions:** Solve each rational equation and check the result. Then choose the letter that corresponds to this result. Some problems may have more than one answer. If there is no solution, then choose E (No Solution). If the result is not among the choices, then choose F (None of these). The first one has been done for you.

Nice try, but you can't just erase the hard parts.

**A** $2$        **B** $-1$

**C** $1$        **D** $\frac{1}{2}$

**E** No Solution        **F** None of These

| | | |
|---|---|---|
| $\dfrac{5x+5}{x+4}=2$ <br><br> $5x+5=2(x+4)$ <br><br> $5x+5=2x+8$ <br><br> $3x=3$ <br><br> $x=1$ **(C)** | $3=\dfrac{8-x}{x+2}$ | $\dfrac{4x-2}{x+2}=4$ |
| $\dfrac{x-3}{2x+3}=5$ | $\dfrac{3}{x}+\dfrac{4}{x+2}=-\dfrac{1}{x}$ | $\dfrac{5}{2}=\dfrac{11+x}{x+5}$ |
| $\dfrac{x+2}{x-4}=\dfrac{2}{x-3}$ | $\dfrac{x-4}{x+1}=-\dfrac{2x+7}{(x+2)(x+1)}$ | $\dfrac{3x^2+5}{7x+5}=2$ |

# Student Activity
## Working with the Smaller Pieces

### Practice with Uniform Motion Problems

We use the formula $d = rt$ to solve motion problems, where $d$ is the distance, $r$ is the rate, and $t$ is the time.

1.  A husband and wife go jogging together.  The husband can run 2 miles an hour faster than the wife.

    a.  If the wife runs at a rate of $x$ miles per hour, what is the rate of the husband? _____

    b.  If they run 10 miles, write an expression that describes the *time* it takes for the wife to run this distance. _____

    c.  If they run 10 miles, write an expression that describes the *time* it takes for the husband to run this distance. _____

2.  Two trains (the Midnight Express and the Night Special) leave a station traveling in opposite directions.  The Midnight Express is 10 mph faster than the Night Special, which travels at 50 mph.  How far apart are the trains after two hours? _____ Every hour the trains are _____ more miles apart.

3.  The same two trains as the previous problem leave a station traveling in the *same* direction.  How far apart are the trains after two hours? _____ Every hour the trains are _____ more miles apart.

### Practice with Shared-work Problems

Work completed, work rate, and time can be related using the formula: $w = rt$

Often the "work completed" is represented by 1 (for 1 job completed).

4.  Suppose Joann can clean her house in 2 hours and Marty can clean the house in 5 hours. If JoAnn and Marty work together to clean the house, will it take
    **a.** less than 2 hours     **b.** between 2 and 5 hours        **c.** more than 5 hours

5.  If it takes you 2 hours to clean one pool, how long will it take you to clean 3 pools of the same size? _____

6.  It normally takes you 2 hours to clean the pool.  If you have help to clean it, will it take you
    **a.** more than 2 hours      **b.** less than 2 hours

7.  Saundra can type 5 pages in 30 minutes.  How long should it take her to type 8 pages? _____

8.  Shell can paint 500 square feet of wall in 40 minutes.  What is his work rate in square feet per minute? _____

# Guided Learning Activity
## Understanding Shared-Work Problems

Work completed = rate of work · time completed or $W = RT$ where

$W$ = fraction of one job that is completed

$R$ = Rate of work = $\dfrac{\text{amount of work}}{\text{time}}$

$T$ = time (in the same units as the rate of work)

Start each problem below by first designating what a "job" is. Then fill in the blank boxes in the table, write an equation that can be used to solve the problem, and find the solution.

---

1. If the community pool is filled from the reserve water tower, it will take 6 hours to fill the pool. If the pool is filled from the city water pipes, it will take 4 hours to fill the pool. How long will it take to fill the pool if water from both the reserve water tower and the city water pipes are used?

| Job: Fill one community pool | Rate | Time | Work completed |
|---|---|---|---|
| Reserve water tower | $\dfrac{1 \text{ pool}}{6 \text{ hours}} = \dfrac{1}{6}$ | $x$ hours | $\dfrac{x}{6}$ |
| City water pipes | | | |

Equation:                                    Solution:

2. It takes Mr. Dupree 5 hours to audit 12 bank files. His associate, Ms. Carmichael can audit 16 files in 4 hours. If they work together, how long will it take them to audit 30 files?

| Job: | Rate | Time | Work completed |
|---|---|---|---|
| Mr. Dupree | | | |
| Ms. Carmichael | | | |

Equation:                                    Solution:

3. Macon Construction can erect 100 square feet of rock facing on an exterior wall in 8 hours. Thomson Masonry can erect 200 square feet of rock facing in 12 hours. If you hire the two crews to work together, but Thomson Masonry starts working 2 hours after Macon Construction, how long will it take them to erect 250 square feet of rock facing?

| Job: | Rate | Time | Work completed |
|---|---|---|---|
| Macon Construction | | | |
| Thomson Masonry | | | |

Equation:                                    Solution:

# Student Activity

## Working with the Language of Proportions

**Directions:** Fill in the table using the language of proportions. Find the units for each ratio and set up the proportion. You do not need to solve the problem. The first one has been done for you.

| Problem | Fill in the Missing Values | | Proportion $\dfrac{A}{B} = \dfrac{C}{D}$ |
|---|---|---|---|
| | Ratio 1 | Ratio 2 | |
| **1.** If 12 doughnuts cost $2.50, how many doughnuts can you buy for $5.00? | $\dfrac{12 \text{ doughnuts}}{\$2.50}$ | $\dfrac{x \text{ doughnuts}}{\$5.00}$ | $\dfrac{12}{2.50} = \dfrac{x}{5.00}$ |
| **2.** A chocolate chip cookie recipe uses 3 cups of flour to make 5 dozen cookies, how much flour is required to make 12 dozen cookies? | $\dfrac{\boxed{\phantom{xx}} \text{ cups flour}}{\boxed{\phantom{xx}} \text{ dozen cookies}}$ | $\dfrac{\boxed{\phantom{xx}} \text{ cups flour}}{\boxed{\phantom{xx}} \text{ dozen cookies}}$ | |
| **3.** If it takes Joel 4.5 hours to run a 26.2 mile marathon, how many minutes does it take him to run 1 mile? | $\dfrac{\boxed{\phantom{xx}} \text{ miles}}{\boxed{\phantom{xx}} \text{ minutes}}$ | $\dfrac{\boxed{\phantom{xx}} \text{ miles}}{\boxed{\phantom{xx}} \text{ minutes}}$ | |
| **4.** If jeans are on sale at three pairs for $50, how much will seven pairs cost? | $\dfrac{\boxed{\phantom{xx}} \text{ pairs}}{\$\boxed{\phantom{xx}}}$ | $\dfrac{\boxed{\phantom{xx}} \text{ pairs}}{\$\boxed{\phantom{xx}}}$ | |
| **5.** If Jamie gets paid $1000 for a 40-hour work week, how many hours will he have to work to earn $800? | $\dfrac{\$\boxed{\phantom{xx}}}{\boxed{\phantom{xx}} \text{ hours}}$ | $\dfrac{\$\boxed{\phantom{xx}}}{\boxed{\phantom{xx}} \text{ hours}}$ | |
| **6.** There are 1160 calories in a 32 oz. chocolate shake from McDonalds. How many calories are in the 24 oz. shake? Source: www.mcdonalds.com | $\dfrac{\boxed{\phantom{xx}} \text{ calories}}{\boxed{\phantom{xx}} \text{ oz.}}$ | $\dfrac{\boxed{\phantom{xx}} \text{ calories}}{\boxed{\phantom{xx}} \text{ oz.}}$ | |
| **7.** According to American Red Cross guidelines, CPR should consist of cycles of 30 chest compressions followed by 2 breaths. If a rescuer has given 270 compressions, how many breaths should she have given? Source: www.redcross.org | $\dfrac{\boxed{\phantom{xx}} \text{ compressions}}{\boxed{\phantom{xx}} \text{ breaths}}$ | $\dfrac{\boxed{\phantom{xx}} \text{ compressions}}{\boxed{\phantom{xx}} \text{ breaths}}$ | |

308

# Student Activity

## Proportion Heteronyms

**Directions:** In writing, there are words that are spelled the same but have different pronunciations and different definitions; these are called heteronyms. Many mathematical expressions and equations look similar but are really very different (almost like mathematical heteronyms). In each set of "proportion heteronyms" below, first **identify** whether you are being given an equation or an expression. If it is an expression, simplify it. If it is an equation, solve it. **Do your work (simplifying or solving) on another sheet of paper.** The first line has been done for you.

|  | | | |
|---|---|---|---|
| **Heteronym Variation I** | $\dfrac{5}{x} = \dfrac{2}{3}$ | Equation. | $x = \dfrac{15}{2}$ |
|  | $\dfrac{5}{x} - \dfrac{2}{3}$ | | |
|  | $\dfrac{5}{x} + \dfrac{2}{3}$ | | |
|  | $\dfrac{5+2}{x+3}$ | | |
|  | $\dfrac{5}{x} - \dfrac{2}{3} = 0$ | | |
|  | $\dfrac{5}{x} \div \dfrac{2}{3}$ | | |

|  | | | |
|---|---|---|---|
| **Heteronym Variation II** | $\dfrac{x+1}{8} + \dfrac{x}{2}$ | | |
|  | $\dfrac{x+1}{8} \div \dfrac{x}{2}$ | | |
|  | $\dfrac{x+1}{8} = \dfrac{x}{2}$ | | |
|  | $x + \dfrac{1}{8} = \dfrac{x}{2}$ | | |
|  | $\dfrac{x}{8} + 1 = \dfrac{x}{2}$ | | |
|  | $\dfrac{x}{8} + 1 + \dfrac{x}{2}$ | | |

# Student Activity
## Equation Lineup

**Directions:** We have solved many different types of equations at this point in our learning. For each equation below, categorize the type of equation, write the strategies and key steps for remembering how to solve it, then solve the equation. The first one has been started for you.

**Equation Types:** Linear, quadratic, rational

---

**1.** $3x^2 - x = 4$

**Type of Equation?** Quadratic

**Strategies and Key Steps**

Get =0 on one side of the equation, then factor to solve.

**Solve it!**

---

**2.** $-7(x + 2) = 1 + 5(x - 1)$

**Type of Equation?**

**Strategies and Key Steps**

**Solve it!**

---

**3.** $\dfrac{x+4}{x-4} = \dfrac{x-1}{x+3}$

**Type of Equation?**

**Strategies and Key Steps**

**Solve it!**

---

**4.** $-\dfrac{3}{8} = \dfrac{1}{4} - \dfrac{5x}{16}$

**Type of Equation?**

**Strategies and Key Steps**

**Solve it!**

**5.** $2x^2 = 3(x+3)$

**Type of Equation?**

**Strategies and Key Steps**

**Solve it!**

---

**6.** $\dfrac{x}{x+4} - \dfrac{1}{x+1} = 1$

**Type of Equation?**

**Strategies and Key Steps**

**Solve it!**

---

**7.** $5(2x+1) = 3(x+4) + 7x$

**Type of Equation?**

**Strategies and Key Steps**

**Solve it!**

---

**8.** $x^2 = 10x$

**Type of Equation?**

**Strategies and Key Steps**

**Solve it!**

# Guided Learning Activity

## Orienting Similar Triangles

Similar triangles are triangles with the same shape, but not necessarily the same size. These pairs of triangles can be easy to spot when the orientation of the triangles is the same.

To the right, the two triangles are similar, and we would write that relationship like this: $\triangle ABC \sim \triangle DEF$.

For two triangles to be similar, two of the angle measures in the triangles must be the same (this forces all three angles to be the same).

**Example:**

The sum of the angles in a triangle is _____.

**Find** the missing angles in triangles $\triangle JKL$ and $\triangle MNP$ to the right.

Are the two triangles similar? _____

Which angles are equal?  $\angle J =$ _____      $\angle K =$ _____      $\angle L =$ _____

Use a different color to highlight each pair of similar angles.

Use a different color to highlight each pair of similar sides.

Which sides are similar?  $JK \sim$ _____      $KL \sim$ _____      $LJ \sim$ _____

Because of these side similarities, we can set up equal ratios of similar sides (fill in the missing denominators):  $\dfrac{JK}{\rule{1cm}{0.4pt}} = \dfrac{KL}{\rule{1cm}{0.4pt}} = \dfrac{LJ}{\rule{1cm}{0.4pt}}$

This can be written as three different proportions. What are they?

$\dfrac{\rule{1cm}{0.4pt}}{\rule{1cm}{0.4pt}} = \dfrac{\rule{1cm}{0.4pt}}{\rule{1cm}{0.4pt}}$         $\dfrac{\rule{1cm}{0.4pt}}{\rule{1cm}{0.4pt}} = \dfrac{\rule{1cm}{0.4pt}}{\rule{1cm}{0.4pt}}$         $\dfrac{\rule{1cm}{0.4pt}}{\rule{1cm}{0.4pt}} = \dfrac{\rule{1cm}{0.4pt}}{\rule{1cm}{0.4pt}}$

| Triangles | Are the triangles similar? If so, state the side similarities. | Find the lengths of the side(s) indicated by variables. |
|---|---|---|
| **1.** | | |
| **2.** | | |
| **3.** | | |
| **4.** | | |

# Student Activity

## Charting the Language of Variation

**Directions:** Fill in the blank spaces in the chart. The first one has been done for you.

| In Words | In Equation Form | Type of Variation (direct, inverse, joint, or combined) |
|---|---|---|
| **1.** The amount of sales tax, $s$, on an item varies <u>directly</u> as the price, $p$, of the item. | $s = kp$ | direct |
| **2.** The time, $t$, it takes to clean the house is _____ proportional to the number of people, $n$, who help clean it. | $t = \dfrac{k}{n}$ | |
| **3.** The area of a circle, $A$, varies directly with as square of the radius, $r$. | | |
| **4.** The time, $t$, it takes to arrive at a destination is _____ proportional to the rate of travel, $r$. | $t = \dfrac{k}{r}$ | |
| **5.** The force of wind, $F$, on a billboard varies jointly as the area of the billboard, $A$, and the square of the wind velocity, $v$. | | |
| **6.** The voltage, $V$, in an electrical circuit with a fixed resistance is _____ proportional to the current, $I$. | $V = kI$ | |
| **7.** The mass, $m$, of an iron solid varies directly as the volume, $V$. | | |
| **8.** The number of supports, $n$, needed to hold up a bridge varies _____ as its length, $\ell$. | $n = k\ell$ | |
| **9.** The price of gasoline, $p$, varies inversely as the number of gallons of crude oil, $g$, that are refined. | | |
| **10.** The pressure, $P$, of a gas is directly proportional to the temperature, $T$, and inversely proportional to the volume, $V$. | | |

# Guided Learning Activity
## Constant of Variation... an Oxymoron

What does the word *constant* mean? _____

In mathematics, what is a *constant*? _____

What does the word *variation* mean? _____

In mathematics, what is a *variable*? _____

When we work with variation problems, we use the letter $k$ to represent the constant of variation. But what does it mean when we say *constant of variation*? Let's investigate.

**Hourly Pay:**

**1.** Fill in the tables for Monte, Clay, and Lisa to tell them how much gross pay (before taxes and other fees are removed) they will take home for working the given number of hours. Then write an equation for each that relates their gross pay, $P$, to the hours worked, $h$.

Monte: $7.50/hour

| Hours | Pay |
|-------|-----|
| 10 | |
| 15 | |
| 20 | |
| 30 | |
| 40 | |

Clay: $22/hour

| Hours | Pay |
|-------|-----|
| 10 | |
| 15 | |
| 20 | |
| 30 | |
| 40 | |

Lisa: $18.25/hour

| Hours | Pay |
|-------|-----|
| 10 | |
| 15 | |
| 20 | |
| 30 | |
| 40 | |

Monte: $P =$ _____    Clay: $P =$ _____    Lisa: $P =$ _____

**2.** In this example, what does $k$ represent? _____

**3.** Write an equation for: "gross pay is directly proportional to hours worked" using $k$ as the constant of variation. _____

   **a.** Lisa makes $365 for working 20 hours. Use this data and the equation above to solve for the constant $k$.

   $k =$ _____    Look familiar?

   **b.** Now write the variation equation with the numeric value of $k$ in place: _____

   **c.** Use this equation to find Lisa's gross pay if she works 25 hours. _____

   **d.** Could you use the equation from part b to find gross pay if Monte works for 25 hours? Why or why not?

**Density:** The density, $\rho$, of an object is its mass per unit volume, that is $\rho = \dfrac{m}{V}$. If the object is made of a uniform material, then the density is constant.

1. Calculate the density (in $g/cm^3$) for the given mass and volume of each material in the tables below.

Balsa Wood:

| Mass | 12 g | 15 g | 24 g | 39 g | 45 g |
|---|---|---|---|---|---|
| Volume | 100 cm$^3$ | 125 cm$^3$ | 200 cm$^3$ | 325 cm$^3$ | 375 cm$^3$ |
| Density | | | | | |

Ice:

| Mass | 229.25 g | 366.8 g | 458.5 g | 495.18 g | 687.75 g |
|---|---|---|---|---|---|
| Volume | 250 cm$^3$ | 400 cm$^3$ | 500 cm$^3$ | 540 cm$^3$ | 750 cm$^3$ |
| Density | | | | | |

2. What is constant for Balsa wood? _____ What is constant for ice? _____

3. Is the density constant for Balsa wood the same as the density constant for ice? _____

Now we'll turn this around.

4. Write an equation for: "mass is directly proportional to volume" using $k$ as the constant of variation. _____

    **a.** The mass of 100 cm$^3$ of balsa wood is 12 g. Use this data and the equation you just wrote to solve for $k$.

        $k = $ _____ Look familiar?

    **b.** Now write the variation equation with the numeric value of $k$ in place: _____

    **c.** Use this equation to find the volume of a 500 g piece of balsa wood.

    **d.** Could you use the equation from part b to find the volume of 500 g of ice? Why or why not?

5. See if you can now explain what it means when we say *constant of variation.*

# Assess Your Understanding
## Rational Expressions and Equations

For each of the following, describe the strategies or key steps that will help you **start** the problem.  You do **not** have to complete the problems.

| | | What will help you to start this problem? |
|---|---|---|
| **1.** | Subtract: $\dfrac{4x}{x+1} - \dfrac{2x+3}{x+1}$ | |
| **2.** | Simplify: $\dfrac{x^2 - 2x - 35}{7x + 35}$ | |
| **3.** | Add: $\dfrac{3}{x+2} + \dfrac{4}{x+3}$ | |
| **4.** | Multiply: $\dfrac{x^2 + 8x}{6x + 18} \cdot \dfrac{12x - 24}{x^2 + 6x - 16}$ | |
| **5.** | Where is $\dfrac{x+3}{x^2 - 16}$ undefined? | |
| **6.** | Add: $\dfrac{4x+5}{3x+5} + \dfrac{2x+5}{3x+5}$ | |
| **7.** | Divide: $\dfrac{x^2 + 4x}{x+8} \div \dfrac{x^3 + 4x^2}{x^2 + 6x - 16}$ | |
| **8.** | Simplify: $\dfrac{x^2 + 9x + 14}{4 - x^2}$ | |

| | | |
|---|---|---|
| **9.** | Subtract: $\dfrac{x^2+3}{x^2-4}-\dfrac{x+5}{x-2}$ | |
| **10.** | Multiply: $\dfrac{3x^2+12x}{x-6}\cdot\dfrac{x^2-4x-12}{x^2+2x}$ | |
| **11.** | If you add the same number to the numerator and denominator of $\dfrac{7}{9}$, the result is $\dfrac{5}{6}$. What is the number? | |
| **12.** | Add: $\dfrac{4}{x+3}+\dfrac{1}{3x}$ | |
| **13.** | Solve: $\dfrac{2}{x+6}+\dfrac{5}{x^2+6x}=1$ | |
| **14.** | Simplify: $\dfrac{\dfrac{2}{3}+\dfrac{5}{x}}{\dfrac{1}{2x}-4}$ | |
| **15.** | On a map, a distance of 1.5 inches represents 1.125 miles. How many miles does 2.2 inches on the map represent? | |

318

# Metacognitive Skills
## Rational Expressions and Equations

Metacognitive skills refer to the ability to judge how well you have learned something and to effectively direct your own learning and studying. This is a self-evaluation tool designed to help you focus your studying and to improve your metacognitive skills with regards to this math class.

Fill the 1st column out **before** you begin studying. Fill the 2nd column out after you study for your test.

Go back to this assessment after your test and circle any of the ratings that you would change – this identifies the "disconnects" between what you **thought** you knew well and what you **actually** knew well.

**Use the scale below to assign a number to each topic.**
5  *I am confident I can do any problems in this category correctly.*
4  *I am confident I can do most of the problems in this category correctly.*
3  *I understand how to do the problems in this category, but I still make a lot of mistakes.*
2  *I feel unsure about how to do these problems.*
1  *I know I don't understand how to do these problems.*

| Topic or Skill | Before Studying | After Studying |
|---|---|---|
| Factoring polynomials. | | |
| Correctly identifying opposite factors. | | |
| Simplifying a rational expression. | | |
| Finding the values where a rational expression is undefined. | | |
| Multiplying rational expressions. | | |
| Dividing rational expressions. | | |
| Simplifying expressions that involve units of measurement. | | |
| Adding or subtracting rational expressions with like denominators. | | |
| Finding the LCD of two (or more) rational expressions. | | |
| Building an equivalent rational expression with a desired denominator. | | |
| Understanding why you need a common denominator when you add or subtract rational expressions. | | |
| Identifying denominators that are opposites. | | |
| Adding or subtracting rational expressions with unlike denominators. | | |
| Identifying a complex fraction. | | |
| Simplifying a complex fraction by simplifying the numerator & denominator and then dividing. | | |
| Simplifying a complex fraction by finding the LCD for all terms in the complex fraction and then multiplying by LCD/LCD. | | |
| Finding the values where a rational *equation* is undefined and knowing why it is important to find these values. | | |
| Clearing the fractions from a rational equation. | | |

*Continued on the next page.*

**Use the scale below to assign a number to each topic.**

5  *I am confident I can do any problems in this category correctly.*
4  *I am confident I can do most of the problems in this category correctly.*
3  *I understand how to do the problems in this category, but I still make a lot of mistakes.*
2  *I feel unsure about how to do these problems.*
1  *I know I don't understand how to do these problems.*

| Topic or Skill | Before Studying | After Studying |
|---|---|---|
| Understanding why you cannot "clear the fractions" from a rational expression. | | |
| Solving a linear equation that may result from a rational equation. | | |
| Solving a quadratic equation that may result from a rational equation. | | |
| Checking the solutions of a rational equation. | | |
| Solving a formula containing rational expressions for one of the variables. | | |
| Setting up a table of information for a uniform motion application problem. | | |
| Setting up a table of information for a shared-work application problem. | | |
| Setting up a table of information for an investment application problem. | | |
| Converting the information in a table to an equation that can be solved. | | |
| Writing ratios from sentences. | | |
| Creating a proportion from an application problem. | | |
| Solving a proportion. | | |
| Understanding the specific properties that allow a proportion equation to be solved differently from other rational equations. | | |
| Knowing the types of triangles that you can apply the Pythagorean Theorem to, and being able to correctly set up the Pythagorean Theorem for those triangles. | | |
| Knowing what makes two triangles similar. | | |
| Setting up a proportion of sides for two similar triangles. | | |
| Understanding what the constant of variation is. | | |
| Converting the language of variation problems to variation equations containing a constant of variation. | | |
| Solving for $k$ in a variation equation. | | |
| Using the variation equation to solve for something else in a problem. | | |

# RAD: Radical Expressions and Equations

# Student Activity
## Nightmare on Radical Street

**Directions:**

Each of the radical expressions listed below is either real or non-real.

If the expression is real and **is** a perfect square, then place it in the appropriate **rational** number box.

If the expression is real and **is not** a perfect square, then place it in the appropriate **irrational** number box.

If the expression is **non-real**, then cross it out to eliminate Scotty's nightmares!

**Do this activity without a calculator!**

*I promise, there are no square roots of negatives hiding under your bed. Those aren't even real !*

*Never again would Scotty make the mistake of eating spicy food before he finished his algebra homework.*

$$\sqrt{\frac{64}{49}} \quad -\sqrt{-1} \quad \sqrt{\frac{1}{2}} \quad \sqrt{\frac{4}{9}} \quad \sqrt{-9} \quad \sqrt{\frac{16}{25}} \quad \sqrt{36} \quad \sqrt{\frac{39}{2}} \quad \sqrt{1} \quad \sqrt{9}$$

$$\sqrt{3} \quad \sqrt{16} \quad \sqrt{6} \quad \sqrt{45} \quad \sqrt{-4} \quad \sqrt{14} \quad \sqrt{0} \quad \sqrt{\frac{60}{2}} \quad \sqrt{-25} \quad \sqrt{50}$$

| Rational Numbers | | Irrational Numbers | |
|---|---|---|---|
| $\boxed{\phantom{xx}} = \dfrac{4}{5}$    $\boxed{\phantom{xx}} = 6$ | | $0 < \boxed{\phantom{xx}} < 1$    $4 < \boxed{\phantom{xx}} < 5$ | |
| $\boxed{\phantom{xx}} = 4$    $\boxed{\phantom{xx}} = 3$ | | $1 < \boxed{\phantom{xx}} < 2$    $5 < \boxed{\phantom{xx}} < 6$ | |
| $\boxed{\phantom{xx}} = \dfrac{2}{3}$    $\boxed{\phantom{xx}} = 0$ | | $2 < \boxed{\phantom{xx}} < 3$    $6 < \boxed{\phantom{xx}} < 7$ | |
| $\boxed{\phantom{xx}} = \dfrac{8}{7}$    $\boxed{\phantom{xx}} = 1$ | | $3 < \boxed{\phantom{xx}} < 4$    $7 < \boxed{\phantom{xx}} < 8$ | |

# Student Activity
## Square Roots Using a Calculator

aOften the $\sqrt{\phantom{x}}$ function is found ABOVE one of the keys, and students will need to use a $\boxed{\text{2nd}}$ button to use the $\sqrt{\phantom{x}}$ function.

Your calculator probably falls into one of two groups:

### Group 1: Radicand first

On these calculators, you would find $\sqrt{9}$ by keying in $\boxed{9}$ then $\boxed{\sqrt{\phantom{x}}}$. It is likely that this will immediately give you the answer 3 without pressing any other keys.

### Group 2: Radical Symbol first

On most graphing calculators and some of the newer scientific calculators, you would do the keystrokes like this: $\boxed{\sqrt{\phantom{x}}}$ and then $\boxed{9}$. Then you probably see $\sqrt{(}9$ in your display. You have to close the parentheses by pressing $\boxed{)}$ and then $\boxed{\text{ENTER}}$ to evaluate.

Now fill out the table below, by finding the proper keystrokes to do the evaluation on your calculator. In the last column, square the calculator result to see if you get the radicand back. The first one has been done for you.

| | Expression | Calculator Keystrokes and result | Check by squaring the result |
|---|---|---|---|
| **a.** | $\sqrt{8}$ | Keystrokes will vary. | $(2.8284)^2 \approx 7.9998 \approx 8$ |
| **b.** | $\sqrt{169}$ | | |
| **c.** | $\sqrt{\dfrac{16}{25}}$ | | |
| **d.** | $\sqrt{\dfrac{1}{2}}$ | | |
| **e.** | $\sqrt{\dfrac{2}{3}}$ | | |
| **f.** | $\sqrt{-4}$ | | |
| **g.** | $\sqrt{\dfrac{4}{9}}$ | | |
| **h.** | $\sqrt{1.5625}$ | | |

# Student Activity

## Match Up on Simplifying Radical Expressions

**Directions:** Match each of the expressions in the squares in the table below with its simplified form from the top. Assume all variables represent positive numbers. If the simplified form is not found among the choices A through D, then choose E (none of these).

**A** $3x^2$

**B** $2x$

**C** $-5x$

**D** $4x^3$

**E** None of these

We haven't been able to determine the root of all evil, however, we have positively identified the radicand...

| | | |
|---|---|---|
| $\sqrt{9x^4}$ | $\sqrt[3]{-125x^3}$ | $\sqrt[4]{81x^8}$ |
| $\sqrt[6]{64x^6}$ | $\sqrt[3]{64x^9}$ | $\sqrt{-25x^2}$ |
| $2x^2 + x^2$ | $\sqrt[3]{27x^6}$ | $\sqrt{\dfrac{36x^2}{9}}$ |
| $\sqrt{9x}$ | $\sqrt{x^2} + x$ | $\sqrt{16x^6}$ |
| $-\sqrt[4]{625x^4}$ | $\sqrt[3]{12x^9}$ | $2 \cdot \sqrt[3]{8x^9}$ |

# Guided Learning Activity
## Radical Graphing

The graph of $f(x) = \sqrt{x}$, is a curve with a definite endpoint.

**1.** Draw the graph of $f(x) = \sqrt{x}$ by plotting points.

| $x$ | $f(x)$ |
|-----|--------|
| 9 | |
| 4 | |
| 1 | |
| 0 | |
| −1 | |
| −4 | |
| −9 | |

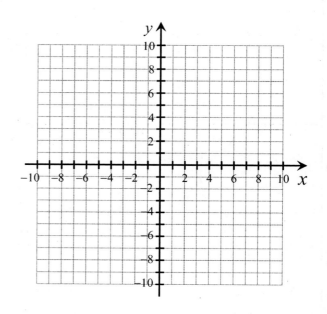

What is the domain of $f$? _____
What is the range of $f$? _____
Why are there no points for $x < 0$?

The graph of $f(x) = \sqrt[3]{x}$, is a curve that has a similar shape to the curve $y = x^3$.

**2.** Draw the graph of $f(x) = \sqrt[3]{x}$ by plotting points.

| $x$ | $f(x)$ |
|-----|--------|
| 8 | |
| 1 | |
| 0 | |
| −1 | |
| −8 | |

What is the domain of $f$? _____
What is the range of $f$? _____

These radical functions can be translated and reflected in the exact same way as $y = |x|$, $y = x^2$, and $y = x^3$.

**3.** Draw the graph of $g(x) = \sqrt{x} - 3$ by shifting the graph of $f(x) = \sqrt{x}$ _____ 3 units.

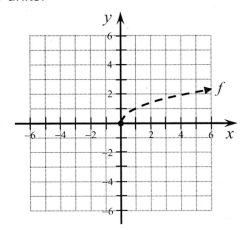

What is the domain of $g$? _____

What is the range of $g$? _____

**5.** Draw the graph of $g(x) = -\sqrt{x}$ by reflecting the graph of $f(x) = \sqrt{x}$ over the ___-axis.

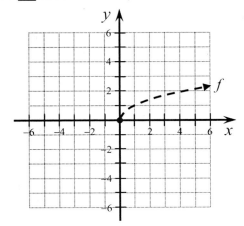

What is the domain of $g$? _____

What is the range of $g$? _____

**4.** Draw the graph of $g(x) = \sqrt[3]{x} + 2$ by shifting the graph of $f(x) = \sqrt[3]{x}$ _____ 2 units.

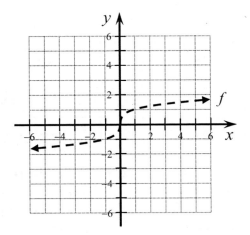

What is the domain of $g$? _____

What is the range of $g$? _____

**6.** Draw the graph of $g(x) = \sqrt[3]{x + 1} - 4$ by shifting the graph of $f(x) = \sqrt[3]{x}$ _____ 1 unit and _____ 4 units.

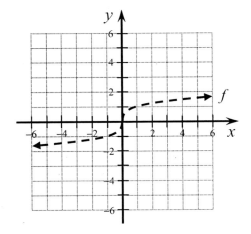

What is the domain of $g$? _____

What is the range of $g$? _____

327

# Student Activity
## Difficult Choices

If the variables can represent any real number and the radical expression represents a real number:

$$\sqrt[n]{x^n} = |x| \text{ if } n \text{ is even.}$$

$$\sqrt[n]{x^n} = x \text{ if } n \text{ is odd.}$$

If the variables only represent positive values, it is not necessary to use absolute values for the even roots.

**Directions:** Simplify each expression and choose the simplified expression from A or B. If the radical expression represents a non-real number (an imaginary number), then choose C (non-real).

| | A | B | C |
|---|---|---|---|
| **1.** Simplify $\sqrt{36x^2}$ if $x$ can be any real number. | $6x$ | $6|x|$ | Non-real |
| **2.** Simplify $\sqrt{-36x^2}$ if $x$ can be any real number. | $6x$ | $-6x$ | Non-real |
| **3.** Simplify $\sqrt{36x^2}$ if $x$ is a positive number. | $6x$ | $6|x|$ | Non-real |
| **4.** Simplify $\sqrt[3]{8x^6}$ if $x$ can be any real number. | $2x^2$ | $|2x^2|$ | Non-real |
| **5.** Simplify $\sqrt[3]{-8x^6}$ if $x$ can be any real number. | $-2x^2$ | $2x^2$ | Non-real |
| **6.** Simplify $\sqrt{49x^2}$ if $x$ can be any real number. | $7x$ | $7|x|$ | Non-real |
| **7.** Simplify $\sqrt{49x^2}$ if $x$ is a positive number. | $7x$ | $7|x|$ | Non-real |
| **8.** Simplify $\sqrt{16x^4}$ if $x$ can be any real number. | $4x^2$ | $4|x^2|$ | Non-real |
| **9.** Simplify $\sqrt{-16x^4}$ if $x$ can be any real number. | $-4x^2$ | $|-4x^2|$ | Non-real |
| **10.** Simplify $\sqrt[3]{64x^3}$ if $x$ can be any real number. | $4x$ | $4|x|$ | Non-real |
| **11.** Simplify $\sqrt[4]{16x^4}$ if $x$ is a positive number. | $2x$ | $2|x|$ | Non-real |
| **12.** Simplify $\sqrt[4]{16x^4}$ if $x$ can be any real number. | $2x$ | $2|x|$ | Non-real |
| **13.** Simplify $\sqrt[4]{-16x^4}$ if $x$ is a positive number. | $2x$ | $2|x|$ | Non-real |
| **14.** Simplify $\sqrt[5]{-32x^5}$ if $x$ can be any real number. | $-2x$ | $2|x|$ | Non-real |
| **15.** Simplify $\sqrt[5]{32x^5}$ if $x$ can be any real number. | $2x$ | $2|x|$ | Non-real |

328

# Student Activity
## Paint by Rational Exponents

**Directions:** Convert each expression according to the directions. Then shade in the box that contains your answer. The first one has been done for you. There's a surprise when you're finished!

**Convert to radical notation:**

**1.** $(xy)^{1/2} = \sqrt{xy}$

**2.** $(25 + x^2)^{1/2}$

**3.** $5x^{1/2}$

**4.** $(-x)^{1/5}$

**5.** $\dfrac{y}{x^{-1/2}}$

**Convert to rational exponent notation:**

**6.** $2\sqrt{xy}$

**7.** $5\sqrt[4]{x^3}$

**8.** $\sqrt{3x^5}$

**9.** $\sqrt[6]{(10x+7)^7}$

**10.** $\sqrt[5]{-ab^4}$

| $3x^{2/5}$ | $\left(3x^5\right)^{1/2}$ | $(3x)^{5/2}$ | $\sqrt{25+x^2}$ | $5+x$ |
|---|---|---|---|---|
| $\dfrac{1}{\sqrt[5]{x}}$ | $\sqrt[5]{-x}$ | $-\sqrt{x^5}$ | $y\sqrt{x}$ | $-\dfrac{y}{\sqrt{x}}$ |
| $(xy)^2$ | $10x+7^{6/7}$ | $10x^{7/6}+7^{7/6}$ | $x\sqrt{y}$ | $a^{1/5}b^{-4/5}$ |
| $5x^{4/3}$ | $\sqrt{5x}$ | $\sqrt{xy}$ | $(2xy)^2$ | $ab^{-4/5}$ |
| $5x^{3/4}$ | $\dfrac{r}{\sqrt{t}}$ | $10x+7^{7/6}$ | $2(xy)^2$ | $\left(-ab^4\right)^{1/5}$ |
| $(5x)^{3/4}$ | $5\sqrt{x}$ | $-t\sqrt{r}$ | $2(xy)^{1/2}$ | $(2xy)^{1/2}$ |
| $(5x)^{\frac{4}{3}}$ | $-r\sqrt{t}$ | $(10x+7)^{7/6}$ | $(10x+7)^{6/7}$ | $2xy^{1/2}$ |

# Student Activity
## Language of the Other Roots

**Directions:** For each set, fill in the missing boxes in the first row with numbers that make the two expressions equal. Write out a statement that describes the relationship between the two expressions in the second row. The first two have been started for you.

| Expression written with root or power | simplified expression |
|---|---|
| Write out a statement in words. | |

| | |
|---|---|
| $\sqrt[3]{125}$ | 5 |
| The cube root of ____ is ____. | |

| | |
|---|---|
| $(\quad)^3$ | 216 |
| The cube of ____ is ____. | |

| | |
|---|---|
| $(\quad)^{\frac{1}{4}}$ | 2 |
| | |

| | |
|---|---|
| $\sqrt[4]{\quad}$ | 9 |
| | |

| | |
|---|---|
| | 27 |
| The cube of ____ is ____. | |

| | |
|---|---|
| $\sqrt[3]{\quad}$ | 7 |
| | |

| | |
|---|---|
| | $\dfrac{1}{16}$ |
| _____ raised to the -2 power is _____. | |

| | |
|---|---|
| $(\quad)^{-2}$ | |
| 3 raised to the ____ power is _____. | |

| | |
|---|---|
| $\dfrac{1}{5^2}$ | |
| | |

| | |
|---|---|
| $1^{-\frac{3}{4}}$ | |
| | |

| | |
|---|---|
| $(-1000)^{\frac{1}{3}}$ | |
| | |

| | |
|---|---|
| $(\quad)^{\frac{3}{2}}$ | |
| _____ raised to the $\dfrac{3}{2}$ is 8. | |

# Student Activity
## Rational Exponents Using a Calculator

By now you should be familiar with the keys on your calculator used to evaluate expressions like $\sqrt{196}$, $\sqrt[5]{3125}$. Your calculator can also evaluate expressions containing rational exponents.

Consider the following expression: $\left(\sqrt[4]{625}\right)^3$

To evaluate this expression with a calculator, you would first need to write an equivalent expression with a rational exponent:

$$\left(\sqrt[4]{625}\right)^3 = 625^{3/4}$$

Most calculators have an exponent key that looks like $\boxed{\wedge}$. When entering a rational exponent, you should always use parentheses around the exponent. Enter each of the following expressions exactly as they are shown:

$625\boxed{\wedge}3/4 = $ _____

$625\boxed{\wedge}(3/4) = $ _____

You should see two *very* different answers! Which one is correct? If you calculate $\sqrt[4]{625}$ then cube the result you get 125. This is the correct answer. Your calculator will always follow the order of operations, so without parentheses around the 3/4, your calculator will cube 625 *then* divide the result by 4. The result is a very large number!

Now try finding these with your calculator. The first row has been done for you.

| Radical Notation | Rational Exponent Notation | Keystrokes to Enter into Calculator | Answer |
|---|---|---|---|
| 1. $\left(\sqrt[3]{64}\right)^2$ | $64^{2/3}$ | $64\boxed{\wedge}(2/3)$ | 16 |
| 2. | $-243^{6/5}$ | | |
| 3. $\dfrac{1}{\sqrt[3]{8^2}}$ | | | |
| 4. | $1024^{-2/5}$ | | |
| 5. $\dfrac{1}{\left(\sqrt[4]{1296}\right)^{-3}}$ | | | |
| 6. | $625^{3/4}$ | | |
| 7. $\left(\sqrt[5]{32}\right)^{-3}$ | | | |

# Student Activity

## Escape the Rational Exponent Matrix

**Directions:** Assume all variables represent positive numbers. Begin at the box marked START. By shading in adjacent pairs of squares that contain equivalent expressions, you will eventually find the path to "escape" this matrix of boxes. The first "step" in the path and a couple steps in the middle have been taken for you.

| START $\left(36b^6\right)^{1/2}$ | $18b^3$ | $9b^3$ | $x+\sqrt{x}$ | $7^{2/3}$ |
|---|---|---|---|---|
| $6b^3$ | $36b^3$ | $3y$ | $\sqrt[4]{3y}$ | $x^{9/5}$ |
| $x^{3/4}\cdot x^2$ | $\sqrt[4]{x^{11}}$ | $\left(3y^{2/3}\right)^3$ | $9y^2$ | $\left(\sqrt{2x}\right)^3$ |
| $\sqrt{x^3}$ | $x^{4/11}$ | $27y^2$ | $\sqrt[3]{3y^2}$ | $2x^{4/3}$ |
| $x^{4/3}$ | $x^{-4/3}$ | $\dfrac{x^3}{x^{7/3}}$ | $\sqrt[3]{x^2}$ | $x^{5/6}\left(x^{1/6}+x^{1/2}\right)$ |
| $x^{1/3}$ | $\dfrac{1}{x^7}$ | $x^7$ | $x^{1/9}$ | $x+\sqrt[3]{x^4}$ |
| $7$ | $\dfrac{x^3}{8}$ | $\left(\dfrac{x^5}{32}\right)^{3/5}$ | $y^{1/20}$ | $\sqrt[4]{\sqrt[5]{y}}$ |
| $\dfrac{1}{y^2}$ | $\sqrt[10]{y^5}$ | $y^2$ | $y^{5/4}$ | $y^{20}$ |
| $3\sqrt{y}$ | $y^{1/2}$ | $49$ | **ESCAPE** $343^{2/3}$ | $\sqrt{y^{400}}$ |

# Student Activity
## Radical Heteronyms

**Directions:** In writing, there are words that are spelled the same but have different pronunciations and different definitions; these are called heteronyms. Many mathematical expressions look similar but are really very different (almost like mathematical heteronyms). Simplify each of the expressions below, paying close attention to the use of parentheses, the mathematical operations, and notation. Assume all variables represent positive numbers.

**1.**

| $9^{-1}$ | $9^{1/2}$ | $9^2$ | $9^{-2}$ | $-9^{-1}$ | $9^{-1/2}$ |
|---|---|---|---|---|---|
|  |  |  |  |  |  |

**2.**

| $\sqrt{25-9}$ | $(25-9)^{1/2}$ | $25^{1/2}-9^{1/2}$ | $(25-9)^{-1}$ | $25^{-1}-9^{-1}$ |
|---|---|---|---|---|
|  |  |  |  |  |

**3.**

| $x^{1/2}x^{1/3}$ | $\left(x^{1/2}\right)^{1/3}$ | $x^{1/2}+x^{1/3}$ | $\dfrac{x^{1/2}}{x^{1/3}}$ | $\left(\dfrac{1}{2}x\right)\left(\dfrac{1}{3}x\right)$ | $\dfrac{1}{3}\left(\dfrac{1}{2}x\right)$ |
|---|---|---|---|---|---|
|  |  |  |  |  |  |

**4.**

| $\left(16a^4\right)^{-2}$ | $\left(16a^4\right)^{1/2}$ | $\left(16a^4\right)^{1/4}$ | $\dfrac{1}{2}\left(16a^4\right)$ | $16\left(a^4\right)^{1/2}$ |
|---|---|---|---|---|
|  |  |  |  |  |

**5.**

| $\left(\dfrac{9}{4}\right)^{-2}$ | $\left(\dfrac{9}{4}\right)^{1/2}$ | $\left(\dfrac{9}{4}\right)^{-1/2}$ | $\left(\dfrac{9}{4}\right)^{-1}$ | $\dfrac{9^{1/2}}{4}$ | $\dfrac{9^{-1/2}}{4}$ |
|---|---|---|---|---|---|
|  |  |  |  |  |  |

# Student Activity

## Factor Trees

**Directions:** Use a factor tree to find the prime factorization for each number. Build branches of the tree using factor pairs. When a prime number is reached in one of the branches, circle it in red to indicate that it is one of the "apples" (prime factors) on the tree. To write the prime-factored form, collect all the "apples" on the tree. The factor tree shown here tells us that the prime factorization of 80 is $2 \cdot 2 \cdot 2 \cdot 2 \cdot 5$. To find $\sqrt{80}$, we find pairs of factors in $\sqrt{2 \cdot 2 \cdot 2 \cdot 2 \cdot 5}$ and bring them outside the radical as a single factor to get $2 \cdot 2\sqrt{5}$ or $4\sqrt{5}$. To make a particularly colorful and "treelike" factor tree below, use brown for all the "branches," green for all the numbers, and red to circle and shade in the prime numbers. In the factor tree below, your "trees" will go in all directions to make the branches. The number 90 has been prime factored for you.

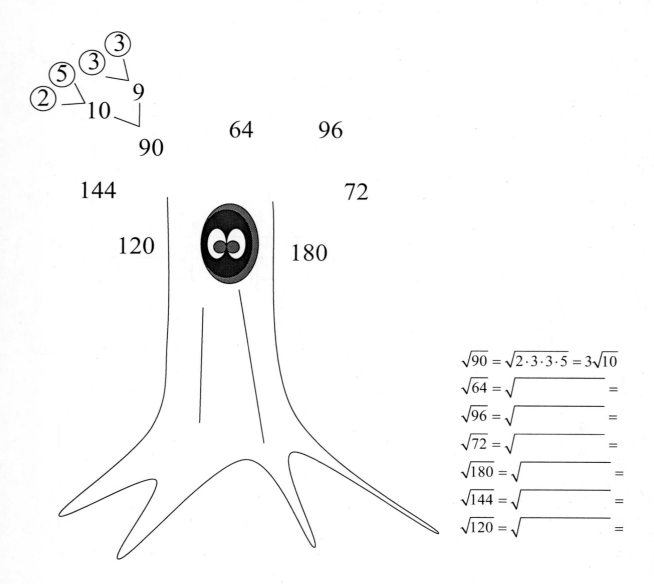

$$\sqrt{90} = \sqrt{2 \cdot 3 \cdot 3 \cdot 5} = 3\sqrt{10}$$

$$\sqrt{64} = \sqrt{\rule{2cm}{0pt}} = \rule{1.5cm}{0pt}$$

$$\sqrt{96} = \sqrt{\rule{2cm}{0pt}} = \rule{1.5cm}{0pt}$$

$$\sqrt{72} = \sqrt{\rule{2cm}{0pt}} = \rule{1.5cm}{0pt}$$

$$\sqrt{180} = \sqrt{\rule{2cm}{0pt}} = \rule{1.5cm}{0pt}$$

$$\sqrt{144} = \sqrt{\rule{2cm}{0pt}} = \rule{1.5cm}{0pt}$$

$$\sqrt{120} = \sqrt{\rule{2cm}{0pt}} = \rule{1.5cm}{0pt}$$

# Guided Learning Activity
## Higher on the Factor Tree

Recall that a **square root** is usually written without an index, so $\sqrt{a} = \sqrt[2]{a}$. In a square root we had to find **pairs** of a factor in order to simplify the radicand.

Similarly, in a **cube root**, like $\sqrt[3]{a}$, we would need to find **triples** of a factor in order to simplify the radicand.

And in a **fourth root**, $\sqrt[4]{a}$, we would need to find **quadruples** of a factor in order to simplify the radicand.

Notice that the index of the radical tells us how many matching factors we need to find in order to simplify.

$$\rightarrow \sqrt[2]{a} \qquad \rightarrow \sqrt[3]{a} \qquad \rightarrow \sqrt[4]{a}$$

One way to search for these pairs, triples, quadruples, etc. is to use factor trees to find the prime factorization for each radicand.

**Example:** Simplify: $\sqrt[3]{162}$.

We look for triples since the index is three.

Using a factor tree, we factor 162.

$$\sqrt[3]{162} = \sqrt[3]{\boxed{3\,3\,3}\, 3 \cdot 2} = 3\sqrt[3]{3 \cdot 2} = 3\sqrt[3]{6}$$

**Example:** Simplify: $\sqrt[4]{80}$

We look for quadruples since the index is four.

Using a factor tree, we factor 80.

$$\sqrt[4]{80} = \sqrt[4]{5\,\boxed{2\,2\,2\,2}} = 2\sqrt[4]{5}$$

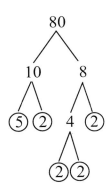

**Directions:** Use factor trees to help simplify the following higher order roots.

**1.** $\sqrt[3]{3,000}$

**4.** $\sqrt[3]{625}$

**2.** $\sqrt[4]{432}$

**5.** $\sqrt[4]{512a^8}$

**3.** $\sqrt[5]{243}$

**6.** $\sqrt[3]{216x^5}$

# Student Activity

## Match Up on Simplifying Radicals

**Directions (READ them):** Simplify each radical expression. Then look at the remaining radicand and choose the letter that corresponds to the remaining radicand.

For example, for $2\sqrt{3}$, $x\sqrt{3}$, $5\sqrt{3}$, and $x^2\sqrt{3}$, you would choose B. If there is no radical left after simplification, choose E.

If none of these choices are correct, choose F. Assume that all variables represent nonnegative numbers.

**A** 2

**B** 3

**C** 5

**D** $x$

**E** No radical left

**F** None of these

Radical Expressionists

| | | | |
|---|---|---|---|
| $\sqrt{72}$ | $\sqrt{80}$ | $\sqrt{81}$ | $\sqrt[3]{375}$ |
| $\sqrt{90}$ | $\sqrt{x^7}$ | $\sqrt{125}$ | $\sqrt[3]{x^4}$ |
| $\sqrt{200}$ | $\sqrt{45}$ | $\sqrt{30}$ | $\sqrt[3]{3000x^3}$ |
| $\sqrt{108}$ | $\sqrt{121}$ | $\sqrt{16x}$ | $\sqrt[3]{1024}$ |

# Student Activity
## Thread of Like Terms II

**Directions:** Simplify each of the expressions that follow.

1.

| | | | |
|---|---|---|---|
| $3x^2 + 5x^2$ | $\dfrac{3}{11} + \dfrac{5}{11}$ | $3\sqrt{x} + 5\sqrt{x}$ | $3\sqrt[3]{3} + 5\sqrt[3]{3}$ |

2.

| | | | |
|---|---|---|---|
| $x^3 - 5x^3$ | $\dfrac{1}{3} - \dfrac{5}{3}$ | $\sqrt{5} - 5\sqrt{5}$ | $\sqrt[3]{3} - 5\sqrt[3]{3}$ |

3.

| | | | |
|---|---|---|---|
| $\dfrac{8}{x+8} - \dfrac{1}{x+8}$ | $8xy - xy$ | $8\sqrt{y} - \sqrt{y}$ | $8\sqrt[3]{2} - \sqrt[3]{2}$ |

4.

| | | | |
|---|---|---|---|
| $4x^2y^2 - 6x^2y^2$ | $\dfrac{4}{5x^2} - \dfrac{6}{5x^2}$ | $4\sqrt{xy} - 6\sqrt{xy}$ | $4\sqrt[3]{15} - 6\sqrt[3]{15}$ |

5. What is the lesson to be learned here about addition and subtraction in algebra and mathematics in general?

6. Now be a little more careful with this group of expressions. If the expression can't be simplified, just say so!

| | | | |
|---|---|---|---|
| $8\sqrt{3} + 2\sqrt{3}$ | $6\sqrt{2} + 6\sqrt{3}$ | $\sqrt{5} + \sqrt{5}$ | $\sqrt{5}\sqrt{5}$ |
| $3\sqrt{2} \cdot \sqrt{2}$ | $3\sqrt{2} + \sqrt{2}$ | $\sqrt[3]{x} - \sqrt[3]{x}$ | $\sqrt{3} + \sqrt{3} + \sqrt{3}$ |
| $3\sqrt[3]{ab} - \sqrt[3]{ab}$ | $-5\sqrt{2} + 5\sqrt{2}$ | $5\sqrt{x} + 5\sqrt{y}$ | $\sqrt{4} \cdot \sqrt{4}$ |

# Student Activity
## Paint by Radicals

**Directions:** Simplify each expression and shade in the corresponding square in the grid below (that contains the correctly simplified version). Assume all variables represent positive numbers. The first one has been done for you. There's a surprise when you're finished!

1. Simplify $\sqrt{72} - \sqrt{50} = 6\sqrt{2} - 5\sqrt{2} = \sqrt{2}$

2. Simplify: $\sqrt{72} + \sqrt{50}$

3. Simplify: $\sqrt{16 + 9}$

4. Simplify: $\sqrt{40} - \sqrt{90}$

5. Simplify: $10\sqrt{20} + \sqrt{5}$

6. Simplify: $\sqrt{300} - 10\sqrt{3}$

7. Simplify: $\sqrt{75} + \sqrt{12}$

8. Simplify: $\sqrt{48y^2} - y\sqrt{27}$

9. Simplify: $\sqrt{75x^2 + 25x^2}$

10. Simplify: $\sqrt{81b^2} + \sqrt{500b}$

11. Simplify: $-13 + \sqrt{4x^2} + 5\sqrt{2} - 5\sqrt{2}$

12. Simplify: $10 + 10\sqrt[3]{5} + 2\sqrt[3]{5} + 2$

13. Simplify: $\sqrt{52b^2 - 3b^2}$

14. Simplify: $\sqrt{900} - \sqrt{80} + \sqrt{363} + \sqrt{80}$

15. Simplify: $\sqrt[3]{54} - \sqrt[3]{64} - 3\sqrt[3]{2}$

| 7 | $21\sqrt{5}$ | 5 | 0 | $11\sqrt{2}$ | $10x$ |
|---|---|---|---|---|---|
| $\sqrt{5}$ | $y\sqrt{3}$ | $\sqrt{3y}$ | $2\sqrt{2xy}$ | $\sqrt{122}$ | $-13 + \sqrt{2} + 2x$ |
| $11b\sqrt{5}$ | $\sqrt{2}$ | $\sqrt{22}$ | $30 + 11\sqrt{3}$ | $-8$ | $2x - 13$ |
| $-\sqrt{10}$ | $7b$ | $24\sqrt[3]{5}$ | $19b\sqrt{10b}$ | $-4$ | $4 - 6\sqrt[3]{2}$ |
| $\sqrt{-10}$ | $12 + 12\sqrt[3]{5}$ | $80 + 11\sqrt{3}$ | $9b + 10\sqrt{5b}$ | $\sqrt{3}$ | $7\sqrt{3}$ |

## Radical Addition and Multiplication Tables

Here is a simple addition table with radical inputs. Fill in the missing boxes, use $\times$ if the inputs cannot be combined with the given operation.

**Addition Table I:**

| + | $\sqrt{2}$ | $\sqrt{3}$ | $\sqrt{5}$ | $\sqrt{6}$ |
|---|---|---|---|---|
| $\sqrt{2}$ | $2\sqrt{2}$ | $\times$ | $\times$ | $\times$ |
| $\sqrt{3}$ | | | | |
| $\sqrt{5}$ | | | | |
| $\sqrt{6}$ | | | | |

Now try this addition table that also involves variables! Again, use $\times$ if you cannot combine the input terms. You may want to simplify some of the radicals before you add. Assume $x > 0$.

**Addition Table II:**

| + | $\sqrt{3x}$ | $\sqrt{12x}$ | $\sqrt{75x}$ | $\sqrt{3x^2}$ | $\sqrt{12x^2}$ | $\sqrt{75x^2}$ |
|---|---|---|---|---|---|---|
| $\sqrt{3x}$ | | | | | | |
| $\sqrt{12x}$ | | | | | | |
| $\sqrt{75x}$ | | | | | | |
| $\sqrt{3x^2}$ | | | | | | |
| $\sqrt{12x^2}$ | | | | | | |
| $\sqrt{75x^2}$ | | | | | | |

Here is a simple multiplication table with radical inputs. Perform the multiplications and simplify each result.

**Multiplication Table I:**

| $\bullet$ | $\sqrt{2}$ | $\sqrt{3}$ | $\sqrt{5}$ | $\sqrt{6}$ |
|---|---|---|---|---|
| $\sqrt{2}$ | $2$ | $\sqrt{6}$ | $\sqrt{10}$ | $2\sqrt{3}$ |
| $\sqrt{3}$ | | | | |
| $\sqrt{5}$ | | | | |
| $\sqrt{6}$ | | | | |

Now try this multiplication table that also involves variables! Make sure to simplify the result of each multiplication. Assume $x > 0$.

**Multiplication Table II:**

| $\bullet$ | $\sqrt{3x}$ | $\sqrt{12x}$ | $\sqrt{3x^2}$ | $\sqrt{12x^2}$ | $\sqrt{3x^3}$ | $\sqrt{12x^3}$ |
|---|---|---|---|---|---|---|
| $\sqrt{3x}$ | | | | | | |
| $\sqrt{12x}$ | | | | | | |
| $\sqrt{3x^2}$ | | | | | | |
| $\sqrt{12x^2}$ | | | | | | |
| $\sqrt{3x^3}$ | | | | | | |
| $\sqrt{12x^3}$ | | | | | | |

# Student Activity
## Tempting Radical Expressions

**Directions:** The expressions on this page might try to tempt you away from the strict mathematical rules you have learned about radical expressions. So, work carefully to simplify each expression and think about every move you make! If an expression cannot be simplified, then say so.

| | | |
|---|---|---|
| **1.** $\sqrt{25-9}$ | **2.** $7\sqrt{3}-\sqrt{3}$ | **3.** $5-2\sqrt{2}$ |
| **4.** $\sqrt{8}+\sqrt{1}$ | **5.** $\sqrt{25}\cdot\sqrt{25}$ | **6.** $\sqrt{5}+\sqrt{5}$ |
| **7.** $\sqrt[3]{8}+1$ | **8.** $\sqrt[4]{4}+12$ | **9.** $\sqrt{169-25}$ |
| **10.** $5\sqrt{x}-\sqrt{x}$ | **11.** $\sqrt{x^2}\cdot\sqrt{x^2}$ | **12.** $\sqrt[3]{3}+\sqrt[3]{3}+\sqrt[3]{3}$ |
| **13.** $\dfrac{\sqrt{3}}{2}+\dfrac{\sqrt{3}}{2}$ | **14.** $\dfrac{\sqrt{3}}{2}\cdot\dfrac{\sqrt{3}}{2}$ | **15.** $\sqrt[3]{\dfrac{8-7}{8}}$ |

## Scrambled Answers:

| | | | | |
|---|---|---|---|---|
| $\sqrt[3]{9}$ | 12 | $6\sqrt{3}$ | $x^2$ | 4 |
| $2\sqrt{5}$ | $\dfrac{1}{2}$ | $\dfrac{3}{4}$ | Cannot be simplified further | 2 |
| $2\sqrt{2}+1$ | $4\sqrt{x}$ | $\sqrt{3}$ | $3\sqrt[3]{3}$ | 25 |

# Student Activity
## Multiplication Madness Match Up

**Directions:** There are so many different ways to combine multiplication and radicals it can be easy to mix up the rules. Carefully work to simplify each of these expressions applying the appropriate rules as you go. Assume all roots are real numbers.

$$\sqrt[n]{a} \cdot \sqrt[n]{b} = \sqrt[n]{a \cdot b} \qquad a\sqrt[n]{b} \cdot c\sqrt[n]{d} = ac\sqrt[n]{bd} \qquad \left(\sqrt[n]{a}\right)^n = a$$

**A** $4x$     **B** $6 + 4\sqrt{2}$     **C** $36 - x$     **D** $18\sqrt{10}$     **E** None of these

| | | | | |
|---|---|---|---|---|
| $\sqrt{2}\left(3\sqrt{2}+4\right)$ | $3\sqrt{2} \cdot 6\sqrt{5}$ | $\left(\sqrt[3]{4x}\right)^3$ | $3\left(2+4\sqrt{2}\right)$ | $\left(\sqrt{36}\right)^2 - \left(\sqrt{x}\right)^2$ |
| $\left(\sqrt[3]{6}\right)^3 - \left(\sqrt[3]{x}\right)^3$ | $\sqrt{2x} \cdot \sqrt{8x}$ | $2\sqrt{5} \cdot 9\sqrt{2}$ | $2\left(3+2\sqrt{2}\right)$ | $4\sqrt{x} \cdot \sqrt{x}$ |
| $\sqrt{2x} \cdot 2\sqrt{2x}$ | $\left(6-\sqrt{x}\right)\left(6+\sqrt{x}\right)$ | $\left(\sqrt{2}+2\right)^2$ | $\sqrt{x}\left(36-\sqrt{x}\right)$ | $\sqrt{5} \cdot 18\sqrt{2}$ |
| $\left(6+x\right)\left(6-x\right)$ | $\left(6-\sqrt{x}\right)^2$ | $\sqrt[3]{64 \cdot x^3}$ | $\left(9\sqrt{10}\right)^2$ | $\left(\sqrt{6}\right)^2 + \sqrt{32}$ |

# Student Activity

## Double the Fun on Radical Expressions

**Directions:** When you simplify expressions that have multiplication or division of radicals, you sometimes have a choice. You could first simplify each radicand and then apply the operation, or you could first combine the radicals with the operation, and then simplify. For each of the expressions below, simplify the radical expression both ways (let $x > 0$). Your simplified expression should be the same for both methods. If they are not, you'll have to go back and look for a mistake. The first one has been done for you.

| | Radical Expression | Simplify the radicands first | Combine the radicals first |
|---|---|---|---|
| 1. | $\dfrac{\sqrt{18x}}{\sqrt{2x^3}}$ | $\dfrac{\sqrt{18x}}{\sqrt{2x^3}} = \dfrac{3\sqrt{2x}}{x\sqrt{2x}} = \dfrac{3\cancel{\sqrt{2x}}^{1}}{x\cancel{\sqrt{2x}}^{1}} = \dfrac{3}{x}$ | $\dfrac{\sqrt{18x}}{\sqrt{2x^3}} = \sqrt{\dfrac{18x}{2x^3}} = \sqrt{\dfrac{9}{x^2}} = \dfrac{3}{x}$ |
| 2. | $\sqrt{75x^2} \cdot \sqrt{48x^4}$ | | |
| 3. | $\dfrac{\sqrt{108x^6}}{\sqrt{12x^4}}$ | | |
| 4. | $\dfrac{\sqrt{625x^4}}{\sqrt{25x^8}}$ | | |
| 5. | $\sqrt{12x} \cdot \sqrt{24x^2}$ | | |
| 6. | $\dfrac{\sqrt[3]{192x^6}}{\sqrt[3]{3x^3}}$ | | |

344

# Student Activity

## Rationalize with the Missing Form of 1

**Directions:** In each radical expression below, we need to rationalize the denominator. We do this rationalization by multiplying by a **form of one** that changes the denominator from a radical expression to a non-radical expression. Fill in the "ones" to rationalize the denominator of each expression and simplify the result.

**Example:** $\dfrac{6}{\sqrt{3}} \cdot \boxed{\phantom{--}}$ becomes $\dfrac{6}{\sqrt{3}} \cdot \boxed{\dfrac{\sqrt{3}}{\sqrt{3}}} = \dfrac{6\sqrt{3}}{3} = 2\sqrt{3}$.

**Hint for problems 6-10: Remember to use a conjugate.**

1. $\dfrac{8}{\sqrt{2}} \cdot \boxed{\phantom{--}}$

2. $\dfrac{5\sqrt{3}}{\sqrt{15}} \cdot \boxed{\phantom{--}}$

3. $\dfrac{\sqrt{5}}{\sqrt{10}} \cdot \boxed{\phantom{--}}$

4. $\sqrt{\dfrac{7}{2}} = \dfrac{\sqrt{\phantom{-}}}{\sqrt{\phantom{-}}} \cdot \boxed{\phantom{--}}$

5. $\dfrac{4}{\sqrt{18x}} = \dfrac{4}{\sqrt{\phantom{-}}} \cdot \boxed{\phantom{--}}$

6. $\dfrac{4}{\sqrt{2}+1} \cdot \boxed{\phantom{--}}$

7. $\dfrac{12}{3-\sqrt{5}} \cdot \boxed{\phantom{--}}$

8. $\dfrac{\sqrt{3}}{-2-\sqrt{3}} \cdot \boxed{\phantom{--}}$

9. $\dfrac{\sqrt{3}-2}{\sqrt{3}+4} \cdot \boxed{\phantom{--}}$

10. $\dfrac{\sqrt{6}}{x+\sqrt{2}} \cdot \boxed{\phantom{--}}$

# Guided Learning Activity
## Rationalizing Higher-order Roots

Believe it or not, rationalizing denominators with square roots is a pretty intuitive procedure. The procedure for rationalizing denominators with higher-order roots is not so obvious. Let's look at a comparison and see what's different.

**Example with square roots:**

To rationalize $\dfrac{6}{\sqrt{3}}$ we multiply by $\dfrac{\sqrt{3}}{\sqrt{3}}$, to get: $\dfrac{6}{\sqrt{3}} \cdot \dfrac{\sqrt{3}}{\sqrt{3}} = \dfrac{6\sqrt{3}}{\sqrt{3^2}} = \dfrac{6\sqrt{3}}{3} = 2\sqrt{3}$.

**Example with cube roots:**

If we followed exactly the same procedure as for square roots, here's what happens:

We multiply $\dfrac{6}{\sqrt[3]{2}}$ by $\dfrac{\sqrt[3]{2}}{\sqrt[3]{2}}$ to get: $\dfrac{6}{\sqrt[3]{2}} \cdot \dfrac{\sqrt[3]{2}}{\sqrt[3]{2}} = \dfrac{6\sqrt[3]{2}}{\sqrt[3]{2^2}} = ?$

The problem is that $\sqrt[3]{2^2}$ doesn't simplify. However, $\sqrt[3]{2^3}$ would simplify, equaling 2.

**Let's try again:**

**1.** Rationalize: $\dfrac{6}{\sqrt[3]{2}}$

It helps to think first of the radical denominator that you're going to aim for. Then you can figure out what to multiply by to get it. Fill in the blanks in the next few problems.

**2.** Rationalize $\dfrac{15}{\sqrt[3]{5}}$ by aiming for $\sqrt[3]{125}$ or $\sqrt[3]{5^3}$ in the denominator.

$$\dfrac{15}{\sqrt[3]{5}} \cdot \dfrac{\sqrt[3]{\phantom{xx}}}{\sqrt[3]{\phantom{xx}}} = \dfrac{}{\sqrt[3]{125}} = \dfrac{}{5} =$$

**3.** Rationalize $\dfrac{6}{\sqrt[4]{27}}$ by aiming for $\sqrt[4]{81}$ or $\sqrt[4]{3^4}$ in the denominator.

$$\dfrac{6}{\sqrt[4]{27}} \cdot \dfrac{\sqrt[4]{\phantom{xx}}}{\sqrt[4]{\phantom{xx}}} = \dfrac{}{\sqrt[4]{81}} = \dfrac{}{3} =$$

**4.** Rationalize $\dfrac{5}{2\sqrt[3]{25}}$ by aiming for $2\sqrt[3]{125}$ or $2\sqrt[3]{5^3}$ in the denominator.

$$\dfrac{5}{2\sqrt[3]{25}} \cdot \dfrac{\sqrt[3]{\phantom{xx}}}{\sqrt[3]{\phantom{xx}}} = \dfrac{}{2\sqrt[3]{125}} = \dfrac{}{2 \cdot 5} =$$

We always want to aim for a radical in the denominator that has a perfect power in the radicand that corresponds with the index of the root.

**Let's practice deciding on a good denominator to aim for:**

**5.** Rationalize $\dfrac{12}{\sqrt[5]{4}}$ by aiming for _____ in the denominator.

**6.** Rationalize $\dfrac{\sqrt[3]{3}}{\sqrt[3]{4}}$ by aiming for _____ in the denominator.

**7.** Rationalize $\dfrac{6x^2}{\sqrt[3]{3x}}$ by aiming for _____ in the denominator.

**8.** Rationalize $\dfrac{\sqrt[4]{2}}{\sqrt[4]{9m}}$ by aiming for _____ in the denominator.

**Finally, we'll put it all together. Rationalize the denominators in the following problems.**

**9.** Rationalize: $\dfrac{40}{\sqrt[3]{10}}$

**10.** Rationalize: $\dfrac{x}{\sqrt[4]{8x^2}}$

**11.** Rationalize: $\sqrt[3]{\dfrac{25}{3}}$  (Hint: first write this with two separate radicals.)

**12.** Rationalize: $\dfrac{\sqrt[4]{6}}{\sqrt[4]{9x^2}}$

# Student Activity
## The Radical Reunion

**Directions:** Now that we've made it through addition, subtraction, multiplication, and division, it seems only fair to bring the whole gang back together for a reunion. Assume all variables represent positive numbers. Simplify each of the following radical expressions. If an expression cannot be simplified, say so.

| $\sqrt{50} + \sqrt{18}$ | $\sqrt{50-18}$ | $\sqrt{50+18}$ | $\sqrt{50} \cdot \sqrt{18}$ | $\dfrac{\sqrt{50}}{\sqrt{18}}$ |
|---|---|---|---|---|
| | | | | |

| $\sqrt{16a^2b} + \sqrt{9a^2b}$ | $\sqrt{16a^2b - 9a^2b}$ | $\sqrt{16a^2b} \cdot \sqrt{9a^2b}$ | $\dfrac{\sqrt{16a^2b}}{\sqrt{9a^2b}}$ | $\left(\sqrt{16a^2b}\right)^2$ |
|---|---|---|---|---|
| | | | | |

| $\sqrt[3]{135xy^4} - \sqrt[3]{40x^4y}$ | $\sqrt[3]{135xy^4 - 40x^4y}$ | $\sqrt[3]{135xy^4} \cdot \sqrt[3]{40x^4y}$ | $\dfrac{\sqrt[3]{135xy^4}}{\sqrt[3]{40x^4y}}$ | $\left(\sqrt[3]{40x^4y}\right)^3$ |
|---|---|---|---|---|
| | | | | |

# Student Activity

## Is it a Solution?

**Directions:** If the number in the square **IS** a solution of the equation, then circle the solution (placing an **O** on the square). If the given number **IS NOT** a solution, then put an **X** on the square.

| | | |
|---|---|---|
| $\sqrt[3]{7-x} = 1-x$ <br><br> $-1$ | $(3-x)^{1/2} = x-3$ <br><br> $3$ | $\sqrt{3-x} = x-3$ <br><br> $0$ |
| $-3x = 3\sqrt{-2x-1}$ <br><br> $3$ | $-3x = 3\sqrt{-2x-1}$ <br><br> $0$ | $\sqrt{x} + \sqrt{x+3} = x-4$ <br><br> $1$ |
| $x = \sqrt[4]{x^3+16}$ <br><br> $-1$ | $\sqrt{3x} + 10 = 10$ <br><br> $0$ | $-2 = (3x-2)^{1/3}$ <br><br> $-2$ |

# Guided Learning Activity
## Radical Isolation

When we solve radical equations, it is important to **isolate the radical** before using the power rule of equality. *But why?* Consider the radical equation $\sqrt{x+3}+4=6$.

**Squaring first:** Attempting to square both sides without isolating the radical leaves another radical term. This is messier looking than what we started with and we're not even done!

$$\sqrt{x+3}+4=6$$
$$\left(\sqrt{x+3}+4\right)^2=\left(6\right)^2$$
$$\left(\sqrt{x+3}+4\right)\left(\sqrt{x+3}+4\right)=36$$
$$x+3+8\sqrt{x+3}+16=36$$

This is worse than the original problem!

**Isolating first:** If we isolate the radical term first, and then square both sides of the equation, we can quickly solve for $x$. Checking the solution shows that $x=1$ is a valid solution to this radical equation.

$$\sqrt{x+3}+4=6$$
$$\sqrt{x+3}=2$$
$$\left(\sqrt{x+3}\right)^2=\left(2\right)^2$$
$$x+3=4$$
$$x=1$$

---

**Now let's try some together!**

1.

| Isolate the radical: | Solve by squaring both sides: | Check: |
|---|---|---|
| $\sqrt{2x+3}+5=8$ | | |

2.

| Isolate the radical: | Solve by squaring both sides: | Check: |
|---|---|---|
| $8\sqrt{x+5}-12=28$ | | |

3.

| Isolate the radical: | Solve by squaring both sides: | Check: |
|---|---|---|
| $5+\sqrt{x+2}=0$ | | |

**4.**

| Isolate the radical: | Solve by cubing both sides: | Check: |
|---|---|---|
| $5 + 2x^{1/3} = 13$ | | |

**5.**

| Isolate a radical: | Solve by cubing both sides: | Check: |
|---|---|---|
| $\sqrt[3]{3a+4} = \sqrt[3]{2a+7}$ | | |

**6.**

| Isolate the radical | Solve by squaring both sides: | Check: |
|---|---|---|
| $\sqrt{x+7} - 5 = x$ | | |

**7.**

| Isolate a radical: | Squaring both sides: | |
|---|---|---|
| $\sqrt{x+4} + \sqrt{x-1} = 5$ | | |
| **Isolate the radical:** | **Solve by Squaring both sides:** | **Check:** |
| | | |

# Student Activity
## Match Up on Solving Radical Equations

**Directions:** Match each of the equations in the squares in the table below with its solution from the top. If the solution is not found among the choices A through D, then choose E (none of these).

**A** 5      **B** 3      **C** −2      **D** No real number solution      **E** None of these

| | | |
|---|---|---|
| $\sqrt{7-x}=2$ | $3+\sqrt{x+6}=5$ | $3\sqrt{x+9}=6$ |
| $(2y-1)^{1/2}=3$ | $8+\sqrt{x-1}=5$ | $6\sqrt[3]{a-4}=6$ |
| $x=2+\sqrt{14-x}$ | $\sqrt[4]{3w}=\sqrt[4]{5+2w}$ | $x+\sqrt[3]{3-12x}=x+3$ |
| $-2\sqrt{3x-1}=10$ | $x-1=(7-x)^{1/2}$ | $x=\sqrt{4x-11}+2$ |

# Student Activity
## Working with the Language of Radicals

Fill in the table below.  The first row has been done for you.

| In words | Equation or Expression? | In Math Notation |
|---|---|---|
| **1.** the square root of the quantity $x+5$ | Expression | $\sqrt{x+5}$ |
| **2.** The quantity $3x+2$ squared is zero. | | |
| **3.** The sum of 5 and the square root of $x$ is 30. | | |
| **4.** | | $5(x+6)^2+8$ |
| **5.** | | $5\sqrt{x+6}+8$ |
| **6.** | | $7\sqrt{x}-7=42$ |
| **7.** twice the square of the quantity $5x+9$ | | |
| **8.** 40 less the square root of $x$ is 35. | | |
| **9.** | | $5\sqrt[3]{50-x}$ |
| **10.** 5 times the cube root of the quantity $50-x$ is 25. | | |

# Student Activity
## Sail Into the Pythagorean Sunset

**Directions:** Almost all of the triangles in the figure below are right triangles. Using the sparse information you are given and the Pythagorean Theorem, work out the lengths of all the missing sides in the sailboat below.  You should be able to work out the sides of the non-right triangles by piecing together the information you find from the right triangles.  It might be helpful to start by finding the measure of $a$.

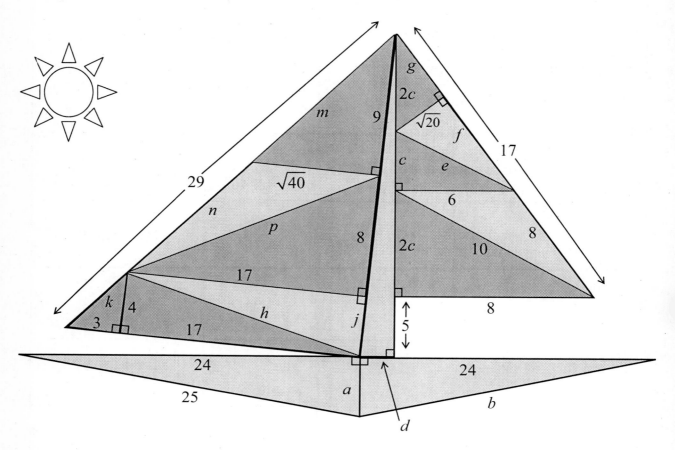

$a =$

$b =$

$c =$

$d =$

$e =$

$f =$

$g =$

$h =$

$j =$

$k =$

$m =$

$n =$

$p =$

# Student Activity
## These Triangles are Just Special

Several triangles have special names. In every diagram, the equal sides are represented with thick lines.

**Equilateral Triangle**
(all sides are equal,
all angles are equal)

**Isosceles Triangle**
(two equal sides,
two equal angles)

**Isosceles Right Triangle**
(one right angle and two equal angles)
**45-45-90 Triangle**

1. Use some geometric reasoning and the Pythagorean Theorem to work out the measure of each missing side for the triangles below.

Based on your work above, complete the following conjecture: In an isosceles right triangle, if the legs have length *l*, then the hypotenuse has length _____.

If we take an equilateral triangle and cut it exactly down the middle, we get two congruent 30-60-90 triangles, demonstrated here:

**Equilateral Triangle**
(all sides are equal,
all angles are equal)

Cut the triangle
down the center.

$AD \cong DB = \frac{1}{2} AB$

**30-60-90
Triangle**

**2.** Use some geometric reasoning and the Pythagorean Theorem to work out the measure of each missing side for the triangles below.

30° 2 60°

30° 6 60°

30° x 60°

Based on your work above, complete the following conjecture: In a 30-60-90 triangle, if the shorter leg has length $a$, then the longer leg has length _____, and the hypotenuse has length _____.

**3.** Now use what you've learned about 30-60-90 and 45-45-90 triangles to work out the measure of the missing sides for the triangles below.

Fairy Tales that didn't make it: Goldilocks and the Special Triangles

... but this one is juuuuust right!

**a.**

45° 7 45°

**b.**

60° 30° $12\sqrt{3}$

**c.**

45° 45° $9\sqrt{2}$

**d.**

30° 60° 4

# Guided Learning Activity

## The Distance Formula

**1.** Find the length of side $AB$ and side $BC$ by measuring the distances on the graphing grid. Then use the Pythagorean Theorem to find the length of side $AC$.

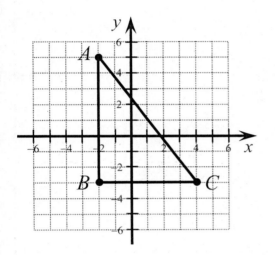

**2.** Find the length of line segment $DE$ using a method similar to problem 1.

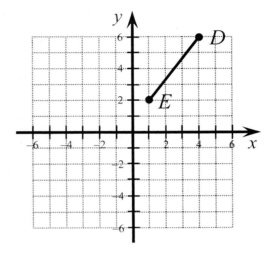

**3.** Find the length of the side $d$ in the triangle below in terms of $x_1, x_2, y_1,$ and $y_2$.

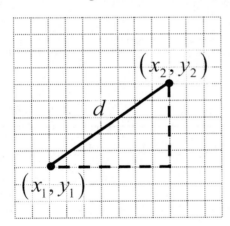

**4.** Now find the distance between $(4, -1)$ and $(7, -5)$ using the formula from Problem 3.

# Assess Your Understanding

## Radical Expressions and Equations

For each of the following, describe the strategies or key steps that will help you **start** the problem. You do **not** have to complete the problems.

| | | What will help you to start this problem? |
|---|---|---|
| **1.** | Multiply: $\left(3+\sqrt{5}\right)\left(3-\sqrt{5}\right)$ | |
| **2.** | Divide: $\dfrac{\sqrt{120x^4y}}{\sqrt{12y^5}}$ | |
| **3.** | Add: $\sqrt{8}+\sqrt{50}$ | |
| **4.** | Simplify: $\sqrt{50}+\sqrt{18}$ | |
| **5.** | Solve: $\sqrt{3x-1}+6=0$ | |
| **6.** | Rationalize the denominator: $\dfrac{12}{6-\sqrt{3}}$ | |
| **7.** | Is $\sqrt{8}$ rational or irrational? | |
| **8.** | Simplify: $\sqrt{120x^5}$ | |

| | | **What will help you to start this problem?** |
|---|---|---|
| **9.** | Evaluate: $4^{3/2}$ | |
| **10.** | Find the distance between $(3,-6)$ and $(2,-4)$. | |
| **11.** | Solve: $\sqrt{3x-1}+6=0$ | |
| **12.** | The smallest side of a 30-60-90 triangle is 6 cm.  Find the lengths of the other two sides. | |
| **13.** | Simplify: $\sqrt[4]{80x^4y^{10}}$ | |
| **14.** | Solve: $3\sqrt{2x-1}=6$ | |
| **15.** | Simplify: $x^{1/3}x^{2/5}$ | |

# Metacognitive Skills
## Radical Expressions and Equations

Metacognitive skills refer to the ability to judge how well you have learned something and to effectively direct your own learning and studying. This is a self-evaluation tool designed to help you focus your studying and to improve your metacognitive skills with regards to this math class.

Fill the 1$^{st}$ column out **before** you begin studying. Fill the 2$^{nd}$ column out after you study for your test.

Go back to this assessment after your test and circle any of the ratings that you would change – this identifies the "disconnects" between what you **thought** you knew well and what you **actually** knew well.

**Use the scale below to assign a number to each topic.**
5  *I am confident I can do any problems in this category correctly.*
4  *I am confident I can do most of the problems in this category correctly.*
3  *I understand how to do the problems in this category, but I still make a lot of mistakes.*
2  *I feel unsure about how to do these problems.*
1  *I know I don't understand how to do these problems.*

| Topic or Skill | Before Studying | After Studying |
|---|---|---|
| Recognizing perfect squares (like 1, 4, 9, 16, 25…). | | |
| Recognizing perfect cubes (like 1, 8, 27, 64, 125…), perfect fourths (like 1, 16, 81, …), and perfect fifths (like 1, 32, 243, …). | | |
| Ability to use the terms radical, radicand, radical symbol, square, and square root, appropriately. | | |
| Understanding why $\sqrt{a}$ is not a real number if $a < 0$. | | |
| Understanding why $\sqrt[n]{a}$ is non-real if $n$ is even and $a < 0$. | | |
| Understanding why $\sqrt[n]{a}$ is a real number as long as $n$ is odd. | | |
| Categorizing a radical as rational or irrational, real or imaginary. | | |
| Using a calculator to approximate an irrational radical. | | |
| Simplifying radical expression involving numbers and variables that are perfect powers (like $\sqrt{16x^2}$). | | |
| Rewriting a radical expression with rational (fractional) exponents or vice versa. | | |
| Knowing which part of a rational (fractional) exponent represents the root and which part is the power. | | |
| Evaluating numbers with rational exponent powers, like $8^{3/2}$ or $9^{-1/2}$. | | |
| Using a calculator to evaluate radicals with higher-order roots or numbers written with rational exponents.. | | |
| Applying exponent rules correctly to simplify expressions involving rational exponents. | | |

*Continued on next page.*

| Topic or Skill | Before Studying | After Studying |
|---|---|---|
| Applying the product and quotient rules for radicals correctly; using these rules to simplify radical expressions like $\sqrt{\dfrac{9x^2}{16}}$ or $\sqrt{4\cdot 25}$. | | |
| Simplifying radical expressions by finding perfect squares or perfect powers inside the radicand (for example: $\sqrt{8}$, $\sqrt[3]{x^4}$, or $\sqrt{50x^3y^2}$ ). | | |
| Adding or subtracting radical expressions by simplifying each radical and using like terms. | | |
| Simplifying radical expressions involving multiplication and/or distribution like $3\sqrt{6}\cdot 2\sqrt{2}$, $2\sqrt{3}\left(4\sqrt{2}-5\sqrt{3}\right)$, or $\left(4+\sqrt{3}\right)\left(5-\sqrt{2}\right)$. | | |
| Simplifying a radical expression that is raised to a power, like $\left(2+\sqrt{3}\right)^2$ or $\left(\sqrt{2x}\right)^2$. | | |
| Rationalizing the denominator when there is a single term involving a square root in the denominator. | | |
| Rationalizing the denominator when there are two terms involving at least one square root. | | |
| Rationalizing the numerator of an expression involving square roots. | | |
| Rationalizing the denominator when there is a single term involving a higher-order root (like a cube or fourth root). | | |
| Solving an equation containing a single radical. | | |
| Solving an equation with a rational exponent power. | | |
| Solving an equation with more than one radical expression. | | |
| Checking the solution to a radical equation. | | |
| Recognizing when a radical equation cannot possibly have a real number solution. | | |
| Solving application problems involving radical equations. | | |
| Knowing when the Pythagorean Theorem can be applied and correctly using it. | | |
| Applying the distance formula correctly. | | |
| Knowing the special properties of 45-45-90 and 30-60-90 triangles. | | |

# CPLX: Complex Numbers

# Student Activity

## Complex Numbers on Trial

You're bluffing! That's not even a REAL number...*

*Editors note: This joke is only funny when $i^2 = -1$.

**Directions:** Each of the numbers below is "on trial" for being purely imaginary, purely real, or complex, and you are the judge.

If the number is purely imaginary, then sentence it to life on the imaginary number line by placing a dot in the appropriate place on the imaginary number line.

If the number is purely real, then sentence it to life on the real number line by placing a dot in the appropriate spot on the real number line.

If the number is not purely real or purely imaginary, then release it in the complex plane so that the real part and imaginary part intersect where the dot is placed.

**A** and **B** have been done for you.

A. $6i$

B. $-4 - 2i$

C. $7$

D. $4 + 3i$

E. $-8i$

F. $-5i + 4$

G. $-7 + 0i$

H. $\sqrt{-9}$

I. $-\sqrt{-25}$

J. $\sqrt{4} + \sqrt{-100}$

K. $-\sqrt{64} - \sqrt{-16}$

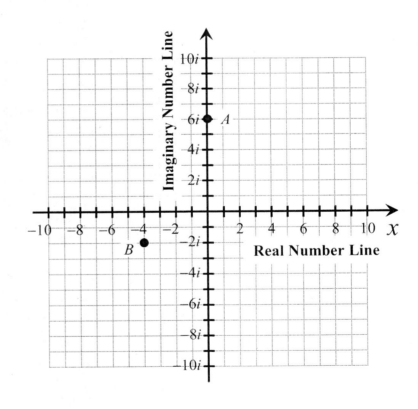

# Student Activity

## Match Up on Complex Numbers

**Directions:** In each box, simplify the complex number expression given. Write the letter of the corresponding solution. If the solution does not appear in choices A – D, choose E (none of these).

**A** $-2$     **B** $-i$     **C** $5-5i$     **D** $5+5i$     **E** None of these

| | |
|---|---|
| $(6+7i)+(-1-2i)$ | $-i(5+5i)$ |
| $5(1-i)$ | $\dfrac{-8-10i}{4+5i}$ |
| $\dfrac{-6-4i}{3+2i}$ | $\dfrac{15+5i}{2-i}$ |

First you tell me there's no Santa, then no Easter Bunny. Now you're telling me some numbers are imaginary? How am I supposed to trust you?

| | | | |
|---|---|---|---|
| $(2-i)(3-i)$ | $5(1-5i)$ | $(10-i)-(5-6i)$ | $(6+7i)-(1+2i)$ |
| $(i^2)(i^2)(i^2)(i)$ | $\dfrac{-10+10i}{2i}$ | $\dfrac{1}{2}(1+i)(-1-i)$ | $(1+i)(-1+i)$ |

# Student Activity
## Rotation Using Complex Numbers

**Example 1:** The complex number $4+3i$ is graphed in the complex plane. Also, you see a circle, centered at the origin, with radius 5. Notice that the point that represents $4+3i$ is a point on the circle.

Multiply $4+3i$ by $i$ and simplify. Graph this point.

Multiply the new number by $i$ and simplify. Graph this point.

Multiply the new number by $i$ and simplify. Graph this point.

Multiply the new number by $i$ and simplify. Notice anything?

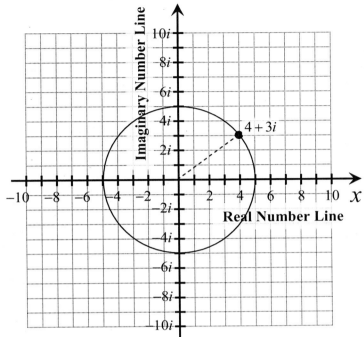

Draw more dashed lines that connect the **origin** to each of the points you have graphed. What do you notice about the lines?

**Example 2:** Repeat this process of graphing and multiplying by $i$ beginning with the number $-6+8i$, which has been graphed for you.

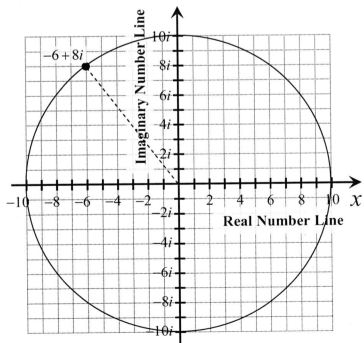

When we multiply by $i$, what affect does that have on the graph?

# Student Activity

## Parallel Expressions

**Directions:** Simplify each expression.

| | | |
|---|---|---|
| $(2+5x)+(4-3x)$ | $(4+6x)-(3+x)$ | $(3+x)(4+x)$ |
| $(2+5\sqrt{3})+(4-3\sqrt{3})$ | $(4+6\sqrt{5})-(3+\sqrt{5})$ | $(3+\sqrt{2})(4+\sqrt{2})$ |
| $(2+5i)+(4-3i)$ | $(4+6i)-(3+i)$ | $(3+i)(4+i)$ |

| | | |
|---|---|---|
| $(3x)^2$ | $(3+x)^2$ | $3(4-5x)$ |
| $(3\sqrt{x})^2$ | $(3+\sqrt{2})^2$ | $3(4-5\sqrt{2})$ |
| $(3i)^2$ | $(3+i)^2$ | $3(4-5i)$ |

| | | |
|---|---|---|
| $\sqrt{20}$ | $2\sqrt{3}\cdot 5\sqrt{3}$ | $\dfrac{4}{\sqrt{2}}$ |
| $\sqrt{-20}$ | $2i\cdot 5i$ | $\dfrac{4}{i}$ |

| | | |
|---|---|---|
| $\dfrac{5}{2\sqrt{6}}$ | $\dfrac{4}{2+\sqrt{3}}$ | $\dfrac{2+\sqrt{3}}{3-\sqrt{3}}$ |
| $\dfrac{5}{2i}$ | $\dfrac{4}{2+i}$ | $\dfrac{2+i}{3-i}$ |

# Assess Your Understanding
## Complex Numbers

For each of the following, describe the strategies or key steps that will help you **start** the problem.  You do **not** have to complete the problems.

| | | What will help you to start this problem? |
|---|---|---|
| **1.** | Multiply.  Write the answer in the form $a+bi$: $(5+2i)(5-2i)$ | |
| **2.** | Simplify: $\sqrt{-9}+\sqrt{-36}$ | |
| **3.** | Subtract: $(3-4i)-(5+9i)$ | |
| **4.** | Simplify: $\dfrac{4}{3+i}$ | |
| **5.** | Simplify: $i^{25}$ | |

# Metacognitive Skills
## Complex Numbers

Metacognitive skills refer to the ability to judge how well you have learned something and to effectively direct your own learning and studying. This is a self-evaluation tool designed to help you focus your studying and to improve your metacognitive skills with regards to this math class.

Fill the 1st column out **before** you begin studying. Fill the 2nd column out after you study for your test.

Go back to this assessment after your test and circle any of the ratings that you would change – this identifies the "disconnects" between what you **thought** you knew well and what you **actually** knew well.

**Use the scale below to assign a number to each topic.**
**5** *I am confident I can do any problems in this category correctly.*
**4** *I am confident I can do most of the problems in this category correctly.*
**3** *I understand how to do the problems in this category, but I still make a lot of mistakes.*
**2** *I feel unsure about how to do these problems.*
**1** *I know I don't understand how to do these problems.*

| Topic or Skill | Before Studying | After Studying |
|---|---|---|
| Understanding why $\sqrt{a}$ is not a real number if $a < 0$. | | |
| Rewriting the square root of a negative number in terms of $i$. | | |
| Adding or subtracting complex numbers and writing the answer in the form $a + bi$. | | |
| Multiplying complex numbers and writing the answer in the form $a + bi$. | | |
| Simplifying expressions involving complex numbers in fractions and writing the answer in the form $a + bi$. | | |
| Simplifying powers of $i$ using the pattern of powers of $i$. | | |

# QUAD: Quadratics

# Student Activity
## Paint by Double-Signs

**Directions:** Simplify each of the expressions below. Then shade in the equivalent expressions in the grid below. When you're done, you should see a picture!

1. $-1 \pm 6$

2. $-3 \pm 5$

3. $\dfrac{5 \pm 10\sqrt{2}}{5}$

4. $\dfrac{6 \pm 2\sqrt{2}}{2}$

5. $\dfrac{-8 \pm 24}{4}$

6. $\dfrac{12 \pm \sqrt{18}}{3}$

7. $\dfrac{-2 \pm \sqrt{25-9}}{2}$

.... and I thought I'd never need
to use math if I became an artist! Ugh!"

| $12 - \sqrt{2}$ | $12 + \sqrt{2}$ | $4 + \sqrt{2}$ | $4 + \sqrt{18}$ | $4 - \sqrt{6}$ |
|---|---|---|---|---|
| $-2$ | $4 - \sqrt{18}$ | $-7$ | $7$ | $3 - 2\sqrt{2}$ |
| $2$ | $4 - \sqrt{2}$ | $-3$ | $-8$ | $3 - \sqrt{2}$ |
| $3 + 2\sqrt{2}$ | $6 + \sqrt{2}$ | $3 + \sqrt{2}$ | $-14$ | $6 - \sqrt{2}$ |
| $4 + \sqrt{6}$ | $-6$ | $1 + 2\sqrt{2}$ | $1 + 10\sqrt{2}$ | $3$ |
| $8$ | $-5$ | $0$ | $1 - 10\sqrt{2}$ | $-4$ |
| $-8$ | $5$ | $1$ | $1 - 2\sqrt{2}$ | $4$ |

# Guided Learning Activity
## The Square Root Property

I am thinking of a number. If I square the number, I get 9. What is the number?

The easy choice here is 3, since $3^2 = 9$.

What **other** number could satisfy the condition? $(\underline{\hspace{1cm}})^2 = 9$

You could also set up the equation $x^2 = 9$ to solve this problem. We have learned how to solve this equation **by factoring**. Go ahead and do that. *Hint: Remember to get =0 on one side first.*

Solve: $x^2 = 9$

We can also solve this equation (often in less steps) by **taking a square root** on both sides of the equation.

Solve: $\quad x^2 = 9$

$\quad\quad \sqrt{x^2} = \sqrt{9}$

$\quad\quad\quad x = 3$

But there is a problem with the process we have written above; we are missing one of the possible solutions! Recall that 9 has two square roots, -3 and 3. However, the notation $\sqrt{9}$ only gives the principal (positive) square root. Thus, we have to solve $x^2 = 9$ by listing both square root solutions, as follows.

Solve: $\quad\quad\quad x^2 = 9$

$\quad\quad x = \sqrt{9} \quad$ or $\quad x = -\sqrt{9}$

$\quad\quad x = 3 \quad\quad | \quad\quad x = -3$

We now have both of the correct solutions using the square root property $\{\pm 3\}$

**The Square Root Property of Equations:**
For any nonnegative real number $c$, if $x^2 = c$, then $x = \sqrt{c}$ or $x = -\sqrt{c}$.

To solve with the Square Root Property, we must **isolate the squared part** of the equation first. Try these together with your instructor.

**Example 1:** Solve $5x^2 - 10 = 70$.

**Example 2:** Solve $\dfrac{(x-5)^2}{2} = 32$.

**Example 3:** Solve $3x^2 - 33 = 0$.

**Example 4:** Solve $4 = (4x-3)^2 - 9$.

**Example 5:** Solve $x^2 + 25 = 0$.

# Student Activity
## Square Isolation Tic-Tac-Toe

The square root property can be used to solve a quadratic equation if
   1) you can isolate the part of the equation that is written as a square.
   2) the only variables in the equation are in the squared part.

**Directions:** If the equation **can** be solved easily with the square root property using the rules above, then isolate the squared part, and circle the equation (placing an O on the square). If the equation cannot be easily solved with the square root property, then place an X on the square.

| | | |
|---|---|---|
| $3(x+2)^2 - 25 = 50$ | $x^2 = 5x - 4$ | $-5(3x-1)^2 = -45$ |
| $3x^2 = 75$ | $24 = (x+7)^2 + 8$ | $x^2 + 8x - 20 = 0$ |
| $16x^2 - 24x = -9$ | $(x-3)^2 + 2x = 0$ | $x^2 + 6x = 2(3x+2)$ |

# Student Activity
## Double the Fun on Quadratic Equations

**Directions:** When you solve some quadratic equations, you can choose to solve by factoring or by using the Square Root Property. **If possible**, solve the following equations both ways. If it is not possible to solve the problem using one of the methods, say so. The first one has been started for you.

| | Quadratic Equation | Solve by Factoring | Solve by Square Root Property |
|---|---|---|---|
| **1.** | $x^2 = 25$ | $x^2 = 25$ <br> $x^2 - 25 = 0$ | $x^2 = 25$ <br> $x = \sqrt{25}$ or $x = -\sqrt{25}$ |
| **2.** | $4x^2 - 81 = 0$ | | |
| **3.** | $(x-3)^2 = 4$ | Hint: Expand the binomial first. | |
| **4.** | $x^2 - 8 = 0$ | | |
| **5.** | $x^2 + 16x + 64 = 25$ | | Hint: Do you see a perfect square? |

# Student Activity
## The Square Egg

**Directions:** Fill in the blanks in each equation to complete the square and write the expression as a perfect square. All of the solutions are provided inside the "egg". When you've crossed out all the solutions, you have "hatched" the egg. The first one has been done for you.

$$x^2 + 4x + \underline{\ 4\ } = \underline{(x+2)^2}$$

$$x^2 + 3x + \underline{\quad} = \underline{\quad}$$

$$x^2 - 4x + \underline{\quad} = \underline{\quad}$$

$$x^2 - 16x + \underline{\quad} = \underline{\quad}$$

$$x^2 + 5x + \underline{\quad} = \underline{\quad}$$

$$x^2 + 8x + \underline{\quad} = \underline{\quad}$$

$$x^2 + 12x + \underline{\quad} = \underline{\quad}$$

$$x^2 - x + \underline{\quad} = \underline{\quad}$$

$$x^2 - 10x + \underline{\quad} = \underline{\quad}$$

$$x^2 + 2x + \underline{\quad} = \underline{\quad}$$

Inside the egg:

$\dfrac{9}{4}$  $\left(x - \dfrac{1}{2}\right)^2$  $\dfrac{25}{4}$

$\dfrac{1}{4}$  16  $(x-2)^2$

64

$\cancel{(x+2)^2}$

25  $\left(x + \dfrac{5}{2}\right)^2$  4

$(x-8)^2$  $(x+4)^2$  $(x+6)^2$

$(x+1)^2$

$(x-5)^2$

$\cancel{4}$  36  1

$\left(x + \dfrac{3}{2}\right)^2$

# Guided Learning Activity

## Learning to Complete

Recall that we can use the square root property to solve quadratic equations, if all of the variables appear within a quantity squared.

**Example 1:** Solve $(x+2)^2 = 9$

Now we will use the process of completing the square together with the square root property to help us solve quadratic equations.

**Example 2:** Solve $x^2 - 8x - 9 = 0$ by completing the square.

a) Get the constant term on the right side of the equation and leave a "blank" to complete the square in on the left side. (Note: Sometimes the equation is already set up this way.)

$$x^2 - 8x \underline{\hspace{2cm}} = 9$$

b) Complete the square by adding the same constant to both sides.

$$x^2 - 8x \underline{+16} = 9 + 16$$

c) Write the left side as a perfect square, combine constant terms on the right side.

$$(x-4)^2 = 25$$

d) Now solve using the square root property.

e) And check your answers.

**Example 3:** Solve $x^2 - 10x + 5 = 0$ by completing the square.

Moonlighting Math T.A. – Epic FALL

**Example 4:** Solve $a^2 = 24(a - 1)$ by completing the square. *Hint: Simplify first!*

**Example 5:** Solve $x^2 + 5x + \dfrac{1}{4} = 0$ by completing the square.

**Example 6:** Solve $w^2 - w + 1 = 0$ by completing the square.

What if the lead coefficient is not 1? We then begin by dividing both sides of the equation by the lead coefficient. Let's work through some of these problems together.

**Example 7:** Solve $4x^2 + 16x - 48 = 0$ by completing the square.

**Example 8:** Solve $3x^2 + 13x - 30 = 0$ by completing the square.

# Student Activity

## Recognizable Foe

**Directions:** Follow the directions for each step and fill in the boxes with the missing pieces as you go. Careful – as you move from line to line, the boxes might contain different expressions! At the end, you should get a formula that you might recognize!

Solve: $\qquad ax^2 + bx + c = 0$

Divide both sides by $a$: $\qquad x^2 + \boxed{\phantom{x}} + \dfrac{c}{a} = \boxed{\phantom{x}}$

Subtract $\boxed{\phantom{x}}$ from both sides: $\quad x^2 + \boxed{\phantom{x}} = -\dfrac{c}{a}$

Complete the square: $\dfrac{1}{2}\left(\boxed{\phantom{x}}\right) = \boxed{\phantom{x}}$ and $\left(\boxed{\phantom{x}}\right)^2 = \boxed{\phantom{x}}$

Add $\boxed{\phantom{x}}$ to both sides: $\qquad x^2 + \dfrac{b}{a}x + \boxed{\phantom{x}} = -\dfrac{c}{a} + \boxed{\phantom{x}}$

Write the left side as a perfect square: $\qquad \left(x + \boxed{\phantom{x}}\right)^2 = -\dfrac{c}{a} + \boxed{\phantom{x}}$

Find a common denominator on the right: $\qquad \left(x + \boxed{\phantom{x}}\right)^2 = -\dfrac{\boxed{\phantom{x}}}{4a^2} + \dfrac{b^2}{4a^2}$

Combine the terms on the right side: $\qquad \left(x + \boxed{\phantom{x}}\right)^2 = \dfrac{\boxed{\phantom{x}} - 4ac}{4a^2}$

Apply the square root property: $\qquad x + \boxed{\phantom{x}} = \pm\sqrt{\dfrac{\boxed{\phantom{x}} - 4ac}{4a^2}}$

Simplify the denominator on the right side: $\qquad x + \boxed{\phantom{x}} = \pm\dfrac{\sqrt{\boxed{\phantom{x}} - 4ac}}{\boxed{\phantom{x}}}$

Subtract $\boxed{\phantom{x}}$ from both sides: $\qquad x = \boxed{\phantom{x}} \pm \dfrac{\sqrt{\boxed{\phantom{x}} - 4ac}}{\boxed{\phantom{x}}}$

Combine terms with like denominators: $\quad x = \dfrac{\boxed{\phantom{x}} \pm \sqrt{\boxed{\phantom{x}} - 4ac}}{\boxed{\phantom{x}}}$

If you're really good… try it on your own on a blank sheet of paper without peeking!

# Student Activity
## Quadratic Formula with a Calculator

It can be tricky to evaluate the quadratic formula using your calculator because of all the order of operations that are implied by the symbols.

For example, suppose we wanted to calculate the value of:

$$\frac{-3 \pm \sqrt{(-3)^2 - 4 \cdot 1 \cdot (-2)}}{2 \cdot 1}.$$

There are many extra parentheses we would have to add to do the calculation directly. First, the numerator and denominator need parentheses, since a fraction bar separates two distinct groups of terms:

$$\frac{\left(-3 \pm \sqrt{(-3)^2 - 4 \cdot 1 \cdot (-2)}\right)}{(2 \cdot 1)}$$

Then the radicand needs a set of parentheses, since the radicand must be simplified before the square root is taken:

$$\frac{\left(-3 \pm \sqrt{\left((-3)^2 - 4 \cdot 1 \cdot (-2)\right)}\right)}{(2 \cdot 1)}$$

Then we need to calculate two expressions because of the double-sign:

$$\frac{\left(-3 - \sqrt{\left((-3)^2 - 4 \cdot 1 \cdot (-2)\right)}\right)}{(2 \cdot 1)} \quad \text{or} \quad \frac{\left(-3 + \sqrt{\left((-3)^2 - 4 \cdot 1 \cdot (-2)\right)}\right)}{(2 \cdot 1)}$$

Entering this into **my** calculator would look like this:

$$\boxed{(}\ \boxed{(-)}\ \boxed{3}\ \boxed{-}\ \boxed{2\text{ND}}\ \boxed{\sqrt{\ }}\ \boxed{(}\ \boxed{(}\ \boxed{(-)}\ \boxed{3}\ \boxed{)}\ \boxed{\wedge}\ \boxed{2}\ \boxed{-}\ \boxed{4}\ \boxed{\times}\ \boxed{1}\ \boxed{\times}\ \boxed{(-)}\ \boxed{2}\ \boxed{)}\ \boxed{)}\ \boxed{\div}\ \boxed{(}\ \boxed{2}\ \boxed{\times}\ \boxed{1}\ \boxed{)}\ \boxed{\text{ENTER}}$$

The point is, **you will want to simplify your quadratic formula** to a more manageable expression **before** you enter it into your calculator.

$$\frac{-3 \pm \sqrt{(-3)^2 - 4 \cdot 1 \cdot (-2)}}{2 \cdot 1} = \frac{-3 \pm \sqrt{9 + 8}}{2} = \frac{-3 \pm \sqrt{17}}{2} \quad \leftarrow \text{This is much more manageable!}$$

Now we calculate the value of the two expressions: (your keystrokes may vary)

$$\frac{\left(-3 - \sqrt{17}\right)}{2} \qquad \boxed{(}\ \boxed{(-)}\ \boxed{3}\ \boxed{-}\ \boxed{2\text{ND}}\ \boxed{\sqrt{\ }}\ \boxed{1}\ \boxed{7}\ \boxed{)}\ \boxed{\div}\ \boxed{2}\ \boxed{\text{ENTER}} \quad \approx -3.56$$

$$\frac{\left(-3 + \sqrt{17}\right)}{2} \qquad \boxed{(}\ \boxed{(-)}\ \boxed{3}\ \boxed{+}\ \boxed{2\text{ND}}\ \boxed{\sqrt{\ }}\ \boxed{1}\ \boxed{7}\ \boxed{)}\ \boxed{\div}\ \boxed{2}\ \boxed{\text{ENTER}} \quad \approx 0.56$$

**NOTE:** On some calculators, you can edit an old entry. For example, on most TI graphing calculators, use ENTRY to pull up the last entry. You can use this to change the $\boxed{-}$ to a $\boxed{+}$ and then press ENTER to get the second answer.

Now try finding these with your calculator.  The first row has been done for you.  Be careful not to round values until the **end** of the problem.  Use a $\approx$ symbol if the calculated result is rounded.  Round to two decimal places if necessary.

| Quadratic Formula Expression | Simplified Expression | Calculated Result |
|---|---|---|
| 1. $\dfrac{-(-4)\pm\sqrt{(-4)^2-4(3)(1)}}{2(3)}$ | | |
| 2. $\dfrac{-5\pm\sqrt{5^2-4(2)(-3)}}{2\cdot2}$ | | |
| 3. $\dfrac{-11\pm\sqrt{11^2-4(-2)(-6)}}{2(-2)}$ | | |
| 4. $\dfrac{-(-3,850)\pm\sqrt{(-3,850)^2-4(550)(-5,500)}}{2(550)}$ | | |
| 5. $\dfrac{-(-4.18)\pm\sqrt{(-4.18)^2-4(1.32)(-1.54)}}{2(1.32)}$ | | |

## Scrambled Answers

| 0.5 | $\approx 0.61$ | $\approx -0.33$ | $-3$ | $\approx 8.22$ |
|---|---|---|---|---|
| 1 | 3.5 | $\approx 0.33$ | $\approx -1.22$ | $\approx 4.89$ |

# Student Activity

## Triple the Fun on Quadratic Equations

**Directions:** For many quadratic equations, you have the choice of solving by using the factoring method, the square root method (completing the square if necessary), or the quadratic formula. Solve each of the quadratic equations by the indicated methods (if possible). If it is **not** possible to use one of the methods, say so. We've given you some hints on the first problem. For each problem, put a ☺ by the method that seemed easiest.

| 1. Solve: $x^2 - 3x - 40 = 0$ | 2. Solve: $x^2 + 2x - 2 = 0$ |
|---|---|
| **Solve by Factoring (if possible):** $$x^2 - 3x - 40 = 0$$ $$(\quad)(\quad) = 0$$ | **Solve by Factoring (if possible):** $$x^2 + 2x - 2 = 0$$ |
| **Solve by the Square Root Property:** (completing the square if necessary) $$x^2 - 3x - 40 = 0$$ $$x^2 - 3x + \underline{\quad} = 40 + \underline{\quad}$$ | **Solve by the Square Root Property:** (completing the square if necessary) $$x^2 + 2x - 2 = 0$$ |
| **Solve using the Quadratic Formula:** $$x^2 - 3x - 40 = 0$$ $$a = \underline{\quad}, \quad b = \underline{\quad}, \quad c = \underline{\quad}$$ | **Solve using the Quadratic Formula:** $$x^2 + 2x - 2 = 0$$ |

**3.  Solve:** $4x^2 - 5x - 6 = 0$

**Solve by Factoring (if possible):**
$4x^2 - 5x - 6 = 0$



**Solve by the Square Root Property:**
(completing the square if necessary)
$4x^2 - 5x - 6 = 0$



**Solve using the Quadratic Formula:**
$4x^2 - 5x - 6 = 0$

**4.  Solve:** $x^2 - 8 = 0$

**Solve by Factoring (if possible):**
$x^2 - 8 = 0$



**Solve by the Square Root Property:**
(completing the square if necessary)
$x^2 - 8 = 0$



**Solve using the Quadratic Formula:**
$x^2 - 8 = 0$

# Student Activity
## Match Up on the Discriminant

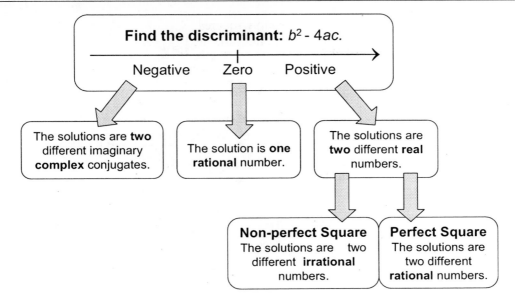

**Find the discriminant:** $b^2 - 4ac$.

Negative — Zero — Positive

The solutions are **two** different imaginary **complex** conjugates.

The solution is **one** **rational** number.

The solutions are **two** different **real** numbers.

**Non-perfect Square** The solutions are two different **irrational** numbers.

**Perfect Square** The solutions are two different **rational** numbers.

**Directions:** In each box, find the discriminant and then classify it.

**A** Two different rational numbers

**B** Two different irrational numbers

**C** One rational number

**D** Two different imaginary numbers

| | | |
|---|---|---|
| $25x^2 - 40x + 16 = 0$ <br> $d = (-40)^2 - 4(25)(16) = 0$ <br><br> **C** | $3x^2 + 14x - 24 = 0$ | $x^2 - 2x + 2 = 0$ |
| $9x^2 = 4$ | $6x^2 - 3 = 0$ | $x^2 + 3 = 5x$ |
| $5(5x^2 + 2x) = -1$ | $2x^2 + 3x = -3$ | $3x^2 + 11x = 5$ |
| $45x^2 + 24 = 67x$ | $144x^2 - 264x + 121 = 0$ | $4x^2 + 36x = -81$ |

# Student Activity
## Strategizing on Quadratic Equations

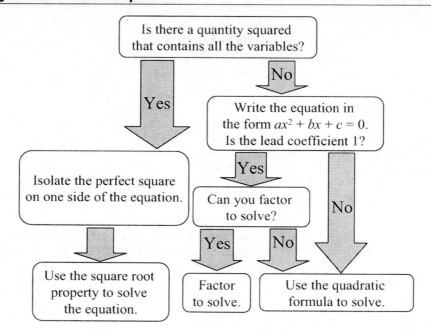

Is there a quantity squared that contains all the variables?

Yes

No

Write the equation in the form $ax^2 + bx + c = 0$. Is the lead coefficient 1?

Isolate the perfect square on one side of the equation.

Yes

Can you factor to solve?

No

Yes

No

Use the square root property to solve the equation.

Factor to solve.

Use the quadratic formula to solve.

**Directions:** For each problem, use the flow chart above to help you make a decision about **how** to solve the quadratic equation (the square root property, factoring, or the quadratic formula). You do not need to solve the problems.

| | Square root property | Factoring | Quadratic formula |
|---|---|---|---|
| **1.** Solve: $x^2 - 8x - 20 = 0$ | | | |
| **2.** Solve: $4(x+3)^2 = 25$ | | | |
| **3.** Solve: $4x^2 = 9$ | | | |
| **4.** Solve: $x^2 + 5x - 3 = 0$ | | | |
| **5.** Solve: $6x^2 + 7x = 20$ | | | |
| **6.** Solve: $x^2 - 6x = 0$ | | | |
| **7.** Solve: $x^2 - x + 6 = 0$ | | | |
| **8.** Solve: $8x(x-2) = 2x + 5$ | | | |
| **9.** Solve: $(x-2)^2 - 2x = 16$ | | | |
| **10.** Solve: $(x-2)^2 = 16$ | | | |

# Student Activity
## Sink the Sub

**Directions:** For each problem, follow these steps:

- Find an appropriate y-substitution to make the equation quadratic in form.
- Solve the quadratic equation.
- Reverse the y-substitution to determine the solution set of the original equation.

The first problem is done for you. Two submarines holding the possible answers for $u = ?$ and $x = ?$ are provided to help you stay on track.

$u = ?$

$u = x^{1/3}$    $u = 2x - 3$    $u = \sqrt{x}$

$u = \sqrt{x}$    $u = \dfrac{1}{x}$    $u = x^2$

$x = ?$

$x = \{-3, 3, -i, i\}$    $x = \{-1, 4\}$

$x = \{36\}$    $x = \{-6, 5\}$

$x = \left\{64, \dfrac{1}{8}\right\}$    $x = \{25, 36\}$

| Solve. | $y = ?$ | Substitute, then solve for y. | Reverse the substitution | Solution set |
|---|---|---|---|---|
| **1.** $x^4 - 8x^2 - 9 = 0$ | $y = x^2$ | $y^2 - 8y - 9 = 0$ <br> $(y-9)(y+1) = 0$ <br> $y = 9, -1$ | $x^2 = 9 \quad x^2 = -1$ <br> $x = \pm 3 \quad x = \pm i$ | $x = \{-3, 3, -i, i\}$ |
| **2.** $x - 11\sqrt{x} + 30 = 0$ | | | | |
| **3.** $x - \sqrt{x} - 30 = 0$ | | | | |
| **4.** $2x^{2/3} - 9x^{1/3} + 4 = 0$ | | | | |
| **5.** $(2x - 3)^2 = 25$ | | | | |
| **6.** $\dfrac{30}{x^2} - \dfrac{1}{x} - 1 = 0$ | | | | |

# Guided Learning Activity

## Everything's Coming Up Parabolas

The graph of $y = ax^2 + bx + c$, where $a \neq 0$, is a quadratic function, and its graph is called a parabola. If $a > 0$, then the parabola opens up. If $a < 0$, then the parabola opens down. The vertex is the maximum or minimum point of the parabola (depending on whether it opens up or down. Each parabola is symmetric about the axis of symmetry.

1. On each parabola below, label the **vertex**, the **axis of symmetry**, whether the parabola **opens up** or **opens down** and whether $a > 0$ or $a < 0$.

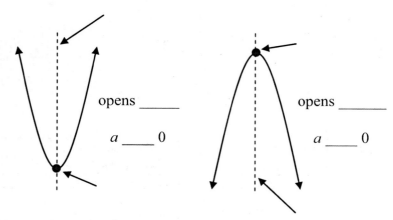

opens _____

$a$ _____ 0

opens _____

$a$ _____ 0

2. Draw the graph of $y = -x^2 + 2x + 3$ by plotting points. On the graph, clearly indicate the point that is the vertex and a dashed line for the axis of symmetry.

| $x$ | $y$ |
|-----|-----|
| $-3$ | |
| $-2$ | |
| $-1$ | |
| $0$ | |
| $1$ | |
| $2$ | |
| $3$ | |

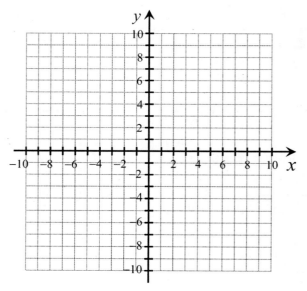

What is the vertex of the parabola? (____, ____)

What is the equation for the axis of symmetry of the parabola? _____

What is the $y$-intercept of the parabola? (____, ____)

What are the $x$-intercepts of the parabola? (____, ____) and (____, ____)

© 2011 Cengage Learning. All Rights Reserved. May not be scanned, copied or duplicated, or posted to a publicly accessible website, in whole or in part.

**3.**

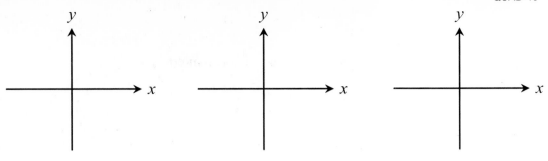

Draw a parabola with **two** *x*-intercepts.

Draw a parabola with **one** *x*-intercept.

Draw a parabola with **no** *x*-intercepts.

Will a parabola ever have **more** than one *y*-intercept?  Why or why not?

Will a parabola ever have **less** than one *y*-intercept?  Why or why not?

To find the *y*-intercept, we find *y* when $x = 0$.

Thus, the **y-intercept** is at a point given by $\left(0, \underline{\phantom{xx}}\right)$.

To find *x*-intercepts, we need to find *x* when $y = 0$.

Likewise, the **x-intercepts** are at point(s) given by $\left(\underline{\phantom{xx}}, 0\right)$.

**4.** Find the *y*- and *x*-intercepts of the graph of $y = x^2 + 4x - 12$.

**5.** Find the *y*- and *x*-intercepts of the graph of $y = 2x^2 + 3$.

The **vertex** of the parabola $y = ax^2 + bx + c$ has an $x$-coordinate of $x = -\dfrac{b}{2a}$.

**6.** Find the vertex of the parabola given by $y = 2x^2 - 8x + 13$.

Find the $x$-coordinate: $\quad x = \dfrac{-b}{2a} = \underline{\quad\quad} = $

Substitute this value of $x$ into the given function to find the $y$-coordinate:

$$y = 2x^2 - 8x + 13$$
$$y = 2(\quad)^2 - 8(\quad) + 13 = \underline{\quad\quad}$$

Vertex: $(\underline{\quad}, \underline{\quad})$

**7.** Find the vertex of the parabola given by $y = -9x^2 - 6x - 8$.

**8.** Put it all together to graph $y = -x^2 + 2x + 8$.

Does the parabola open up or down? $\underline{\quad\quad}$

Find the $y$-intercept: $(0, \underline{\quad})$

Find the $x$-intercept(s), if there are any:
$(\underline{\quad}, 0) \quad (\underline{\quad}, 0)$

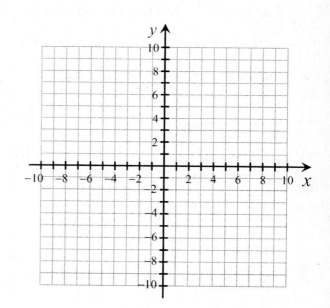

Find the vertex: $(\underline{\quad}, \underline{\quad})$

Write the equation for the axis of symmetry: $\underline{\quad\quad\quad}$

# Student Activity
## Finding the Vertex and Intercepts Using a Calculator

**Example:** Consider the quadratic equation $y = 8x^2 - 18x - 5$. First we will find all the characteristics of the graph using our **algebra** skills.

**a.** Find the vertex: ( ____ , ____ )

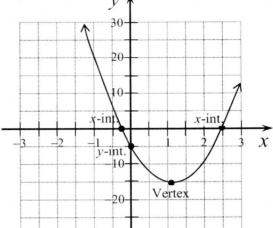

**b.** Find the $x$-intercepts:

( ____ , 0) and ( ____ , 0)

**c.** Find the $y$-intercept: $(0, \_\_\_\_)$

---

Now we'll perform the same analysis using a **graphing calculator**.

☐ Using the $\boxed{y=}$ button or menu, graph the function $y = 8x^2 - 18x - 5$.

☐ Now use $\boxed{\text{WINDOW}}$ to find the window that best displays the parabola. Make sure you can see the vertex of the parabola in your graphing window.

## Maximum or Minimum

Because the vertex of the parabola **is** the maximum or minimum point of the parabola, we can use this feature of the calculator to find the vertex.

☐ Find the graphing menu that has an option to find the MAX or MIN. You may find this in a MATH menu in the graphing panel, or under the CALC function. We will refer to this menu as the Graph Calculations Menu.

☐ Write down how you get to the Graph Calculations Menu on **your** calculator:

☐ Select the option to find the minimum of the graph (since in this case, the vertex is the minimum point).

☐ Your calculator will probably prompt you for the *Left Bound* or *Lower Bound*. Move the cursor until it is somewhere on the left of the vertex and then press ENTER . Repeat this process for the *Right Bound* or *Upper Bound*. Note: If you don't want to use the cursor to scroll left or right, you may just enter an *x*-value to be the bound.

☐ At this point, your calculator may prompt you to *Guess*. If this happens, you can move the cursor close to the vertex and press ENTER (note that you don't *have* to guess, but you do have to press enter).

☐ Your calculator should now display the coordinates of the vertex: (____, ____).

### y-intercept

☐ Go back to the Graph Calculations Menu. This time, select *Value* on the menu. Note: You can also use TRACE to do this.

☐ To find the *y*-intercept, we let $x = 0$. When prompted, enter 0 for the *x*-value.

☐ Your calculator should now display the coordinates of the *y*-intercept: $(0, \underline{\quad})$

### x-intercept(s)

☐ Go back to the Graph Calculations Menu. This time, select *Zero* on the menu. It is called *Zero* because we let *y* be zero when we want to find the *x*-intercepts.

☐ Your calculator will probably prompt you for the *Left Bound* or *Lower Bound*. Move the cursor until it is somewhere on the left of the vertex and then press ENTER . Repeat this process for the *Right Bound* or *Upper Bound*.

☐ At this point, your calculator may prompt you to *Guess*. If this happens, you can move the cursor close to the *x*-intercept and press ENTER (note that you don't *have* to guess, but you do have to press enter).

☐ Your calculator should now display the coordinates of the *x*-intercept: (____, ____).

☐ Repeat this process to find the coordinates of the second *x*-intercept: (____, ____).

### Possible Errors:

ERR: BOUND        You guessed a value that was not within the bounds you set.

ERR: INVALID      You tried to have the calculator find something with an *x*-value outside the viewing window.

Now use your calculator to find the vertex and the $x-$ and $y$-intercepts of the following quadratic equations.

**1.** $y = 5x^2 + 15x + 10$

       Vertex: (_____, _____)
       $y$-intercept: (_____, _____)
       $x$-intercept(s): (_____, _____)  (_____, _____)

**2.** $y = -4x^2 + 20x - 24$

       Vertex: (_____, _____)
       $y$-intercept: (_____, _____)
       $x$-intercept(s): (_____, _____)  (_____, _____)

**3.** $y = x^2 - 4x + 14$

       Vertex: (_____, _____)
       $y$-intercept: (_____, _____)
       $x$-intercept(s): (_____, _____)  (_____, _____)

**4.** $y = \frac{1}{16}x^2$

       Vertex: (_____, _____)
       $y$-intercept: (_____, _____)
       $x$-intercept(s): (_____, _____)  (_____, _____)

# Student Activity
## Parabola Calisthenics

**Directions:** On each graph you are given the graph of the parabola $y = x^2$, the most basic quadratic function. You will be asked to finish a table of values to draw the graphs of other parabolas. If possible, use different colored pencils or pens to fill in each table and draw the corresponding graph.

1. Draw the graphs of $y_1$ and $y_2$ by filling in the missing values and plotting the points on the coordinate axes provided.

   $y_1 = 2x^2$         $y_2 = 10x^2$

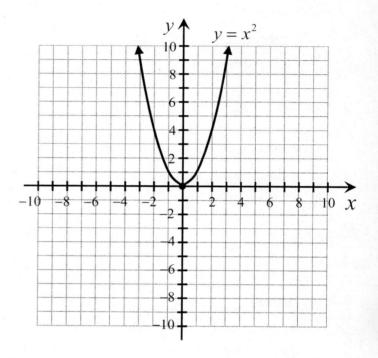

| $x$ | $y_1$ |
|-----|-------|
| $-2$ | 8 |
| $-1$ | |
| 0 | 0 |
| 1 | 2 |
| 2 | |

| $x$ | $y_3$ |
|-----|-------|
| $-1$ | |
| $-\frac{1}{2}$ | 2.5 |
| 0 | |
| $\frac{1}{2}$ | |
| 1 | 10 |

2. How do the graphs of $y_1$ and $y_2$ compare to the basic graph of $y = x^2$?

3. Circle the equations for the graphs that should fall between $y_1 = 2x^2$ and $y_2 = 10x^2$:

   $y = 5x^2$

   $y = 12x^2$

   $y = 9x^2$

   $y = 1.5x^2$

**4.** Draw the graphs of $y_3$ and $y_4$ by filling in the missing values and plotting the points on the coordinate axes provided.

$$y_3 = \frac{1}{2}x^2 \qquad y_4 = \frac{1}{10}x^2$$

| $x$ | $y_2$ |
|---|---|
| $-3$ | 4.5 |
| $-2$ | |
| $-1$ | |
| 0 | 0 |
| 1 | 0.5 |
| 2 | |
| 3 | |

| $x$ | $y_4$ |
|---|---|
| $-5$ | 2.5 |
| $-3$ | |
| $-1$ | 0.1 |
| 0 | 0 |
| 1 | |
| 3 | 0.9 |
| 5 | |

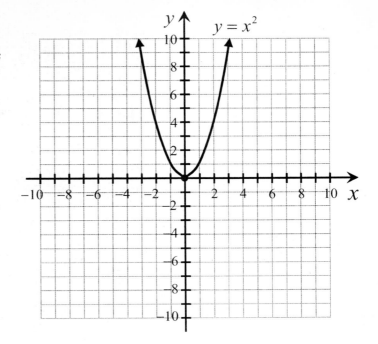

**5.** How do the graphs of $y_4$ and $y_5$ compare to the basic graph of $y = x^2$?

**6.** Circle the equations for the graphs that should fall between $y_3 = \frac{1}{2}x^2$ and $y_4 = \frac{1}{10}x^2$ :

$$y = \frac{3}{4}x^2 \qquad y = \frac{1}{4}x^2 \qquad y = \frac{1}{5}x^2 \qquad y = \frac{1}{20}x^2$$

**7.** To draw $y = a \cdot x^2$ if $a > 1$, how do you transform the graph of $y = x^2$?

**8.** What happens to the graph of $y = a \cdot x^2$ if $a$ is negative?

# Student Activity
## Parabola Transformations

**1.** Draw the desired transformation of $y = x^2$ on each graphing grid below

    **a.** Reflect the graph about the *x*-axis, then shift the graph up 3 units.

    **b.** Shift the graph up 3 units, then reflect the graph about the *x*-axis.

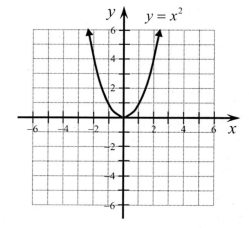

    **c.** Do you get the same result for both graphs? _____ Calculate a few points on the graph of $y = -x^2 + 3$ to see which graph (**a** or **b**) is the correct graph.

**2.** Let's examine some functions that are of the form $y = ax^2$.

    $y = 5x^2$ is narrower than $y = x^2$.

    $y = \dfrac{1}{3}x^2$ is wider than $y = x^2$.

    $y = -x^2$ is the graph of $y = x^2$ reflected about the *x*-axis.

In all these cases, the stretch, compression, or reflection is caused by _____ $x^2$ by a constant.

**3.** Let's examine some functions that include translations:

    Up 3 units: $y = x^2 + 3$          Down 3 units: $y = x^2 - 3$

    Right 3 units: $y = (x - 3)^2$      Left 3 units: $y = (x + 3)^2$

In all these cases, the translation is caused by _____ or _____ of a constant.

When we **graph**, the order of operations still places multiplication before addition or subtraction.

---

**Graphing Order of Operations:**

**1.** Draw the graph of $y = ax^2$.

**2.** Identify the translations of $y = ax^2$ (up or down, left or right) and draw the final graph.

---

398

Yeah, she may look good, but can she function in a relationship?

**4.** Graph $y = -(x+1)^2 + 4$.

   **a.** Draw the graph of $y = -x^2$ with a dashed line.

   **b.** Identify any translations. Use the graph from part **a** to sketch the graph that includes these changes with a solid line.

   **c.** What is the vertex of the parabola?

      (___,___) Does the vertex represent the maximum or minimum value of the graph? _____ What is the maximum or minimum **value** of the graph? _____

   **d.** What is the axis of symmetry of the graph? _____

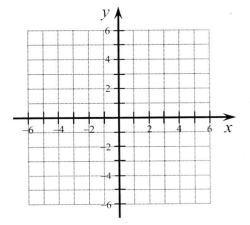

**5.** Graph $y = \frac{1}{2}(x+2)^2 - 3$.

   **a.** Draw the graph of $y = \frac{1}{2}x^2$ with a dashed line.

   **b.** Identify any translations. Use the graph from part **a** to sketch the graph that includes these changes with a solid line.

   **c.** What is the vertex of the parabola?

      (___,___) Does the vertex represent the maximum or minimum value of the graph? _____ What is the maximum or minimum **value** of the graph? _____

   **d.** What is the axis of symmetry of the graph? _____

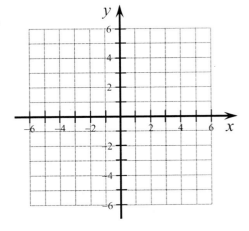

# Student Activity
## Graphical Meaning of Complex Zeros

The *x*-intercepts of a graph can also be called the *zeros* (where the *y*-values are zero). When we try to find the *x*-intercepts of a quadratic function, and the result is a pair of complex numbers, does that give us any useful information for graphing? It turns out that the answer is **yes!**

**Example 1:** $y = x^2 - 4x + 29$

a) Solve $x^2 - 4x + 29 = 0$ using the quadratic formula, as if you were finding the *x*-intercepts of the graph. The solution will be a pair of complex numbers.

b) If the zeros are given by $p \pm qi$, then for this problem, $p = $ _____ and $q = $ _____ . We will also need the value of the leading coefficient: $a = $ _____ .

c) Now we will graph the parabola, using only the values of $p$, $q$, and $a$.

- First we graph the vertex. The *x*-coordinate of the vertex is $p$ and the *y*-coordinate of the vertex is $aq^2$. Write the vertex: $(\_\_,\_\_)$. Then graph the vertex.

- We also need to locate the point $(p, 2aq^2)$. Since this is just a point of reference, just make a small **X** on the graph at this location: $(\_\_,\_\_)$.

- From the reference point marked with an **X**, move $q$ units to the left and plot the point, and move $q$ units to the right and plot the point. Write the coordinates of these two points: $(\_\_,\_\_)$ and $(\_\_,\_\_)$.

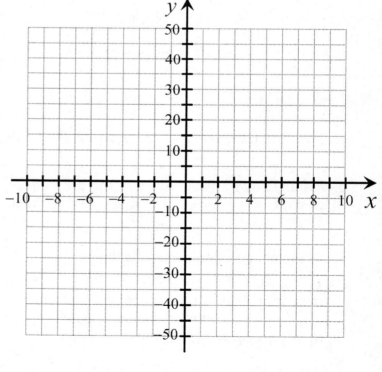

- You should now have three points to use to sketch the graph of the parabola.

**Example 2:** $y = -2x^2 + 4x - 20$

a) Solve $-2x^2 + 4x - 20 = 0$ using the quadratic formula, as if you were finding the *x*-intercepts of the graph.  The solution will be a pair of complex numbers.

b) If the zeros are given by $p \pm qi$, then for this problem, $p =$ _____ and $q =$ _____.  We will also need the value of the leading coefficient:  $a =$ _____ .

c) Now we will graph the parabola.

- Graph the vertex: $(p, aq^2) = ($ _____ , _____ $)$.

- Locate
  $(p, 2aq^2) = ($ _____ , _____ $)$
  with a small **X**.

- From the **X**, plot points $q$ units to the left and $q$ units to the right.  The points are
  ( _____ , _____ ) and
  ( _____ , _____ ).

- You should now have three points to use to sketch the graph of the parabola.

**Connection to complex zeros:**

a) Below each graph, write the complex zeros for the graph in the form $p \pm qi$.  **Then calculate** $p \pm q$ and write those two values below the graphs as well.

b) Draw a horizontal dashed line on each graph that goes through the vertex of the parabola.  Then draw a perfect reflection of the parabola over this dashed line.  We'll call this new parabola the *imaginary parabola* for lack of a better name.

c) Where does the *imaginary parabola* cross the *x*-axis?  In other words, what are its zeros?  _____ and _____  Surprise!

## Quadratics

For each of the following, describe the strategies or key steps that will help you **start** the problem.  You do **not** have to complete the problems.

| | | **What will help you to start this problem?** |
|---|---|---|
| **1.** | Complete the perfect square trinomial: $x^2 - 14x +$ _____ | |
| **2.** | Solve $x^2 + 6x + 5 = 0$ by completing the square. | |
| **3.** | Solve $x^2 - 10 = 0$ using the quadratic formula. | |
| **4.** | Solve: $16(x-3)^2 = 9$ | |
| **5.** | Solve $15x^2 = 4 - 17x$ using the quadratic formula. | |
| **6.** | Solve: $x - 8\sqrt{x} + 12 = 0$ | |
| **7.** | Solve $x^2 + 10 = 7x$ by factoring. | |

| | | **What will help you to start this problem?** |
|---|---|---|
| **8.** | How many solutions does $2x^2 + x - 21 = 0$ have? Are the solutions rational, irrational, or complex numbers? | |
| **9.** | Solve: $x^2 - 8x = 20$ | |
| **10.** | Does the graph of $y = 3 - x^2$ open up or down? | |
| **11.** | Solve: $x^4 - 2x^2 - 35 = 0$ | |
| **12.** | Find the maximum value of $y = -x^2 + 2x + 3$. | |
| **13.** | Find the vertex of the graph of $y = 3x^2 - 3x$.. | |
| **14.** | Write $y = x^2 - 6x + 8$ in standard form. | |
| **15.** | Find the $x$-intercept(s) of the graph of $y = 3x^2 - 3x - 6$. | |

# Metacognitive Skills
## Quadratics

Metacognitive skills refer to the ability to judge how well you have learned something and to effectively direct your own learning and studying. This is a self-evaluation tool designed to help you focus your studying and to improve your metacognitive skills with regards to this math class.

Fill the 1$^{st}$ column out **before** you begin studying. Fill the 2$^{nd}$ column out after you study for your test.

Go back to this assessment after your test and circle any of the ratings that you would change – this identifies the "disconnects" between what you **thought** you knew well and what you **actually** knew well.

**Use the scale below to assign a number to each topic.**
5   *I am confident I can do any problems in this category correctly.*
4   *I am confident I can do most of the problems in this category correctly.*
3   *I understand how to do the problems in this category, but I still make a lot of mistakes.*
2   *I feel unsure about how to do these problems.*
1   *I know I don't understand how to do these problems.*

| Topic or Skill | Before Studying | After Studying |
|---|---|---|
| Solving a quadratic equation using factoring. | | |
| Solving a quadratic equation using the square root property. | | |
| Using the square root property and understanding where the double-sign comes from. | | |
| Simplifying an expression written with a double-sign ( $\pm$ ). | | |
| Completing a perfect square trinomial. | | |
| Solving a quadratic equation by completing the square. | | |
| Knowing the quadratic formula. | | |
| Solving a quadratic equation by using the quadratic formula. | | |
| Finding the discriminant for a quadratic equation. | | |
| Using the discriminant to determine the type and number of solutions. | | |
| Looking at a quadratic equation and choosing the most appropriate method to solve the equation (factoring, square root property, completing the square, or the quadratic formula). | | |
| Recognizing an equation that is quadratic in form. | | |
| Solving an equation that is quadratic in form by making an appropriate $u$-substitution. | | |
| Knowing the general shape of the graph of a quadratic function, including whether the parabola opens up or down. | | |
| Finding the vertex of a parabola by first finding the $x$-coordinate. | | |
| Finding the $x$- and $y$-intercepts of a parabola. | | |
| Graphing a quadratic function of the form $y = ax^2$. | | |
| Graphing a quadratic function in standard form: $y = a(x - h)^2 + k$. | | |
| Finding the vertex if the quadratic function is written in standard form. | | |
| Writing a quadratic function in standard form by completing the square. | | |

# EXPF: Exponential Functions

# Student Activity

## Expressionist Art

**Directions:** For each expression, simplify as much as possible. Shade in the squares containing the answers in the grid below. It helps to write the problem number in the box that you shade in case you make a mistake. The first problem has been done for you!

The Expressionists

$2^{2+\sqrt{3}}$

This must be an early piece, because it hasn't been simplified like the later pieces by this artist.

**1.** $5^{\sqrt{2}} \cdot 5^{\sqrt{2}} = 5^{\sqrt{2}+\sqrt{2}} = 5^{2\sqrt{2}}$

**2.** $\left(8^{\sqrt{3}}\right)^{\sqrt{3}}$

**3.** $a^4 \cdot a^{-\sqrt{4}}$

**4.** $a^{\sqrt{20}} \cdot a^{\sqrt{5}}$

**5.** $\sqrt{2}^{\sqrt{2}} \cdot \sqrt{2}^{2\sqrt{2}}$

**6.** $x^{\sqrt{32}} \div x^{\sqrt{8}}$

**7.** $\left(a^{\sqrt{5}} b^{\sqrt{3}}\right)^{\sqrt{15}}$

**8.** $\left(x^{\sqrt{y}}\right)^{\sqrt{16y}}$

**9.** $x^{\sqrt{y}} \cdot x^{\sqrt{16y}}$

**10.** $\left(ab\right)^{\sqrt{6}} \div ab^{\sqrt{6}}$

**11.** $\sqrt{9}^{2\sqrt{3}} \cdot 3^{\sqrt{3}}$

**12.** $\left(y^{\sqrt{x}}\right)^{\sqrt{x}} \cdot y^{-x}$

**13.** $\left(4^{\sqrt{9}}\right)^{1/3} \cdot \left(2^{\sqrt{3}}\right)$

| $ab$ | $a^{\sqrt{2}}$ | $2^{3\sqrt{2}}$ | | $a^9$ | $xy$ | $y^x$ |
|---|---|---|---|---|---|---|
| $x^{4\sqrt{y}}$ | $a^2$ | $2^{\frac{\sqrt{2}}{2}}$ | | $64$ | $512$ | $2^2 + 2^{\sqrt{3}}$ |
| $a^{5\sqrt{3}} b^{3\sqrt{5}}$ | $5^{\sqrt{2}}$ | $2^{\frac{3\sqrt{2}}{2}}$ | | $1$ | $x^{2\sqrt{2}}$ | $2^{2+\sqrt{3}}$ |
| $x^{5\sqrt{y}}$ | $5^{2\sqrt{2}}$ ①| $a^{\sqrt{6}-1}$ | | $a^{\sqrt{6}}-1$ | $a^{3\sqrt{5}}$ | $a^{5\sqrt{3}}$ |
| $3^{3\sqrt{3}}$ | $25$ | $x^{4y}$ | | $x^{4^y}$ | $a^{\sqrt{15}}$ | $2$ |
| $3^{5\sqrt{3}}$ | $25^{2\sqrt{2}}$ | $x^{y^4}$ | | $a$ | $\sqrt{2}^{\sqrt{2}}$ | $\sqrt{2}$ |

# Guided Learning Activity
## Graphing Exponential Functions

**Example 1:** Graph $f(x) = 2^x$.

a. Begin by filling in the table of solutions.

| $x$ | $f(x) = 2^x$ | $g(x) = 2^x + 2$ | $h(x) = 2^{x+2}$ |
|---|---|---|---|
| $-3$ | | | |
| $-2$ | | | |
| $-1$ | $2^{-1} = \frac{1}{2}$ | | |
| $0$ | | | |
| $1$ | | | |
| $2$ | | | |
| $3$ | | | |

b. Plot the points on the axes provided. Then draw a smooth curve.

c. Notice that as $x$ approaches $-\infty$, the $y$-values get closer and closer to zero. This is called an asymptote – a line that is approached by the graph. **Draw a dashed line** to represent the asymptote at $y = 0$.

d. What is the domain of $f(x) = 2^x$? _____ The range? _____

e. Fill in the table of solutions for $g(x) = 2^x + 2$, how do the $y$-values of $g$ change from the $y$-values for $f$? _____

f. The graph of $g(x) = 2^x + 2$ will be the graph of $f$ shifted _____ units _____.

g. What is the equation of the asymptote of $g$? _____

h. What is the domain of $g$? _____ The range? _____

i. Fill in the table of solutions for $h(x) = 2^{x+2}$, how do the $y$-values of $h$ change from the $y$-values for $f$? _____

j. The graph of $h(x) = 2^{x+2}$ will be the graph of $f$ shifted _____ units _____.

k. What is the equation of the asymptote of $h$? _____

l. What is the domain of $h$? _____ The range? _____

**Example 2:** Graph $f(x) = 3^x$.

a. Begin by filling in the table of solutions.

| $x$ | $f(x) = 3^x$ | $g(x) = \left(\dfrac{1}{3}\right)^x$ | $h(x) = -3^x$ |
|---|---|---|---|
| $-3$ | | | |
| $-2$ | | | |
| $-1$ | | | |
| $0$ | | | |
| $1$ | | | |
| $2$ | | | |
| $3$ | | | |

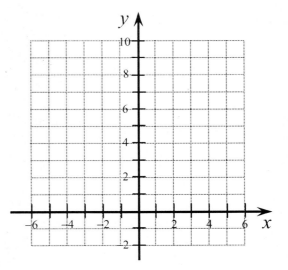

b. Plot the points on the axes provided. Then draw a smooth curve.

c. Again, as $x$ approaches $-\infty$, the $y$- values get closer and closer to zero. **Draw a dashed line** to represent the asymptote at $y = 0$.

d. What is the domain of $f$? _____ The range? _____

e. Fill in the table of solutions for $g(x) = \left(\frac{1}{3}\right)^x$ and graph it. How are the graphs of $f$ and $g$ related? _____

Carbon Emissions

It's not just an exponential curve, it's an Al-Gore-ithmic curve.

f. What is the domain of $g$? _____ The range? _____

g. Write $3^{-x}$ without negative exponents: _____ How is this related to the expression in the function $g$?

h. Fill in the table of solutions for $h(x) = -3^x$ and graph it. How are the graphs of $f$ and $h$ related? _____

i. What is the domain of $h$? _____ The range? _____

# Student Activity
## Graphing Exponential Functions with a Calculator

Let's first examine a graph by carefully finding and plotting a table of solutions.

**By hand:**  $f(x) = \left(\dfrac{3}{2}\right)^x$

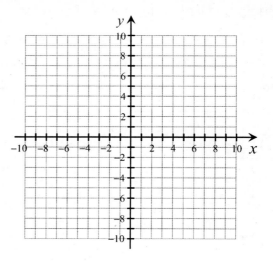

| $x$ | $f(x)$ |
|-----|--------|
| $-3$ | |
| $-2$ | |
| $-1$ | |
| $0$ | |
| $1$ | |
| $2$ | |
| $3$ | |

Graph the points and draw a smooth curve to represent the function.  Use a dashed line to represent the asymptote.

What is the domain of $f$? _____  What is the range of $f$? _____

Is the function increasing or decreasing? _____

What is the equation of the asymptote? _____

**With Calculator:**  Use your calculator to graph $y = \left(\dfrac{3}{2}\right)^x$ .  Your calculator notation might look something like this: y = ( 3 / 2 ) ^ x .

- Use the $\boxed{\text{WINDOW}}$ to look at a window that is the same size as the graph you drew in the previous problem.

- Use the $\boxed{\text{TRACE}}$ feature to verify the values in the table above.  For most graphing calculators, you can enter each $x$-value and press $\boxed{\text{ENTER}}$ to see the corresponding $y$-value.

- Use the $\boxed{\text{TABLE}}$ feature to see what happens as $x$ approaches $-\infty$.  That is, as we look at $x$ values further and further to the left, what happens to the $y$-values?  This is the kind of behavior you will look for to determine the location of a horizontal asymptote.

1. Use your calculator to graph $f(x) = -10^x$. Make sure to use the negative button (not the minus). No parentheses are necessary to enter this expression.

    **a.** Set the window so that $x$ is $[-5,5]$ with a scale of 1 and $y$ is $[-1000,1000]$ with a scale of 100.

    **b.** Use the $\boxed{\text{TRACE}}$ feature to fill in the table of values below.

| $x$ | $f(x)$ |
|---|---|
| $-4$ | |
| $-1$ | |
| $0$ | |
| $\frac{1}{2}$ | |
| $2$ | |

    **c.** Is the graph of $f$ increasing or decreasing? _____

    **d.** Use the $\boxed{\text{TABLE}}$ feature to look for the horizontal asymptote. Hint: You should look at what happens as $x$ approaches $\infty$ or $-\infty$. What is the equation of the horizontal asymptote? _____

2. Use your calculator to graph $g(x) = 3 - 2^{x+1}$. Make sure to use the minus button between 3 and $2^{x+1}$. You will need to put $x+1$ in parentheses.

    **a.** Set the window so that $x$ is $[-5,5]$ with a scale of 1 and $y$ is $[-20,20]$ with a scale of 2.

    **b.** Use the $\boxed{\text{TRACE}}$ feature to fill in the table of values below.

| $x$ | $g(x)$ |
|---|---|
| $-3$ | |
| $-2$ | |
| $-1$ | |
| $0$ | |
| $1$ | |
| $2$ | |

    **c.** Is the graph of $f$ increasing or decreasing? _____

    **d.** Use the $\boxed{\text{TABLE}}$ feature to look for the horizontal asymptote. Hint: You should look at what happens as $x$ approaches $\infty$ or $-\infty$. What is the equation of the horizontal asymptote? _____

411

# Student Activity
## Interesting Compounding

Paula is going to invest $10,000 in an account that pays 6% annual interest for 20 years. Use the compound interest formula to set up and answer each question below. Make sure you don't round decimals until you have reached the final answer.

| Interest is compounded … | $n =$ | $nt =$ | Write the expression: $A = P\left(1+\dfrac{r}{n}\right)^{nt}$ | $A =$ |
|---|---|---|---|---|
| Annually | | | | |
| Quarterly | 4 | 80 | $A = 10{,}000\left(1+\dfrac{.06}{4}\right)^{80}$ | $32,906.63 |
| Monthly | | | | |
| Weekly | | | | |
| Daily | | | | |
| Hourly | | | | |
| Four times an hour | | | | |

**Can Paula just keep making money at a greater rate if the number of compoundings per year were to increase?**

# Student Activity
## Discovering e

Now we look at the expression $\left(1+\dfrac{x}{k}\right)^{k}$ ... Calculate the value of the expression for each of the blanks in the table. Round each result to the nearest hundredth.

| | $x = 1$ | $x = 2$ | $x = 3$ |
|---|---|---|---|
| $k = 10$ | | | |
| $k = 50$ | | $\left(1+\dfrac{2}{50}\right)^{50} \approx 7.11$ | |
| $k = 100$ | | | |
| $k = 500$ | | | |
| $k = 1,000$ | | | |
| Now evaluate these expressions. | $(2.7183)^{1} \approx$ | $(2.7183)^{2} \approx$ | $(2.7183)^{3} \approx$ |

So, as $k \to \infty$, the expression $\left(1+\dfrac{x}{k}\right)^{k}$ equals what? _____

413

# Student Activity
## Exponential Workout

**Directions:** For each expression, simplify and then find and shade the box with its equivalent form in the grid.

**1.** $\left(e^{\sqrt{2}}\right)^{\sqrt{2}}$

**2.** $\dfrac{2}{\sqrt{2}}$

**3.** $\dfrac{10^{3\sqrt{4}}}{10^{2\sqrt{4}}}$

**4.** $e^{-1}$

**5.** $\dfrac{6^{7}}{6^{3}\cdot 36}$

**6.** $\left(\dfrac{1}{3}\right)^{-1}$

**7.** $\left(e^{2}\right)^{3/2}$

**8.** $\left(\sqrt{13}\right)^{4}$

**9.** $7^{36}7^{-34}$

**10.** $3460\left(\dfrac{1}{50}\right)^{2}$

**11.** $\dfrac{1}{\left(\left|-2\right|\right)^{-99}}$

**12.** $1-\dfrac{e^{0}}{9}$

**13.** $\left(\dfrac{1}{e}\right)^{-1}$

**14.** $\left(5^{2\sqrt{5}}\right)^{3}$

**15.** $\left(9^{4}\right)^{1/2}$

**16.** $0.001\left(10^{3}\right)$

**17.** $10^{x}\cdot 10^{3x}$

**18.** $\left(4e\right)^{1/2}$

**19.** $\dfrac{6^{\sqrt{3}+2\sqrt{2}}}{6^{\sqrt{8}}}$

**20.** $2\left(256\right)^{-1}2^{8}$

**21.** $3^{x}-\dfrac{1}{3^{-x}}$

**22.** $\dfrac{e^{2}}{e^{-3}}$

| 1 | $e$ | $2\sqrt{e}$ | $2^{2\sqrt{2}}$ | $4^{4x}$ | $27^{12}$ | 14 | 99 | 36 | | 2 |
|---|---|---|---|---|---|---|---|---|---|---|
| 3 | -2 | $9^{\sqrt{2}}$ | | $e^{2}$ | | 49 | | $6^{\sqrt{3}}$ | 34 | $e^{3}$ |
| | $5^{6\sqrt{5}}$ | 0 | $\dfrac{1}{2}$ | 5 | $\dfrac{8}{9}$ | $\dfrac{1}{4}$ | $\dfrac{2}{3}$ | 81 | 1.384 | |
| $2^{99}$ | 10 | | | $\dfrac{1}{e}$ | $\dfrac{6^{4}}{36}$ | $10^{4x}$ | | 169 | $3^{2-x}$ | 9 |
| $\sqrt{2}$ | 100 | $8^{x}$ | 9 | $-\dfrac{1}{3}$ | | -4 | $e^{5}$ | $2^{2\sqrt{2}}$ | | |

In this activity we will explore where the function $y = e^x$ fits in with the family of exponential curves we already know.

**1.** Graph $f(x) = 2^x$ and $g(x) = 3^x$ on the axes below.

**2.** Fill in the table of solutions.

| $x$ | $h(x) = e^x$ |
|-----|--------------|
| $-2$ | |
| $-1$ | $\approx 0.37$ |
| $0$ | |
| $1$ | |
| $2$ | |

**3.** Plot the solutions from the table on the same graph as the functions in the first problem. Draw a smooth curve to represent $h(x) = e^x$.

**4.** What is the numerical approximation for $e$? _____

**5.** Why is the resulting graph not surprising?

No going back to second,
no going on to third...
I'm stuck at e !

415

# Student Activity

## Modeling with Exponential Equations

**Table 1 Directions:** Fill in the missing information in the table below. The first one has been done for you.

**Table 2 Directions:** Set up an equation to solve each of the problems.

$A = A_0 e^{kt}$

$A$ = amount after time $t$

$A_0$ = initial amount

$k$ = growth or decay rate

*It's a Mother's Day present from my family.*
*It represents my kids ... and my grandkids ... and my great grandkids ...*
*my great great grandkids ... um ... what month is it?*
*... oh yeah, and my great, great, great grand kids*

| Exponential model | Description | Growth Rate ( $r$ ) | Initial Amount ( $P$ ) |
|---|---|---|---|
| $A = 10,000e^{-0.02t}$ | The population of a city is currently 10,000, but the population is decreasing by 2% per year. | −0.02 (decreasing) | 10,000 |
| | Chris deposits $5,000 into an account that pays 7.5% annual interest, compounded continuously. | | |
| | The population of deer on an island is 250 and is increasing at a rate of 4% per year. | | |
| | The amount of radioactivity in a soil sample is $R_0$, and the radioactivity decreases at a rate of 18% per year. | | |

| Equation | Description | Define the Missing Variable |
|---|---|---|
| | Jill's savings account just reached $10,000. The account pays 7.2% interest, compounded continuously. Jill made one deposit into the account 8 years ago. What was the initial deposit? | $P$ = the initial deposit |
| | The population of Tigerlily is 4,234. The city council projects a decline of 5% annually due to poor economic conditions. In how many years will the population of Tigerlily be 4,000? | |
| | The number of bacteria in a petri dish is 500 at the beginning of an experiment. At the end of 1.5 hours, the population has doubled. What is the hourly growth rate? | |

# Assess Your Understanding
## Exponential Functions

For each of the following, describe the strategies or key steps that will help you **start** the problem. You do **not** have to complete the problems.

| | | What will help you to start this problem? |
|---|---|---|
| **1.** | Simplify: $\left(2^{\sqrt{3}}\right)^{2\sqrt{3}}$ | |
| **2.** | A deposit of $5,000 is left in an account that pays 10.5% annual interest, compounded continuously. How much will be in the account after 8 years? | |
| **3.** | Graph $f(x) = 5^x$ and write the equation for the asymptote. | |
| **4.** | Simplify: $5 - \left(2\right)^{-1}$ | |
| **5.** | An account has $6,000 in it after 5 years. The interest rate on the account is 2.4% compounded monthly. How much was initially deposited? | |
| **6.** | The initial population of Dogstar is 450, and it is increasing at an annual rate of 8%. What will the population be in 5 years? | |
| **7.** | Graph $f(x) = e^x - 5$ and write the equation for the asymptote. | |
| **8.** | Simplify: $2\left(3\right)^{-2}$ | |
| **9.** | Graph $f(x) = \left(\dfrac{3}{4}\right)^x + 1$ and write the equation for the asymptote. | |
| **10.** | How can you graph $f(x) = 2^{-x}$ given the graph of $y = 2^x$? | |

# Metacognitive Skills
## Exponential Functions

Metacognitive skills refer to the ability to judge how well you have learned something and to effectively direct your own learning and studying. This is a self-evaluation tool designed to help you focus your studying and to improve your metacognitive skills with regards to this math class.

Fill the $1^{st}$ column out **before** you begin studying. Fill the $2^{nd}$ column out after you study for your test.

Go back to this assessment after your test and circle any of the ratings that you would change – this identifies the "disconnects" between what you **thought** you knew well and what you **actually** knew well.

**Use the scale below to assign a number to each topic.**
**5** *I am confident I can do any problems in this category correctly.*
**4** *I am confident I can do most of the problems in this category correctly.*
**3** *I understand how to do the problems in this category, but I still make a lot of mistakes.*
**2** *I feel unsure about how to do these problems.*
**1** *I know I don't understand how to do these problems.*

| Topic or Skill | Before Studying | After Studying |
|---|---|---|
| Correctly using exponent rules to simplify expressions. | | |
| Evaluating exponential expressions with irrational exponents. | | |
| Graphing exponential functions with translations. | | |
| Graphing basic exponential functions, including $y = e^x$. | | |
| Identifying the asymptote of the graph of an exponential function, its domain and its range. | | |
| Solving application problems involving the compound interest formula or the continuously compounded interest formula. | | |
| Evaluating exponential expressions involving base-$e$. | | |
| Writing an exponential model to represent a real-world growth or decay problem. | | |

# LOG: Logarithms and More

# Student Activity

## Why Logarithms?

**Directions:** Solve each of the equations. For each equation, note the operation that is performed on $x$ and what the inverse of that operation is. There may be problems that you have not been taught how to solve. If you see one, say so.

| | |
|---|---|
| **1.** Solve: $x + 3 = 12$ <br> $x + 3 - 3 = 12 - 3$ <br> $x = 9$ | Operation on $x$: *Addition of 3* |
| | Inverse Operation on $x$: *Subtraction of 3* |
| **2.** Solve: $3x = 12$ | Operation on $x$: |
| | Inverse Operation on $x$: |
| **3.** Solve: $x^3 = 12$ | Operation on $x$: |
| | Inverse Operation on $x$: |
| **4.** Solve: $3^x = 12$ | Operation on $x$: |
| | Inverse Operation on $x$: |
| **5.** Solve: $\sqrt{x} = 12$ | Operation on $x$: |
| | Inverse Operation on $x$: |
| **6.** Solve: $\dfrac{x}{2} = 12$ | Operation on $x$: |
| | Inverse Operation on $x$: |
| **7.** Solve: $x^{3/4} = 12$ | Operation on $x$: |
| | Inverse Operation on $x$: |
| **8.** Solve: $2^x = 12$ | Operation on $x$: |
| | Inverse Operation on $x$: |

Which problems are we stuck on? Why?

# Student Activity
## Paint by Equivalent Equations

**Directions:** Write each logarithmic equation as its corresponding exponential equation. Write each exponential equation as its corresponding logarithmic equation. Shade in your answers in the grid to reveal the picture. The first one has been done for you.

**1.** $5^3 = 125$
$\log_5 125 = 3$

**2.** $2^x = 8$

**3.** $\log x = 2$

**4.** $\log_4 64 = 3$

**5.** $10^x = \dfrac{1}{100}$

**6.** $16^x = 2$

**7.** $\log_3 x = -1$

**8.** $\log 1 = 0$

**9.** $4^{-x} = 2$

**10.** $\left(\dfrac{1}{2}\right)^x = 4$

**11.** $x^2 = 9$

**12.** $\log 1000 = x$

**13.** $\log_x 25 = 2$

**14.** $x^0 = 1$

**15.** $\left(\dfrac{1}{9}\right)^x = 3$

| $\log_x 9 = 2$ | $\log_2 9 = x$ | $\log_x 3 = \dfrac{1}{9}$ | $\log_5 125 = 3$ | $\log_1 x = 0$ | $\log \dfrac{1}{100} = x$ |
|---|---|---|---|---|---|
| $10^2 = x$ | $2^{10} = x$ | $x = \log_8 2$ | $\log_2 4 = -x$ | $\log_{1/2} 4 = x$ | $\log x = \dfrac{1}{100}$ |
| $\log_{1/9} 3 = x$ | $\log_x 1 = 0$ | $\log_2 8 = x$ | $\log_4 2 = -x$ | $\log_4 x = 4$ | $10^0 = 1$ |
| $3^{-1} = x$ | $x^{-1} = 3$ | $4^3 = 64$ | $3^4 = 81$ | $\log_4 \dfrac{1}{2} = x$ | $\log_3 x = \dfrac{1}{9}$ |
| $10^x = 1000$ | $x^2 = 25$ | $\log_{16} 2 = x$ | $\log_2 16 = x$ | $25^x = 2$ | $2^x = 25$ |

# Guided Learning Activity
## Unraveling Exponentials

Let $f(x) = b^x$, where $b > 0$ and $b \neq 1$. Since $f$ is a one-to-one function, we define notation for the inverse: $f^{-1}(x) = \log_b(x)$.

For example: $f(x) = 10^x$ and $f^{-1}(x) = \log_{10} x$ are inverses.

$f(x) = 4^x$ and $f^{-1}(x) = \log_4 x$ are inverses.

By the definition of inverses, $f\left(f^{-1}(x)\right) = x$, and so $b^{\log_b(x)} = x$.

Also by the definition of inverses, $f^{-1}\left(f(x)\right) = x$, and so $\log_b\left(b^x\right) = x$.

**Part I:** Write the corresponding function, inverse, or property to fill in the table below:

| Function | Inverse | $b^{\log_b(x)} = x$ | $\log_b\left(b^x\right) = x$ |
|---|---|---|---|
| $f(x) = 3^x$ | $f^{-1}(x) = \log_3(x)$ | $3^{\log_3(x)} = x$ | $\log_3\left(3^x\right) = x$ |
| $f(x) = 5^x$ | | | |
| | | $10^{\log_{10}(x)} = x$ | |
| | $f^{-1}(x) = \log_2(x)$ | | |
| | | | $\log_e\left(e^x\right) = x$ |
| $f(x) = (1/2)^x$ | | | |

> **Inverse Properties of Logarithms**
> $b^{\log_b(x)} = x$ and $\log_b\left(b^x\right) = x$

**Part II:** Circle the expressions below that simplify to be $x$:

$4^{\log_2(x)}$ $\qquad$ $4^{\log_4(x)}$ $\qquad$ $4\log_4(x)$ $\qquad$ $4^{\log(4x)}$ $\qquad$ $\log_4(4x)$ $\qquad$ $\log_4\left(4^x\right)$

$\log_6\left(6^{2x}\right)$ $\qquad$ $\log_6(6+x)$ $\qquad$ $\log_6(6)$ $\qquad$ $\log_6\left(x^6\right)$ $\qquad$ $6^{\log_6(x)}$ $\qquad$ $6^{\log_x(6)}$

Just like we would never write $2^x$ as $^x2$, we also would never write $\log_2(x)$ as $(x)\log_2$.

$2^{(\ )}$ must operate on a number that follows the 2.

$\log_2(\ )$ must operate on a number that follows the $\log_2$.

**Part III:** Circle the expressions with correct mathematical notation below:

$3^x$ $\qquad\qquad$ $x^3$ $\qquad\qquad$ $^3x$ $\qquad\qquad$ $^x3$ $\qquad\qquad$ $x\log_3$

$\log_3 x$ $\qquad\qquad$ $\log_x 3$ $\qquad\qquad$ $3\log_x$ $\qquad\qquad$ $3^{\log_x}$ $\qquad\qquad$ $\log_3 3$

---

Sometimes, we can simplify expressions involving logs or exponentials by rewriting part of the expression. For example, $\log_9 81$ can be rewritten because 81 is the same as $9^2$.

$$\log_9 81 = \log_9(81) = \log_9(9^2) = 2$$

**Part IV:** Use $\log_b b^x = x$ to finish these problems off with a little bit of clever reasoning:

$\log_2 8 = \log_2(\quad) =$

$\log_4 16 = \log_4(\quad) =$

$\log_5\left(\dfrac{1}{5}\right) = \log_5(\quad) =$

$\log_3 \sqrt{3} = \log_3(\quad) =$

$\log_2\left(\dfrac{1}{4}\right) = \log_2(\quad) =$

# Student Activity
## Match Up on Log and Exponential Expressions

**Directions:** Match each of the expressions in the squares in the table below with its simplified form from the top. If the simplified form is not found among the choices A through D, then choose E (none of these). The first one is done for you.

**A** 2 **B** $\frac{1}{2}$ **C** $-2$ **D** 1 **E** 0 **F** None of these

| | | | |
|---|---|---|---|
| $\log_5 \sqrt{5}$<br><br>$\log_5 5^{1/2} = \frac{1}{2}$<br>B | $2^{\log_2 3}$ | $\log \frac{1}{100}$ | $10^{\log 10}$ |
| $2^{\log 1}$ | $\log_9 3$ | $\log_3 9$ | $10^{\log 2}$ |
| $10^{\log_2 10}$ | $\log_{1/2} \left( \frac{1}{4} \right)$ | $\log 1$ | $\log_4 2$ |
| $\log_2 1$ | $\log_4 16$ | $\log_5 5$ | $2^{\log_5 5}$ |
| $\left( \log_5 2 \right)^0$ | $5^{\log_5 2} - \log_2 16$ | $\log 100$ | $\log_9 \frac{1}{81}$ |

# Student Activity
## Log Graphing

**1.** Graph $f(x) = 2^x$ on the axes below using the table of values provided. Use a dashed line to represent the asymptote of the graph of $f$.

| $x$ | $f(x) = 2^x$ |
|-----|--------------|
| −3 | 0.125 |
| −2 | 0.25 |
| −1 | 0.5 |
| 0 | 1 |
| 1 | 2 |
| 2 | 4 |
| 3 | 8 |

| $x$ | $f^{-1}(x) = \log_2 x$ |
|-----|------------------------|
| | |
| | |
| | |
| | |
| | |
| | |
| | |

**2.** The inverse function is written $f^{-1}(x) = \log_2 x$. We can make a table of solutions for $f^{-1}$ by swapping the $x$- and $y$-coordinates of all the solutions. Fill in the table of solutions for $f^{-1}(x) = \log_2 x$.

**3.** Plot the points on the same graph as $f(x) = 2^x$. Then draw a smooth curve.

**4.** What is the equation for the asymptote of $f(x)$? _____
for $f^{-1}(x)$? _____

**5.** Write the domain and range for each function:

$f$: Domain _____

Range _____

$f^{-1}$: Domain _____

Range _____

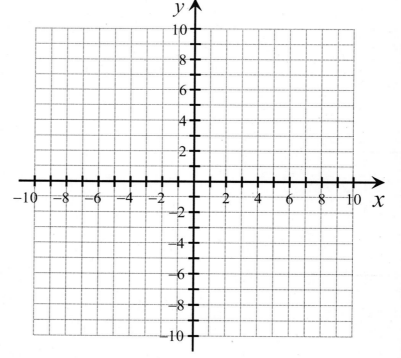

**6.** If it is always true that the point $(0,1)$ is on the graph of $y = b^x$, then what point should always be on the graph of $y = \log_b x$? _____

**7.** If you wanted to graph $y = \log_5 x$, what would be an easy way to do it?

# Guided Learning Activity
**Inverse Power**

> We have **defined** the inverse of the operation $b^{(\ )}$ ($b$ to the ...) to be $\log_b(\ )$ (log base b of ...).

As with all the properties of equality, if we perform an operation on one side of an equation, we must also perform the same operation on the other side of the equation.

**Example 1:** To solve $2^x = 12$, we need to undo the operation $2^{(\ )}$. So we need to use $\log_2(\ )$ on both sides:

$$2^x = 12$$
$$\log_2(2^x) = \log_2(12)$$
$$x = \log_2(12)$$

> Using your calculator, you can evaluate any logarithm using the $\boxed{\text{LOG}}$ key and the
> **Change of Base Formula:** $\log_b(x) = \dfrac{\log(x)}{\log(b)}$

Coach want's you to steal 3rd.

When the logarithm is written without a base it is called a **common log** and stands for $\log_{10}(\ )$.

So, to evaluate $\log_2(12)$, we calculate $\dfrac{\log(12)}{\log(2)}$

and get $\approx 3.585$. We can check this answer by finding $2^{3.585}$, which is $\approx 12.0003 \approx 12$.

**Example 2:** To solve $\log_3 x = 2$, we need to undo the operation $\log_3(\ )$.
So we need to use $3^{(\ )}$ on both sides: $\log_3 x = 2$
$$3^{\log_3 x} = 3^2$$
$$x = 9$$

**Example 3:** To solve $\log x = 1.75$, we need to undo the operation $\log_{10}(\ )$.
So we need to use $10^{(\ )}$ on both sides: $\log x = 1.75$
$$10^{\log_{10} x} = 10^{1.75}$$
$$x \approx 56.23$$

**Directions:** Decide which operation to use on both sides of the equation in order to solve for $x$. Then write out the step of applying the inverse operation. The first two have been done for you. Round decimals to the nearest thousandth.

| | Equation to be solved | Perform this inverse operation on both sides of the equation | First Step | Solve for $x$ |
|---|---|---|---|---|
| **1.** | $3^x = 12$ | $\log_3(\ )$ <br> Take a log-base-3 on both sides. | $\log_3(3^x) = \log_3(12)$ | $x = \dfrac{\log 12}{\log 3} \approx 2.262$ |
| **2.** | $\log_8 x = -1$ | $8^{(\ )}$ <br> Do 8-to-the ... on both sides. | $8^{\log_8 x} = 8^{-1}$ | $x = \dfrac{1}{8}$ |
| **3.** | $\log_3 x = 5$ | | | |
| **4.** | $5^x = 9$ | | | |
| **5.** | $(1/2)^x = 8$ | | | |
| **6.** | $\log_6 x = \dfrac{1}{2}$ | | | |
| **7.** | $0.432 = \log_{10} x$ | | | |
| **8.** | $10 = 4^x$ | | | |
| **9.** | $10^x = 140$ | | | |
| **10.** | $100 = 2^x$ | | | |

# Student Activity

## Split the Log by Solving Equations

**Directions:** Solve each of the equations below and cross out the solutions as you go (if necessary, round to two decimal places). Every solution should be found in the log below. Round decimals to the nearest hundredth. The first one is done for you.

**1.** $3^x = 8$

**2.** $\log x = 2$

**3.** $5^x = 20$

**4.** $\log_3 x = 5$

**5.** $30 = 10^x$

**6.** $5.4 = \log x$

**7.** $9 = 2^x$

**8.** $\log_5 x = 2.7$

**9.** $10^x = 1000$

**10.** $0.002 = \log_2 x$

1.89    251,189    1.001

100    77.13

243    3.17    1.48

3    1.86

# Student Activity
## Isolation Island

**Directions:** The first step to solving each of the equations below is to isolate the log or exponential part of the equation. Do this and you should find the simplified equation on the island.

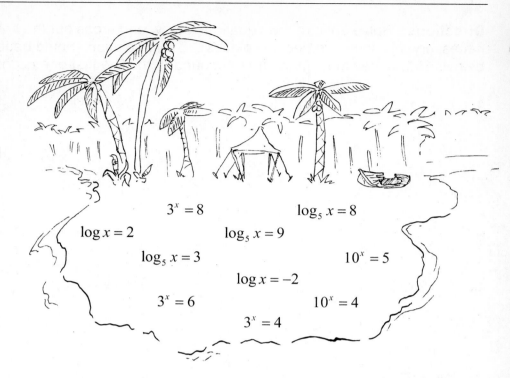

$$3^x = 8$$
$$\log_5 x = 8$$
$$\log x = 2$$
$$\log_5 x = 9$$
$$\log_5 x = 3$$
$$10^x = 5$$
$$\log x = -2$$
$$3^x = 6$$
$$10^x = 4$$
$$3^x = 4$$

1. Solve: $2 + \log_5 x = 10$

2. Solve: $4 \cdot 3^x = 24$

3. Solve: $3 - \log x = 5$

4. Solve: $10^x - 2 = 3$

5. Solve: $26 = 6 \cdot 3^x + 2$

6. Solve: $4 + 3\log_5 x = 13$

7. Solve: $5 - 3^x = -3$

8. Solve: $9\log x - 4 = 14$

9. Solve: $10^2 + 10^x = 104$

10. Solve: $\log_5 x + \log_5 1 = 9$

# Student Activity
## Shifty Logarithms

Complicated-looking variants of the $\ln(x)$ and $e^x$ functions are often just simple **translations** of the function: that is, the original function shifted vertically or horizontally along the *y* or *x* axis, respectively.

Graph each set of functions below on the grid provided, and describe the translation(s).

1. $f(x) = e^x$

2. $g(x) = e^x + 2$

   Graph (2) is the same as $f(x)$

   shifted _____ 2 units.

3. $h(x) = e^{x+2}$

   Graph (3) is the same as $f(x)$

   shifted _____ 2 units.

4. $p(x) = e^{x+2} + 2$

   Graph (4) is the same as $f(x)$

   shifted _____ and _____ 2

   units.

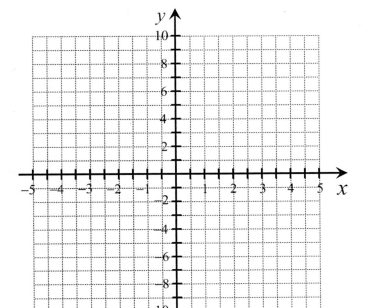

5. $f(x) = \ln(x)$

6. $g(x) = \ln(x-1)$

   $g(x)$ is the same as $f(x)$

   shifted _____ 1 unit.

7. $h(x) = 3 + \ln(x)$

   $h(x)$ is the same as $f(x)$

   shifted _____ 3 units

8. $p(x) = 3 + \ln(x+1)$

   $p(x)$ is the same as $f(x)$

   shifted up ___ units and left ___

   units.

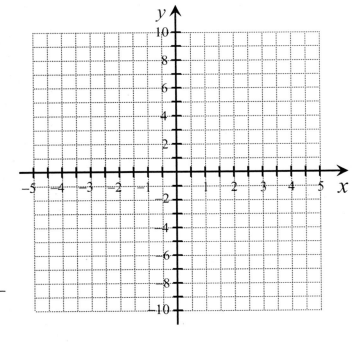

# Student Activity
## Compounding the Problem

**Directions:** First set up an equation to help you solve for the missing information in the table below. Then solve each equation for the missing variable using the space provided at the bottom of the page. If $k$ is infinity, use the formula for continuously compounded interest.

| | $A$ | $P$ | $k$ | $r$ | $t$ | Equation |
|---|---|---|---|---|---|---|
| **1.** | | $2,000 | 12 | 8% | 15 | |
| **2.** | $9,500 | | 4 | 5.9% | 8 | |
| **3.** | $6,000 | | $\infty$ | 6.4% | 5 | |
| **4.** | $8,000 | $6,000 | $\infty$ | 8.2% | | |
| **5.** | $12,000 | $10,000 | 4 | | 6 | |
| **6.** | $1,250 | $1,000 | 12 | 7.15% | | |
| **7.** | $2P_0$ | $P_0$ | $\infty$ | | 10 | |
| **8.** | $3A_0$ | $A_0$ | 52 | 10.4% | | |

## Persnickety Log Notation

**Directions:** Fill in the missing pieces to demonstrate your prowess in using logarithm language and notation!

| | In words | Abbreviated Notation (ln$x$ or log$x$) | Notation with the base indicated |
|---|---|---|---|
| 1. | natural log of $x$ | $\ln x$ | $\log_e x$ |
| 2. | log of the sum of 4 and $x$ | | |
| 3. | | $\ln\left(\dfrac{x}{y}\right)$ | |
| 4. | | | $\log_e(x-2)$ |
| 5. | 10 to the log of 4 | | |
| 6. | | $\ln e^{-3}$ | |
| 7. | | | $e^{\log_{10} x}$ |
| 8. | | $\log(5x)$ | |
| 9. | 2 to the negative log of $x$ | | |
| 10. | the quantity 10 to the natural log of $x$, squared | | |

I need a chainsaw, not a calculator.

# Student Activity

## Interesting Comparison

**1.** Jake invests $5,000 in Trustworthy Bank, which pays an annual interest rate of 5%, compounded **quarterly**, find the amount in his bank account **every ten years** by completing the table below:

| $k =$ | $t$ | $k\,t$ | Write the equation for the interest calculation: | $A$ |
|---|---|---|---|---|
| | 10 | | | |
| | 20 | 80 | $A = 5{,}000\left(1 + \dfrac{0.05}{4}\right)^{80}$ | $13,507.42 |
| | 30 | | | |
| | 40 | | | |

**2.** Jake's friend Bob tells him that his bank, Dependable & Reliable Trust, pays interest compounded continuously. Jake is skeptical but decides to make the calculations to compare the difference between the two. Complete the table below for the same investment at the same interest rate, only this time compounding continuously. Should Jake switch?

| $t$ | $k\,t$ | Write the equation for the interest calculation: | $A$ |
|---|---|---|---|
| 10 | | | |
| 20 | | | |
| 30 | | | |
| 40 | | | |

**3.** If Jake invests at Dependable & Reliable Trust, how long will it take for his investment to double?

# Student Activity
## Simply Logarithmic

**Directions: Simplify each expression** and shade in the equivalent simplified expression. **Solve each equation** and shade the solution. Only shade one square for each problem. The first one has been done for you.

1. $\log_3 3^2 = 2$

2. $6^{\log_6 4}$

3. $\ln e^9$

4. Solve: $2^{\log_2 2x} = 5$

5. $1 + \log 10 + 1$

6. Solve: $10^{\log 16x} = 16$

7. $\log_x 7 = 1$

8. $\log(\log 10)$

9. $\frac{1}{2}\ln e^{4x}$

10. $e^{\ln e}$

11. $5\log_8 8$

12. $\log_9 \dfrac{1}{729}$

13. $3\log_3 9$

14. $4x + \ln e^{-x}$

15. $\ln 1 + \log_3 3^{-1} + \log_3 3$

16. $\log_2 8^{y/3}$

17. $10^{\log e^2}$

18. Solve: $3 = \log_x 1000$

19. $\left(\log_3 9\right)^{-1}$

20. $\log_2 \dfrac{1}{2} + \log_4 \dfrac{1}{4}$

| | | | | | | | | | | |
|---|---|---|---|---|---|---|---|---|---|---|
| *e* | $-4$ | $1+\log 11$ | $112$ | $\log_3 4$ | $100$ | $e-2$ | | $5$ | | $1$ |
| | $11$ | | $8$ | $\frac{1}{2}$ | $3$ | $2$ | $-4$ | $y$ | $\frac{7}{3}$ | $e^2$ |
| $9$ | $-8$ | $\frac{1}{4}$ | $-9$ | | $-10$ | $-1$ | $-99$ | $0$ | $4$ | $-3$ |
| $2x$ | | $6$ | $e^{1/6}$ | $7$ | $10$ | | $\ln 9$ | | $\frac{2}{5}$ | $\frac{5}{2}$ |
| $100$ | $x-1$ | | $\frac{1}{5}$ | | $8$ | $e^{2x}$ | $12$ | $3x$ | $-2$ | |

# Guided Learning Activity

**Become a Believer**

**Directions:** Follow the directions for each step and fill in the boxes with the missing pieces as you go. Careful – as you move from line to line, the boxes might contain different expressions! At the end, you should get a formula that you might recognize!

## Derive the Product Rule for Logarithms:

Let $M = b^x$ and $N = b^y$.

$MN = \boxed{\phantom{xx}} \cdot \boxed{\phantom{xx}}$   Solve $M = b^x$ for $x$.   Solve $N = b^y$ for $y$.

$MN = b^{\boxed{\phantom{xx}}}$   $x = \boxed{\phantom{xxxxx}}$   $y = \boxed{\phantom{xxxxx}}$

Now we take $\log_b (\ )$ on both sides of the equation.

$$\log_b (MN) = \log_b \left( \boxed{\phantom{xxxx}} \right)$$

Simplify on the right side:

$$\log_b (MN) = \boxed{\phantom{xxx}}$$

Replace $x$ and $y$ with the equivalent expressions found above:

$$\log_b (MN) = \boxed{\phantom{xxxxx}} + \boxed{\phantom{xxxxx}}$$

## Derive the Quotient Rule for Logarithms:

Let $M = b^x$ and $N = b^y$. From above, we know: $x = \boxed{\phantom{xxxxx}}$ and $y = \boxed{\phantom{xxxxx}}$.

$$\frac{M}{N} = \frac{\boxed{\phantom{x}}}{\boxed{\phantom{x}}} \quad \rightarrow \quad \frac{M}{N} = b^{\boxed{\phantom{xxx}}}$$

Now we take $\log_b (\ )$ on both sides of the equation.

$$\log_b \left( \frac{M}{N} \right) = \log_b \left( \boxed{\phantom{xxxx}} \right)$$

Simplify on the right side:

$$\log_b \left( \frac{M}{N} \right) = \boxed{\phantom{xxx}}$$

Replace $x$ and $y$ with the equivalent expressions found above:

$$\log_b \left( \frac{M}{N} \right) = \boxed{\phantom{xxxxx}} - \boxed{\phantom{xxxxx}}$$

# Student Activity
## Pesky Product and Quotient Rules

*Ever since I got this new log splitter the savings on my heating bill have been exponential!*

A **product inside** the logarithm = A **sum outside** the logarithms.

$$\log(MN) = \log M + \log N \qquad \ln(MN) = \ln M + \ln N$$

A **quotient inside** the logarithm = A **difference outside** the logarithms.

$$\log\left(\frac{M}{N}\right) = \log M - \log N \qquad \ln\left(\frac{M}{N}\right) = \ln M - \ln N$$

**Step 1:** Use the product and/or quotient rules for logarithms to write each logarithm in expanded form, then simplify if possible. Write your answer in the second shaded column, and be careful not to leave exponents or radical symbols in your answer.

**Step 2:** Substitute $x = 2$ and $y = 3$ in both expressions, and use your calculator to verify that you have simplified correctly.

| Original Expression | Simplified Expression | Substitute $x=2$ and $y=3$ for original expression | Substitute $x=2$ and $y=3$ for simplified expression | Match? |
|---|---|---|---|---|
| $\log 9x$ | $\log 9 + \log x$ | $\log 9 \cdot 2 \approx 1.26$ | $\log 9 + \log 2 \approx 1.26$ | ✓ |
| $\log \dfrac{3x}{4}$ | | | | |
| $\ln xy^3$ | | | | |
| $\log \dfrac{10}{xy}$ | | | | |
| $\ln \dfrac{10^3}{2x}$ | | | | |
| $\log \sqrt{xy}$ | | | | |

# Student Activity
## Escape the Properties Matrix

**Directions:** Begin at the box marked START. Without a calculator, look for pairs of expressions that are equivalent. Shade the matched pairs to reveal the path to escape the matrix! If you get really stuck, use your calculator to check the equivalence of the pairs you have already shaded.

Example: $\ln 5 + \ln 7 = \ln(5 \cdot 7) = \ln 35$

| START $\ln 5 + \ln 7$ | $\ln 35$ | $\ln \dfrac{1}{4}$ | $\ln 2 + \ln 8$ | $\dfrac{\log 3}{\log 5}$ |
|---|---|---|---|---|
| $\ln 12$ | $\log \dfrac{5}{20}$ | $\log 5 - \log 20$ | $\ln 2^8$ | $\log_3 5$ |
| $\ln 3 + \ln 9$ | $\dfrac{\log 5}{\log 20}$ | $\log(5 - 20)$ | $8 \ln 2$ | $5 \ln 20$ |
| $\log e^3$ | $\log 3^2$ | $\ln 1$ | $\log 1$ | $\ln 5^{20}$ |
| $3$ | $2 + \log 3$ | $\log 100 + \log 3$ | $\log 103$ | $e^{\ln 4}$ |
| $\ln 2$ | $\ln 5 - \ln 10$ | $-\ln 5$ | $\log e$ | $\log e^4$ |
| $\log 50 - \log 5$ | $\ln \dfrac{1}{2}$ | $\log 2$ | $\ln e$ | $\ln \dfrac{1}{5}$ |
| $\log 45$ | $\log 4 + \log 4$ | $2 \log 4$ | $\log 10$ | $-\ln 5$ |
| $\log 4 + \log 5$ | $\log 4^4$ | $\ln 3 + \ln 12$ | $\ln 15$ | **ESCAPE** the Properties Matrix |

# Student Activity
## Escape the Logjam

**Directions:** Begin at the box marked START. Either solve the equation or simplify the expression. Look for the result in the movement grid to the right. If your result is not in the movement grid, that's a BAD sign. For example, the answer to $3^x = 7$ is $x \approx 1.7712$, so we move to the right. Next solve $\log(3 - x) = 3$.

| Move Left | Move Right | Move Down | |
|---|---|---|---|
| 0.0025 | ~~1.7712~~ | 3.4190 | 1.0986 |
| 23.1331 | 0.4474 | 0.7906 | 8 |
| | 2.4022 | 2 | −2 |
| | 0.2091 | −997 | 10 |
| | 4.3774 | | |

| START $3^x = 7$ $\rightarrow \rightarrow \rightarrow$ | $\log(3 - x) = 3$ | $e^0 + e^x$ |
|---|---|---|
| $4^x = 16$ | $4^{x-2} = 27$ | $\log_5 25 = x$ |
| $e^{-2.4t} = 14.2$ | $e^{2.8x} = 3.5$ | $\log_2(x - 7) + \log_2 x = 3$ |
| $e^{3x} = 60$ | $\log 8 - \log x = 2$ | $2^{x+2} = 3^x$ |
| $3^{x^2 - 3x + 4} = 81$ | $2^x = 16$ | $\log 3x = \log 6$ |
| $\log 5 - \log 2x = 0.5$ | $5^x = 7^{x-4}$ | $\log \dfrac{2x}{5} = -3$ |
| $e^{2x} = 9$ | $\ln x = 5$ | $\log_2 5x - \log_2 3 = 4$ |
| $3^x - 11 = 3$ | $5^{x+1} = 7$ | $\log x^{900} = 900$ |
| $\dfrac{1}{3}\log(3x + 5) = \log x$ | $\ln 5x = 4$ | $6^{x+4} = 36$ |
| $3^{9y-3} = 1$ | $z + \ln e^{-z}$ | **ESCAPE** the logjam |

# Student Activity
## Popular Applications

Population growth is a process that we can describe using the exponential growth model $P = P_0 e^{kt}$, where $P$ is the population at some time $t$, $P_0$ is the initial population at $t = 0$, and k depends on the rate of growth.

Assume that a city San Populato had an initial population of 100, and that the city doubled its population in the first 5 years.

**1.** First, use this information to find k *(round to four decimal places)*.

$$P =$$
$$P_0 =$$
$$t =$$

$$k \approx \underline{\hspace{2in}}$$

**2.** Write the population growth model that gives the population at any given time.

**3.** Use the model to calculate the population of San Populato after 30 years.

**4.** Calculate the time needed for San Populato's population to grow to 1000 people.

**5. Put it all together:** In another remote part of the Country, the poor city of Lesgo is struggling to survive. A series of natural disasters is causing a steady flow of disgruntled citizens to leave the city. In fact, the situation is so bad that Lesgo lost half of its population in one year. If evacuation continues at its current negative exponential rate, and the city's initial population was about 1000 persons, approximately how much longer does the city have before only one lonely person is left?

# Student Activity
## Decaying Applications

Living organisms consume a variety of carbon isotopes in their supply of nutrients, including $^{14}C$. The percentage of carbon that is $^{14}C$ in living organisms remains constant because it is always being renewed. However, when a living organism dies, it no longer consumes fresh carbon. At this point, the percentage of carbon that is made up of the isotope $^{14}C$ begins to decrease as the $^{14}C$ in the organism decays.
By measuring the amount of Carbon-14 present in a dead organism, and comparing it with the percentage that would be in a live organism, scientists and archeologists can determine the date when it died. The half-life of Carbon-14 is 5,700 years; this means after 5,700 years, there will only be half of whatever amount of $^{14}C$ the sample had initially.

Radioactive decay is a process that we can describe using the exponential decay model $A = A_0 \cdot 2^{-t/h}$, where $A$ is the amount of radioactive material present at time $t$, $A_0$ is the amount of radioactive material present at $t = 0$, and $h$ is the material's half-life.

1. Write a model for the decay of Carbon-14.

2. If 2/3 of the initial estimated amount of Carbon-14 is present in a sample, how many years old is the sample? Hint: Replace $A$ with $\frac{2}{3}A_0$ and solve for $t$.

3. If 75% of the initial estimated amount of Carbon-14 has **decayed**, how many years old is the sample? Hint: Don't replace $A$ with $0.75A_0$.

4. A professor is examining a museum skeleton that he suspects might be a fake, so he takes a measurement of the Carbon-14 remaining in one of the suspect bones of the specimen. From this measurement, he determines that 65% of the original $^{14}C$ remains in the sample. The skeleton is estimated to be 5,000 years or older. Perform your own calculation of the age of the skeleton, and then decide if the skeleton might be a fake.

© 2011 Cengage Learning. All Rights Reserved. May not be scanned, copied or duplicated, or posted to a publicly accessible website, in whole or in part.

# Assess Your Understanding

## Logarithms and More

For each of the following, describe the strategies or key steps that will help you **start** the problem. You do **not** have to complete the problems.

| | | What will help you to start this problem? |
|---|---|---|
| **1.** | Write $\log_2 x + \log_2 5$ as a single logarithm. | |
| **2.** | Simplify: $10^{\log(4x)}$ | |
| **3.** | Expand the expression $\ln\left(\dfrac{e}{x}\right)^2$ and then simplify. | |
| **4.** | Simplify: $\ln\dfrac{1}{e^3}$ | |
| **5.** | Solve: $3 + 2\log x = 9$ | |
| **6.** | Solve: $5^x = 2^{x+3}$ | |
| **7.** | If $f(x) = 3^x$, what is $f^{-1}(x)$? | |
| **8.** | Solve: $800 = 1000e^{-0.04x}$ | |
| **9.** | How are the graphs of $f(x) = \log x$ and $g(x) = \log(x+3)$ related? | |
| **10.** | Solve: $\log_3 x + \log_3 6 = 4$ | |

# Metacognitive Skills
## Logarithms and More

Metacognitive skills refer to the ability to judge how well you have learned something and to effectively direct your own learning and studying. This is a self-evaluation tool designed to help you focus your studying and to improve your metacognitive skills with regards to this math class.

Fill the 1<sup>st</sup> column out **before** you begin studying. Fill the 2<sup>nd</sup> column out after you study for your test.

Go back to this assessment after your test and circle any of the ratings that you would change – this identifies the "disconnects" between what you **thought** you knew well and what you **actually** knew well.

**Use the scale below to assign a number to each topic.**
**5**  *I am confident I can do any problems in this category correctly.*
**4**  *I am confident I can do most of the problems in this category correctly.*
**3**  *I understand how to do the problems in this category, but I still make a lot of mistakes.*
**2**  *I feel unsure about how to do these problems.*
**1**  *I know I don't understand how to do these problems.*

| Topic or Skill | Before Studying | After Studying |
|---|---|---|
| Understanding why and when it is necessary to use logarithms. | | |
| Simplifying logarithmic expressions. | | |
| Simplifying expressions in the form $\log_b b^x$ or $b^{\log_b x}$. | | |
| Graphing basic logarithmic functions, including $y = \ln x$. | | |
| Graphing logarithmic functions with translations. | | |
| Identifying the asymptote of the graph of a logarithmic function, its domain and its range. | | |
| Solving a simple exponential or logarithmic equation. | | |
| Applying the product, quotient, and power rules for logarithms to expand a logarithmic expression. | | |
| Applying the product, quotient, and power rules for logarithms to write an expanded log expression as a single logarithm. | | |
| Solving a logarithmic equation that requires multiple steps or uses the application of log properties. | | |
| Solving an exponential equation that requires multiple steps. | | |
| Solving a logarithmic equation where the product or quotient rule for logarithms needs to be used first. | | |
| Solving application problems involving population growth or decay. | | |
| Solving application problems involving the Richter scale or sound. | | |
| Solving application problems involving pH. | | |
| Solving application problems involving radioactive decay. | | |
| Solving application problems involving continuously compounded interest, including problems with doubling times. | | |

# CNC: Conics

# Student Activity
## Constructing a Circle

Constructing a Circle

**Supplies needed for one construction:**
- Bulletin Board
- Tacks (3)
- Unlined paper (the bigger the better)
- String (length approx half the width of the paper)
- Crayon

**Definition of a Circle:** A **circle** is the *set of all points* in a *plane* that are *equidistant* from a *fixed point* called its **center**. The fixed distance is called the **radius** of the circle.

## Construct a Circle:

1. Use two tacks to affix your paper to the bulleting board.
2. Tie a loop on one end of the string around the crayon (near the tip).
3. Use the last tack to affix the other end of the string to the piece of paper (near the center of the paper).
4. Move the crayon to trace a curve, keeping the string taut.
5. Remove your drawing from the bulletin board.

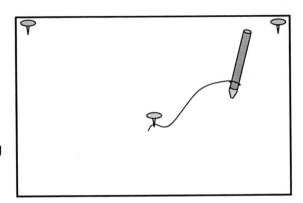

## Add labels:

- The crayon traced the *set of all points* equidistant from the center. In other words, the crayon traced a circle. Label your diagram **Circle**.
- The *fixed point* or *center* is represented by the hole where the center tack was. Label this point with the word **Center** and the point **(h, k)**.
- The *fixed distance* or *radius* is represented by the taut string. Draw a straight line from the center to the drawn circle. Label this line **radius** and the point where the line intersects the circle with **(x, y)**.

447

# Student Activity

**Triple Play on Circles**

**Instructions:** In the table below, you have been given some of the information on a circle: the graph, the equation in standard form, or a description of a circle. Fill in the missing information to complete the table (and the triple play).

**Game 1**

**Equation:**

$$(x-4)^2 + (y-6)^2 = 36$$

**Description:**

**Game 2**

**Equation:**

**Description:**

The circle passes through (2,1).

**Game 3**

**Equation:**

**Description:**

Circle with center at (3, -2) and radius 7.

**Game 4**

**Equation:**

**Description:**

# Student Activity
## Constructing a Parabola

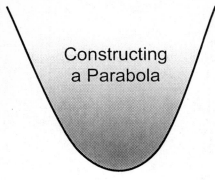

**Supplies needed for one construction:**
- A dark marker
- Straightedge
- Unlined paper for each student

**Definition of a Parabola:** A **parabola** is the set of all points in a plane that are equidistant from a fixed point, called the **focus**, and a fixed line, called the **directrix**.

### Construct a Parabola:

1. Towards the edge of the blank sheet of paper, use the straightedge and the dark marker to draw a line, and then draw a point off of the line. The line and point need to be dark enough that you can see them through the back side of the paper.

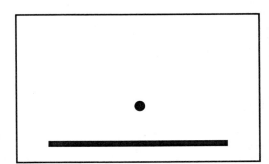

2. Fold the paper so that the point is on the line. Then make sure there is a good crease on the fold of the paper.
3. Unfold and refold the paper so that the point is in a different place on the line. Again, make sure there is a good crease on the fold.
4. Repeat the folding and creasing until you can see a curve formed by the creases in the paper.

### Add labels:

- The *set of all points equidistant from a fixed point and a fixed line* are seen in the shape of a *parabola* made by the creases of the folds. Draw in the curve made by the creases and label this curve **parabola**.

- Label the fixed point with **focus**.

- Label the fixed line with **directrix**.

- Draw a dashed line between the focus and the directrix so that the line is perpendicular to the directrix (this is the shortest line you can draw between the focus and the directrix. Extend the line past the point so that it marks the line of symmetry on the parabola. Label this line **axis of symmetry.**

- Darken the point where the dashed line intersects the parabola curve. Label this point the **vertex** and also label it with the point (***h, k***).

# Student Activity
## Most Wanted: The Axis of Symmetry

MOST WANTED

VERTEX    Y-Intercept    DIRECTRIX

$$X = \frac{1}{4}(y-2)^2 + 3$$

It's a clue to the whereabouts of our Arch Enemy – The Axis of Symmetry

**Directions:** In each "file" that follows, you have been given some of the clues about a parabola. Fill in the missing information to complete the file.

**File One**

**Graph:**

**Equation:**

**Description:**

(6, 2)

## File Two

**Graph:**

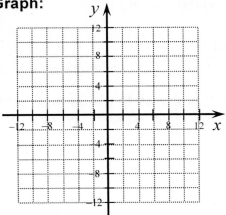

**Equation:**

**Description:**

Vertex at $(4,3)$, $a = 3$, and axis of symmetry at $x = 4$.

## File Three

**Graph:**

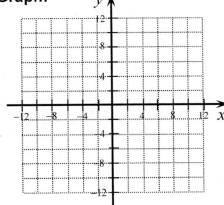

**Equation:**

$$x = \frac{1}{4}(y-2)^2 + 3$$

**Description:**

## File Four

**Graph:**

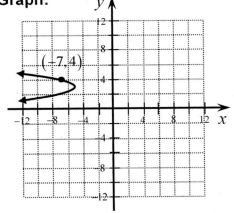

$(-7,4)$

**Equation:**

**Description:**

Vertex at $(4,3)$, $a = 3$, and axis of symmetry at $x = 4$.

# Guided Learning Activity
## Completing the Square for Conics

| General Form of a Circle | Standard Form of a Circle |
|---|---|
| $x^2 + Bx + y^2 + Dy + E = 0$ | $(x-h)^2 + (y-k)^2 = r^2$ |

Complete the Square →

Multiply and simplify ←

We can convert from the standard form of the equation of a circle to the general form by multiplying (squaring) the binomials, simplifying, and setting the equation equal to zero

$$(x-2)(x-2) + (y-3)(y-3) = 25$$

◁ Multiply and simplify   $(x-2)^2 + (y-3)^2 = 5^2$

$$x^2 - 4x + 4 + y^2 - 6y + 9 = 25$$

$$x^2 - 4x + y^2 - 6y - 12 = 0$$

But what if we want to convert the equation that is in general form back to the standard form? We can do it by _completing the square_.

$$x^2 + y^2 + 10x - 12y + 52 = 0$$

Complete the Square ▷   $x^2 + 10x + \underline{\phantom{25}} + y^2 - 12y + \underline{\phantom{36}} = -52$

$$x^2 + 10x + \underline{25} + y^2 - 12y + \underline{36} = -52 + 25 + 36$$

$$(x+5)^2 + (y-6)^2 = 9$$

**Directions:** Fill in the missing spaces in the table below by either completing the square, or by multiplying and simplifying to get the proper form of the circle equation.

| | Circle Equation in General Form | Circle Equation in Standard Form |
|---|---|---|
| 1. | $x^2 + y^2 + 2x - 14y + 34 = 0$ | |
| 2. | | $(x+4)^2 + (y-5)^2 = 4$ |
| 3. | $x^2 + y^2 - 4x - 10y - 7 = 0$ | |
| 4. | | $x^2 + (y+3)^2 = 25$ |

| Parabola General Form | Complete the Square | Parabola Standard Form |
|---|---|---|
| $y = ax^2 + bx + c$ | | $y = a(x-h)^2 + k$ |
| $x = ay^2 + by + c$ | Multiply and simplify | $x = a(y-k)^2 + h$ |

Again, we can convert from the standard form of a parabola to the general form by multiplying (squaring) the binomial and simplifying.

$$y = 3(x^2 - 4x + 4) + 5$$

Multiply and simplify $\quad y = 3(x-2)^2 + 5$

$$y = 3x^2 - 12x + 12 + 5$$

$$y = 3x^2 - 12x + 17$$

To convert from the general form to the standard form, we complete the square.

$$y = 3x^2 - 12x + 17 \quad \boxed{\text{Complete the Square}}$$

$$y = 3x^2 - 12x + 17$$

$$y = 3\left(x^2 - 4x + \underline{\phantom{xx}} - \underline{\phantom{xx}}\right) + 17$$

$$y = 3\left(x^2 - 4x + \underline{4} - \underline{4}\right) + 17$$

$$y = 3\left(x^2 - 4x + 4\right) + 17 - 12$$

$$y = 3(x-2)^2 + 5$$

| | Parabola Equation in General Form | Parabola Equation in Standard Form |
|---|---|---|
| **1.** | $y = 2x^2 - 20x + 53$ | |
| **2.** | | $y = -2(x+4)^2 - 1$ |
| **3.** | $x = y^2 + 12y + 41$ | |
| **4.** | | $x = -4(x-3)^2 + 2$ |

# Student Activity
## Constructing an Ellipse

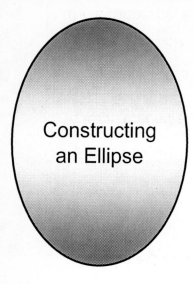

Constructing an Ellipse

**Supplies needed for one construction:**
- Bulletin Board
- Tacks (4)
- Unlined paper (the bigger the better)
- String (length less than the length of the paper)
- Felt tip pen (ballpoint will puncture paper on bulletin board)
- Straightedge

**Definition of an Ellipse:** An **ellipse** is the *set of all points* in a plane for which the *sum of the distances from two fixed points* is a constant. The fixed points are called the **foci** of the ellipse. The **center** is at the midpoint of the two foci. The line segment that forms the longest distance across the ellipse is called the **major axis**. The line segment that forms the smallest distance across the ellipse is called the **minor axis**. The endpoints of the major axis are called **vertices**.

### Construct an Ellipse:
1. Use two tacks to affix your paper to the bulletin board.
2. Use the other two tacks to affix the ends of the string to the piece of paper. Place these to the left and right of the center of the paper (as shown). Leave some slack In the string (do not pull it taut).
3. Use the felt-tip pen to pull the string taut on the upper half of the paper (as shown). Move the pen to trace a curve, keeping the string taut.
4. After drawing half of the ellipse you will have to pick up the pencil and repeat the previous step on the lower half of the paper.
5. Remove your drawing from the bulletin board.

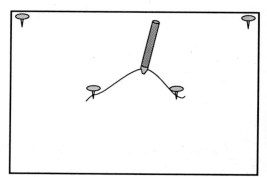

### Add labels:

- The pen traced the *set of all points* in a plane for which the sum of the distances from two fixed points is a constant (the length of the string remained constant). Label your diagram with **Ellipse**.

- The points where the string was held down by a tack were the foci of the ellipse. Label each of these points with the word **focus**.

- Mark the point that is exactly halfway between the two foci and label it the **center**.

- Draw a line segment through the foci and center that extends to the ellipse curve. Label this line segment the **major axis**.

- Draw a line segment through the center that is perpendicular to the major axis that extends to the ellipse curve. Label this line segment the **minor axis**.

- Label each point at the end of the major axis with **vertex**.

# Guided Learning Activity

## Graphing an Ellipse

**To graph an ellipse from a standard equation:** $\dfrac{(x-h)^2}{a^2}+\dfrac{(y-k)^2}{b^2}=1$

1. Find the center, $(h,k)$, and plot it with a small $\times$.
2. Use $a$ to find two points on the ellipse to the left and right of the center.
3. Use $b$ to find two points on the ellipse above and below the center.
4. Draw a smooth curve to represent the ellipse.

**Example:** Graph $\dfrac{(x+2)^2}{36}+\dfrac{(y+4)^2}{64}=1$

1. The center is at $(-2,-4)$. We plot that with a small $\times$.
2. Since $a^2=36$, we deduce $a=6$. Plot a point 6 units to the left of $(-2,-4)$ and another point 6 units to the right of $(-2,-4)$.
3. Since $b^2=64$, we deduce $b=8$. Plot a point 8 units above $(-2,-4)$ and another point 8 units below $(-2,-4)$.
4. Draw the smooth curve of the ellipse.

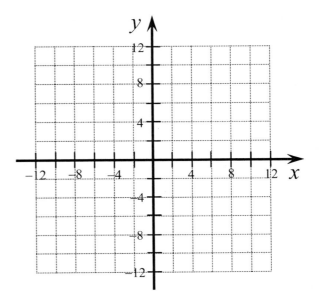

**Now try these!**

**1.** Graph: $\dfrac{x^2}{25}+\dfrac{(y+5)^2}{4}=1$

**2.** Graph: $\dfrac{(x-3)^2}{49}+\dfrac{(y+2)^2}{16}=1$

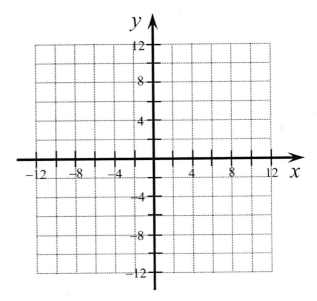

455

# Student Activity
## Triple Play on Ellipses

CNC-10

**Instructions:** In the table below, you have been given some of the information on an ellipse: the graph, the equation in standard form, or a description of an ellipse. Fill in the missing information to complete the table (and the triple play).

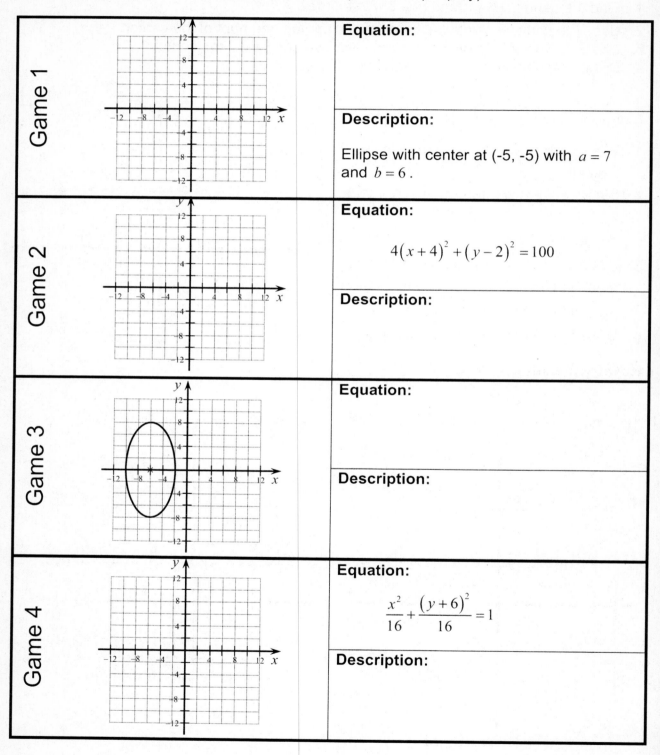

456
© 2011 Cengage Learning. All Rights Reserved. May not be scanned, copied or duplicated, or posted to a publicly accessible website, in whole or in part.

# Student Activity
## Constructing a Hyperbola

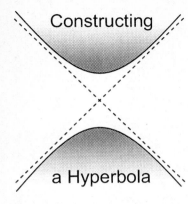

Constructing

a Hyperbola

**Supplies Needed:**
- Unlined paper
- Dark marker or felt-tip pen (if tracing)
- Ruler (a ruler with millimeter markings will work best)
- Hyperbola template

**Definition of a Hyperbola:** A **Hyperbola** is the set of all points in a plane for which the difference of the distances from two fixed points is a constant.

## Construct a Hyperbola:
1. Trace the template of the circle and center on your unlined paper (or use the one provided by your instructor.
2. Choose one of the points on the circle (the ones with the crosshairs) and fold the paper so that the crosshairs fall on point **P**. Crease well.
3. Choose the next point with crosshairs and fold the paper so that these crosshairs fall on point **P**. Crease well.
4. Repeat until you have folded every point with crosshairs to fall on **P**.
5. The creases in the paper will form two curved shapes. We call this a hyperbola. Sketch in the curves along the creases with a pencil.

## Add labels:
- The *set of all points in a plane for which the difference of the distances from two fixed points is a constant* is the curve you have sketched (two curves actually). Label this drawing with **hyperbola**.

- The *two fixed points*, called the foci, are the center of the circle and the point drawn outside the circle. Label the two points (C and P) with the word **focus**.

- Pick a point on the hyperbola and label it **P₁**. Use the ruler to draw a straight line from **C** to **P₁**. Then measure the distance and label the line segment with its distance. Draw a straight line from the **P** to **P₁**. Again, measure the distance and label the line segment with its distance.

- Repeat this procedure for a second point labeled **P₂**.

- For **P₁**, calculate the difference between the two distances. For **P₂**, calculate the difference between the two distances. Are the differences constant?

**Template:** Constructing a Hyperbola

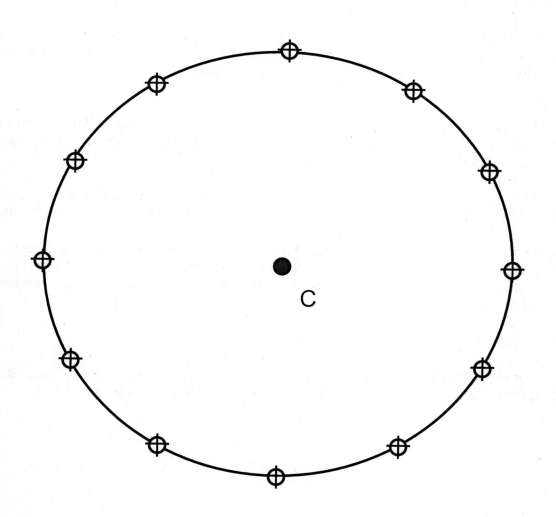

# Guided Learning Activity

## Graphing a Hyperbola

Hyperbola centered at $(h,k)$ opening left and right: $\dfrac{(x-h)^2}{a^2}-\dfrac{(y-k)^2}{b^2}=1$

**To graph this hyperbola:**

1. Plot the **center**, $(h,k)$ with a small $\times$.
2. Draw the **central rectangle**.
   a. From the center, move $a$ units left and plot a point. Then move $a$ units right and plot a point. This marks the left and right sides of the rectangle. Also, these points mark the **vertices** of the hyperbola.
   b. From the center, move $b$ units up and plot a point. Then move $b$ units down and plot a point. This marks the top and bottom sides of the rectangle.
   c. Draw in the rectangle with a light dashed line.
3. Draw the **asymptotes** with dashed lines. The asymptotes are formed by the diagonals of the central rectangle.
4. Find the $x$- and $y$-**intercepts** (if they exist). Plot them.
5. Use the plotted points and the asymptotes to sketch a graph of the hyperbola.

Notice that $a$ is under the variable expression involving $x$ and it indicates the horizontal movement to create the central rectangle. Likewise, $b$ is under the variable expression involving $y$ and it indicates the vertical movement for the central rectangle.

**Example:** Graph the hyperbola:

$\dfrac{x^2}{36}-\dfrac{(y-1)^2}{1}=1$

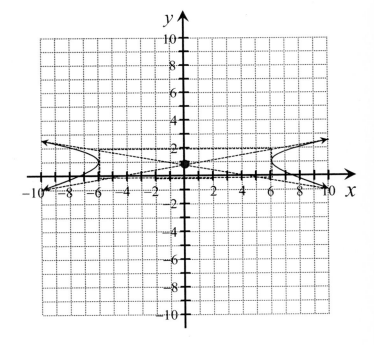

1. The center is at $(0,1)$. Mark this with a small $\times$.

2. From the equation, we see that $a=6$ and $b=1$. Use these to draw the central rectangle.

3. Draw the asymptotes with dashed lines.

4. Find the $x$- and $y$-intercepts and plot these points.

5. Sketch the hyperbola.

**Challenge:**

Use your knowledge of linear equations to write the equations of the asymptotes!

## Now try these!

1. Graph: $\dfrac{(x-1)^2}{9} - \dfrac{(y+2)^2}{16} = 1$

   a. Center: _____

   b. $a =$ ____    $b =$ ____

   c. Vertices: _____ and _____

   d. $x$-intercepts:

   e. $y$-intercepts:

---

Hyperbola centered at $(h,k)$ opening **up and down**: $\dfrac{(y-k)^2}{a^2} - \dfrac{(x-h)^2}{b^2} = 1$

Everything is the same **except** that $a$ now indicates the central rectangle by moving up and down from the center, and $b$ is the number of units left and right of the center.

Notice that $a$ is under the variable expression involving $y$ and it indicates the vertical movement to create the central rectangle. Likewise, $b$ is under the variable expression involving $x$ and it indicates the horizontal movement for the central rectangle.

2. Graph: $\dfrac{y^2}{4} - \dfrac{(x-3)^2}{9} = 1$

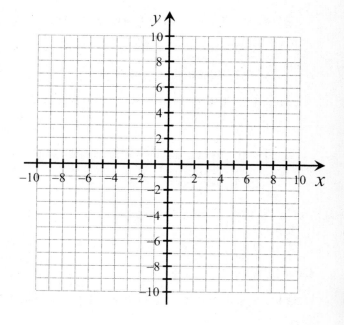

   a. Center: _____

   b. $a =$ ____    $b =$ ____

   c. Vertices: _____ and _____

   d. $x$-intercepts:

   e. $y$-intercepts:

# Student Activity
## Ice Cream Conics

**Directions:** Name the type of conic from the equation (circle, parabola, ellipse, or hyperbola), and then graph it!

Ice Cream Conics

I scream, you scream, we all scream....

**1.** $y = -2(x-3)^2 + 7$

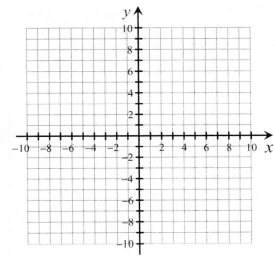

**2.** $4(x-4)^2 + 25(y+6)^2 = 100$

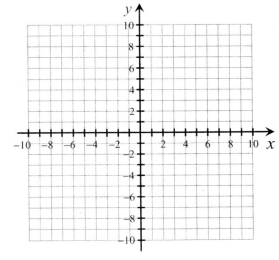

**3.** $\dfrac{x^2}{16} - \dfrac{y^2}{36} = 1$

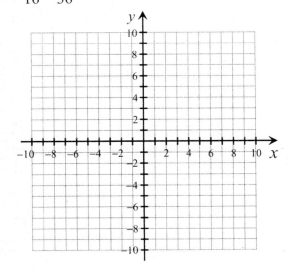

**4.** $(x+1)^2 + (y-5)^2 = 9$

**Name that Conic!**

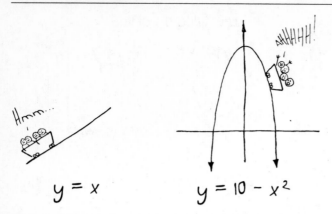

$$y = x$$

$$y = 10 - x^2$$

... Conics are the BEST!!!!!!

**Match-up:** Match each of the equations in the squares of the grid below with the conic classification. You may have to rearrange or simplify equations to determine their classification.

**A** Circle
**B** Ellipse
**C** Parabola opening up or down
**D** Parabola opening right or left

**E** Hyperbola opening right and left
**F** Hyperbola opening up and down
**G** Something else

| | | |
|---|---|---|
| $\dfrac{x^2}{25} - \dfrac{y^2}{9} = 1$ | $x = y^2 + 1$ | $x^2 + 4y^2 = 4$ |
| $x^2 + 3(y-3)^2 = 9$ | $x^2 + (y-3)^2 = 1$ | $x^2 + (y-3)^2 = 9$ |
| $y - (x-4)^2 = 5$ | $y = \dfrac{3}{x}$ | $9(y-4)^2 - 4(x+1)^2 = 36$ |
| $x + (y+9)^2 = 3$ | $x^2 + y^2 = 1$ | $x^2 + 4x - y^2 + 2y = 6$ |
| $x^2 + 5x + 3 - (x^2 + 2x) = y$ | $25x^2 + 100x - 4y^2 = 100$ | $\dfrac{x^2}{4} + \dfrac{y^2}{4} = 1$ |

# Guided Learning Activity
## Solving Nonlinear Systems Algebraically

**Example 1:** Solve: $\begin{cases} y = -x + 2 \\ y = -\dfrac{1}{2}(x-2)^2 + 4 \end{cases}$

First graph the equations to get an idea of where the points of intersection are.

$y = -x + 2$ is the equation of a _____.

$y = -\dfrac{1}{2}(x-2)^2 + 4$ is the equation of a _____.

Estimate the solutions from the graph: (___, ___) and (___, ___)

**Solve by substitution:**

Substitute $y = -x + 2$ into the 2nd equation

and solve for $x$:

$$-x + 2 = -\frac{1}{2}(x-2)^2 + 4$$

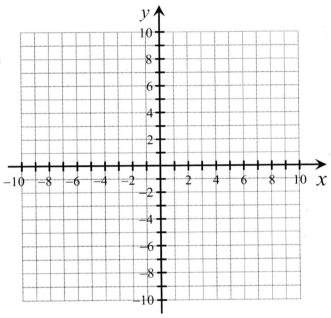

To find the solutions, substitute each value of $x$ into the original equation and evaluate

for $y$: When $x =$ ___, $y =$ ____.　　　　When $x =$ ___, $y =$ ____.

Algebraically, we find the solutions to the system are (___, ___) and (___, ___).

**Example 2:** Solve $\begin{cases} x^2 - y^2 = 1 \\ x^2 + y^2 = 49 \end{cases}$

First graph the equations to get an idea of where the points of intersection are.

$x^2 - y^2 = 1$ is the equation of a _____.

$x^2 + y^2 = 49$ is the equation of a _____.

Estimate the solutions from the graph: (___,___), (___,___), (___,___) and (___,___)

**Solve by Elimination:**

Notice that the $y^2$-terms can be

eliminated by adding the equations:

$$x^2 - y^2 = 1$$
$$\underline{x^2 + y^2 = 49}$$
$$2x^2 \quad = 50$$

Solve this equation for $x$:
(HINT: There are two solutions.)

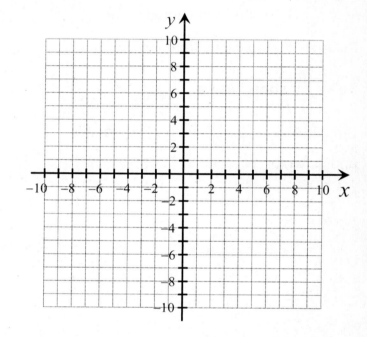

To find $y$, substitute each value of $x$ into one of the original equations.

(HINT: Again, there are two solutions to each of these.)

When $x =$ ____, $y =$ ____.          When $x =$ ____, $y =$ ____.

Algebraically, we find the solutions to the system are (___,___), (___,___), (___,___),

and (___,___).

# Student Activity
## Nonlinear System Detective

**Instructions:** Each graph shows a system of two equations where one of the equations is nonlinear. Write the equations then solve the system algebraically.

**1.**

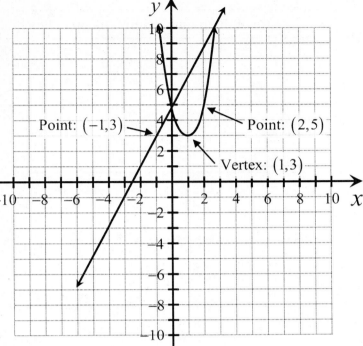

Point: $(-1,3)$

Point: $(2,5)$

Vertex: $(1,3)$

**Equations:**

**Solve:**

**2.**

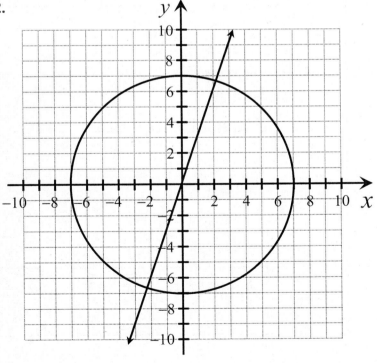

**Equations:**

**Solve:**

# Assess Your Understanding

## Conics

For each of the following, describe the strategies or key steps that will help you **start** the problem. You do **not** have to complete the problems.

| | | **What will help you to start this problem?** |
|---|---|---|
| **1.** | Find the vertex of $y = 2(x-4)^2 + 5$. | |
| **2.** | Graph: $x = (y-1)^2 + 2$ | |
| **3.** | Graph: $9(x-1)^2 + 25(y+2)^2 = 225$ | |
| **4.** | Write the circle $x^2 + 2x + y^2 - 10y = 23$ in standard form. | |
| **5.** | An ellipse is centered at $(-10, 4)$ with $a = 9$ and $b = 3$. Write the equation for the ellipse in standard form. | |
| **6.** | A circle is centered at $(0, 7)$ with radius $\sqrt{3}$. Write the equation for the circle in standard form. | |
| **7.** | Graph: $xy = 4$ | |
| **8.** | Graph: $\dfrac{(y-1)^2}{9} - \dfrac{x^2}{4} = 1$ | |
| **9.** | Write the equation of the hyperbola $9x^2 - 25y^2 = 225$ in standard form. | |
| **10.** | Write the equation of the parabola $x = y^2 - 14y + 54$ in standard form. | |

# Metacognitive Skills
## Conics

Metacognitive skills refer to the ability to judge how well you have learned something and to effectively direct your own learning and studying. This is a self-evaluation tool designed to help you focus your studying and to improve your metacognitive skills with regards to this math class.

Fill the 1st column out **before** you begin studying. Fill the 2nd column out after you study for your test.

Go back to this assessment after your test and circle any of the ratings that you would change – this identifies the "disconnects" between what you **thought** you knew well and what you **actually** knew well.

**Use the scale below to assign a number to each topic.**
5  *I am confident I can do any problems in this category correctly.*
4  *I am confident I can do most of the problems in this category correctly.*
3  *I understand how to do the problems in this category, but I still make a lot of mistakes.*
2  *I feel unsure about how to do these problems.*
1  *I know I don't understand how to do these problems.*

| Topic or Skill | Before Studying | After Studying |
|---|---|---|
| Knowing the equations for the standard form of a circle and ellipse. | | |
| Identifying the center of a circle from its equation in standard form. | | |
| Graphing a circle or ellipse from the standard equations. | | |
| Completing the square to build a squared binomial. | | |
| Converting from the standard equation of a circle or ellipse to the general equation. | | |
| Converting from the general equation of a circle or ellipse to the standard equation. | | |
| Describing a circle or ellipse from its graph. | | |
| Knowing the equations for the standard form of a parabola. | | |
| Identifying whether a parabola will open up, down, left, or right by examining its equation. | | |
| Graphing a parabola from the standard equation. | | |
| Knowing the equations for the standard form of a hyperbola. | | |
| Identifying whether a hyperbola will open left and right, or up and down. | | |
| Graphing a hyperbola from its equation in standard form. | | |
| Identifying the type of conic by inspecting the equation for the conic. | | |
| Identifying $xy = k$ as an equation for a hyperbola. | | |
| Estimating the solution(s) to a nonlinear system of equations by graphing the system. | | |
| Solving a nonlinear system of equations by substitution. | | |

# SEQ: Sequences and More

# Guided Learning Activity:
## Pascal's Triangle

$$(a+b)^0 = \longrightarrow 1$$

$$(a+b)^1 = \longrightarrow 1a + 1b$$

$$(a+b)^2 = \longrightarrow 1a^2 + 2ab + 1b^2$$

$$(a+b)^3 = \longrightarrow 1a^3 + 3a^2b + 3ab^2 + 1b^3$$

$$(a+b)^4 = \longrightarrow 1a^4 + 4a^3b + 6a^2b^2 + 4ab^3 + 1b^4$$

$$(a+b)^5 = \longrightarrow 1a^5 + 5a^4b + 10a^3b^2 + 10a^2b^3 + 5ab^4 + 1b^5$$

$$(a+b)^6 = 1a^6 + 6a^5b + 15a^4b^2 + 20a^3b^3 + 15a^2b^4 + 6ab^5 + 1b^6$$

**Let's make some observations about the binomial expansions for $(a+b)^n$:**

1. Relate the number of terms in each row to the power on the binomial.

2. In any row expansion, what is the degree of each term? How does that relate to the power on the binomial?

See? It's just a basic relationship between the number of gifts and the number of days...

After Halloween, Billy's favorite holiday was definitely Christmas.

3. What is special about the first and last terms?

4. How can you predict the powers on $a$ and $b$ for each term in an expansion?

5. How are the coefficients in one row related to the coefficients in the next row?

Now let's examine the coefficients of each expansion. This array of numbers (forming a triangle) is called Pascal's triangle.

$$(a+b)^0 = \phantom{xxxxxxxxxxxxxxxxxxxxxxxxxx} 1$$

$$(a+b)^1 = \phantom{xxxxxxxxxxxxxxxxxxx} 1a + 1b$$

$$(a+b)^2 = \phantom{xxxxxxxxxxxxxx} 1a^2 + 2ab + 1b^2$$

$$(a+b)^3 = \phantom{xxxxxxxxx} 1a^3 + 3a^2b + 3ab^2 + 1b^3$$

$$(a+b)^4 = \phantom{xxxxx} 1a^4 + 4a^3b + 6a^2b^2 + 4ab^3 + 1b^4$$

$$(a+b)^5 = 1a^5 + 5a^4b + 10a^3b^2 + 10a^2b^3 + 5ab^4 + 1b^5$$

$$(a+b)^6 = 1a^6 + 6a^5b + 15a^4b^2 + 20a^3b^3 + 15a^2b^4 + 6ab^5 + 1b^6$$

**6.** Fill in the missing coefficients for the expansion of $(a+b)^7$ and $(a+b)^8$.

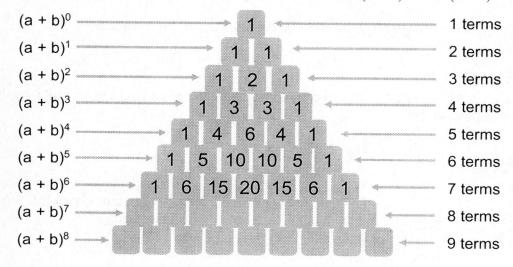

| $(a+b)^0$ | | | | | | 1 | | | | | | 1 terms |
| $(a+b)^1$ | | | | | 1 | | 1 | | | | | 2 terms |
| $(a+b)^2$ | | | | 1 | | 2 | | 1 | | | | 3 terms |
| $(a+b)^3$ | | | 1 | | 3 | | 3 | | 1 | | | 4 terms |
| $(a+b)^4$ | | 1 | | 4 | | 6 | | 4 | | 1 | | 5 terms |
| $(a+b)^5$ | 1 | | 5 | | 10 | | 10 | | 5 | | 1 | 6 terms |
| $(a+b)^6$ | 1 | 6 | 15 | 20 | 15 | 6 | 1 | | | | | 7 terms |
| $(a+b)^7$ | | | | | | | | | | | | 8 terms |
| $(a+b)^8$ | | | | | | | | | | | | 9 terms |

**Use these patterns to fill in the terms of each expansion for problems 7-10:**

**7.** $(a+b)^7 = $ _____ + _____ + _____ + _____ + _____ + _____ + _____ + _____

**8.** $(x+y)^5 = $ _____ + _____ + _____ + _____ + _____ + _____

**9.** $(a+2)^4 = $ _____ + _____ + _____ + _____ + _____

**10.** $(a+(-b))^6 = $ _____ + _____ + _____ + _____ + _____ + _____ + _____

# Student Activity
## Paint by Factorials

**Directions:** Evaluate each of the factorial expressions by simplifying first.  Then shade in the box in the grid that corresponds to the simplified expression.

**Example:** $\dfrac{8!}{5!} = \dfrac{8 \cdot 7 \cdot 6 \cdot 5 \cdot 4 \cdot 3 \cdot 2 \cdot 1}{5 \cdot 4 \cdot 3 \cdot 2 \cdot 1} = \dfrac{8 \cdot 7 \cdot 6 \cdot \cancel{5} \cdot \cancel{4} \cdot \cancel{3} \cdot \cancel{2} \cdot \cancel{1}}{\cancel{5} \cdot \cancel{4} \cdot \cancel{3} \cdot \cancel{2} \cdot \cancel{1}} = = 8 \cdot 7 \cdot 6 = 336$

**1.** Evaluate: $5!$

**2.** Evaluate: $\dfrac{100!}{99!}$

**3.** Evaluate: $\dfrac{99!}{100!}$

**4.** Evaluate: $\dfrac{4!}{0!}$

**5.** Evaluate: $\dfrac{10!}{8!(10-8)!}$

**6.** Evaluate: $4! \cdot 3!$

**7.** Evaluate: $\dfrac{9!}{6!(9-6)!}$

**8.** Evaluate: $\dfrac{2! \cdot 3!}{6!}$

**9.** Evaluate: $\dfrac{12!}{9! \cdot 3!}$

**10.** Evaluate: $\dfrac{9! \cdot 8!}{7! \cdot 10!}$

| | | | |
|---|---|---|---|
| 1 | 100 | 24 | 3 |
| 5 | $\dfrac{1}{100}$ | $\dfrac{1}{60}$ | $\dfrac{7}{2}$ |
| 0 | 220 | 120 | 2 |
| $\dfrac{1}{3}$ | $\dfrac{4}{5}$ | 144 | $12!$ |
| $\dfrac{1}{5}$ | $\dfrac{1}{6}$ | $\dfrac{1}{12}$ | 12 |
| $10!$ | 84 | 45 | undefined |

# Student Activity
## Calculators and Factorials

For this activity you will need to find the factorial function on your calculator. First, check above the keys on your calculator for the exclamation point that symbolizes the factorial function (you may have to use a 2nd or SHIFT first). Sometimes the factorial function is buried in some kind of MATH menu. If you are unable to locate the factorial function in any of these places, you can try looking in the catalog for your calculator. The catalog should give you a list of all of your calculator's available functions. Scroll through this list until you find **!** or **Factorial**, and then select the option.

Once you have found the factorial function on your calculator, use it to evaluate the following factorial expressions:

**1.** Evaluate: $10!$

**3.** Evaluate: $5!3!$

**2.** Evaluate: $\dfrac{25!}{19!}$

**4.** Evaluate: $8!0!$

Your calculator is able to evaluate some factorial expressions that result in **very** large numbers. However, even calculators run out of space eventually.

**5.** What does your calculator do if you try to evaluate $\dfrac{500!}{495!}$ ?

Most calculators are unable to calculate either $500!$ or $495!$. The numbers are simply too large! In this case, we need to first simplify the factorial by reducing common factors. **Then** we can use our calculator to evaluate the simplified expression.

$$\frac{500!}{495!} = \frac{500 \cdot 499 \cdots 496 \cdot 495 \cdot 494 \cdots 3 \cdot 2 \cdot 1}{495 \cdot 494 \cdots 3 \cdot 2 \cdot 1}$$

$$= \frac{500 \cdot 499 \cdots 496 \cdot \cancel{495}^{1} \cdot \cancel{494}^{1} \cdots \cancel{3}^{1} \cdot \cancel{2}^{1} \cdot \cancel{1}^{1}}{\cancel{495}^{1} \cdot \cancel{494}^{1} \cdots \cancel{3}^{1} \cdot \cancel{2}^{1} \cdot \cancel{1}^{1}}$$

$$= 500 \cdot 499 \cdot 498 \cdot 497 \cdot 496$$

$$= 30,629,362,512,400$$

Now try evaluating these factorial expressions using the same technique:

**6.** Evaluate: $\dfrac{315!}{310!}$

**7.** Evaluate: $\dfrac{200!}{198!}$

# Student Activity
## Mathemagic Trick

**Directions:** Mathemagicians often use tricks learned from binomial expansions to speed up their "magic" calculations. In this activity, we're going to learn how to speedily square a 2-digit binomial. First, put away your calculators – you're not going to amaze your friends unless you can do this in your head!

We know the expansion for a binomial squared: $(a+b)^2 = a^2 + 2ab + b^2$

Let's use this to find $26^2$:

**Step 1:** Break the two-digit number into its expanded form.

Example: $26 = 20 + 6 = a + b$

**Step 2:** Find the squared terms $a^2$ and $b^2$, add them and hold them in your memory.

Example: $20^2 = 400 \quad 6^2 = 36 \quad 400 + 36 = 436$

**Step 3:** Find the middle term, $2ab$ (multiply $a$ and $b$, then double it).

Example: $20 \cdot 6 = 120 \quad 2 \cdot 120 = 240$

**Step 4:** Add the middle term to the number in your memory.

Example: $436 + 240 = 676$

**Do the next three problems on paper to see if you've got the hang of it:**

**1.** $41^2$                  **2.** $84^2$                  **3.** $73^2$

**Now try these in your head.**

**4.** $28^2$                  **5.** $19^2$                  **6.** $65^2$

# Student Activity:
## A Sequence of Tables

...84, 88, 92, 96, 100, 104, 108, ...

I think he's counting to n=100 with the sequence $a_n = 4n$.

**Directions:** Fill in the missing information in the table below to demonstrate your knowledge of arithmetic sequences.

Billy loved playing hide and seek ... especially when he got to be it.

| Level 1: Sequences in General | | | | |
|---|---|---|---|---|
| Sequence of numbers | # of terms | State this term | General Term | Find this term |
| $-1, 4, 9, 14...$ | | $a_3 =$ | $a_n = 5n - 6$ | $a_{10} =$ |
| $0, 5, 10, 15, 20, 25, 30$ | | $a_4 =$ | $a_n =$ | $a_6 =$ |
| | 5 | $a_2 =$ | $a_n = \dfrac{(-1)^n}{5^n}$ | $a_5 =$ |
| | Infinite | $a_5 =$ | $a_n = 2^n - 1$ | $a_8 =$ |

| Level 2: Infinite Arithmetic Sequences | | | | |
|---|---|---|---|---|
| Sequence of numbers | $d =$ | $a_1 =$ | $a_n =$ | $a_{20} =$ |
| $6, 12, 18, 24,...$ | | | | |
| | | | $a_n = -2 + 5n$ | |
| | 5 | 8 | | |
| | 3 | | | $a_{20} = 66$ |

| Level 3: Arithmetic Series | | | | | |
|---|---|---|---|---|---|
| Series of numbers | $d =$ | $a_1 =$ | $a_n =$ | Find this term | Find this sum |
| $4 + 12 + 20 + ...$ | | | | $a_{20} =$ | $S_{20} =$ |
| $50 + 55 + 60 + ...$ | | | | $a_{12} =$ | $S_{12} =$ |
| | 3 | $-2$ | | $a_{10} =$ | $S_{10} =$ |
| | | | $a_n = 4 + 3n$ | $a_{100} =$ | $S_{100} =$ |

# Student Activity
## Discovering the Arithmetic Series Sum Formula

**Directions:** Fill in the missing steps to discover the formula for the sum of a finite arithmetic series.

**Example 1:**

$$S_8 = 1 + 7 + 13 + 19 + 25 + 31 + 37 + 43$$

$$S_8 = 43 + 37 + 31 + 25 + 19 + 13 + 7 + 1$$

Add: $\quad 2S_8 = 44 + 44 + 44 + 44 + 44 + 44 + 44 + 44$

Rewrite: $\quad 2S_8 = 8 \cdot 44$

Solve for $S_8$: $\quad S_8 = \dfrac{8 \cdot 44}{2} = 176$

**Example 2:** (you try this one)

$$S_5 = -2 + 4 + 10 + 16 + 22$$

$$S_5 = \rule{6cm}{0.4pt}$$

Add:

Rewrite:

Solve for $S_5$:

**Now, in general:**

$$S_n = \quad a_1 \quad + \quad \left[a_1 + d\right] \quad + \ldots + \left[a_1 + (n-2)d\right] + \left[a_1 + (n-1)d\right]$$

$$S_n = \left[a_1 + (n-1)d\right] + \left[a_1 + (n-2)d\right] + \ldots + \quad \left[a_1 + d\right] \quad + \quad a_1$$

Add: $\quad 2S_n = [\qquad\quad] + [\qquad\quad] + \ldots + [\qquad\quad] + [\qquad\quad]$

Rewrite: $\quad 2S_n =$

Solve: $\quad S_n =$

**Now show how this is the same as** $S_n = \dfrac{n(a_1 + a_n)}{2}$:

# Student Activity
## Another Sequence of Tables

**Directions:** Fill in the missing information in the table below to demonstrate your knowledge of geometric sequences and series.

| Level 1: Infinite Geometric Sequences | | | | |
|---|---|---|---|---|
| Sequence of numbers | $r =$ | $a_1 =$ | $a_n = a_1 \cdot r^{n-1}$ | $a_8 =$ |
| $4, 2, 1, \frac{1}{2}, ..$ | | | | |
| | 5 | $\frac{4}{25}$ | | |
| | | | $a_n = 12\left(\frac{1}{3}\right)^{n-1}$ | |
| | | $\frac{5}{64}$ | | 1280 |

| Level 2: Finite Geometric Series | | | | | |
|---|---|---|---|---|---|
| Series of numbers | $r =$ | $a_1 =$ | $a_n = a_1 \cdot r^{n-1}$ | Find this term | Find this sum |
| $24, 8, \frac{8}{3}, ...$ | | | | $a_6 =$ | $S_6 =$ |
| | 10 | 3 | | $a_5 =$ | $S_5 =$ |
| | | | $a_n = 6 \cdot \left(\frac{1}{2}\right)^{n-1}$ | $a_7 =$ | $S_7 =$ |
| | $\frac{1}{100}$ | | | $a_8 = \frac{32}{25}$ | $S_8 =$ |

| Level 3: Infinite Geometric Series | | | | | |
|---|---|---|---|---|---|
| Series of numbers | $r =$ | $a_1 =$ | $a_n = a_1 \cdot r^{n-1}$ | Find this partial sum | Find the infinite sum (if possible) |
| $20, 10, 5, ...$ | | | | $S_5 =$ | $S =$ |
| | $\frac{3}{2}$ | 1 | | $S_5 =$ | $S =$ |
| | | | $a_n = 2\left(\frac{1}{3}\right)^{n-1}$ | $S_5 =$ | $S =$ |
| | | 8 | | $S_5 =$ | $S =$ |

# Student Activity:
## Discovering the Geometric Series Sum Formula

**Directions:** Fill in the missing steps to discover the formula for the sum of a finite geometric series.

**Example 1:**

$$S_6 = 25 + 5 + 1 + \frac{1}{5} + \frac{1}{25} + \frac{1}{125}$$

Multiply by $r$:   $S_6 \cdot \frac{1}{5} = \quad 5 + 1 + \frac{1}{5} + \frac{1}{25} + \frac{1}{125} + \frac{1}{625}$

Subtract:   $\dfrac{4}{5} S_6 = 25 - \dfrac{1}{625}$

Solve:   $S_6 = \dfrac{5}{4}\left( 25 - \dfrac{1}{625} \right)$

**Example 2:** (you try this one)

$$S_5 = \frac{1}{6} + \frac{1}{3} + \frac{2}{3} + \frac{4}{3} + \frac{8}{3}$$

Multiply by $r$:   $=$ 

_____

Subtract:

Solve:

**Now, in general:** (assuming $r \neq 1$)

$$S_n = a_1 + a_1 r + a_1 r^2 + \ldots + a_1 r^{n-2} + a_1 r^{n-1}$$
$$S_n \cdot r = \quad a_1 r + a_1 r^2 + \ldots + a_1 r^{n-2} + a_1 r^{n-1} + a_1 r^n$$

Subtract:

Some factoring:

Solve:   $S_n =$

**Directions:** Fill in the missing steps to discover the formula for the sum of an infinite geometric series.

### Example 3:

$$S = 25 + 5 + 1 + \frac{1}{5} + \frac{1}{25} + \frac{1}{125} + \ldots$$

Multiply by $r$: $\quad S \cdot \frac{1}{5} = \qquad 5 + 1 + \frac{1}{5} + \frac{1}{25} + \frac{1}{125} + \ldots$

Subtract: $\quad \dfrac{4}{5}S_6 = 25$

Solve: $\qquad S_6 = \dfrac{5}{4}(25)$

### Example 4: (you try this one)

$$S = \frac{8}{3} + \frac{4}{3} + \frac{2}{3} + \frac{1}{3} + \frac{1}{6} \ldots$$

Multiply by $r$: $\qquad =$

Subtract:

Solve:

### Now, in general:

$$S_n = a_1 + a_1 r + a_1 r^2 + \ldots + a_1 r^{n-2} + a_1 r^{n-1} + a_1 r^n + \ldots$$
$$S_n \cdot r = \qquad a_1 r + a_1 r^2 + \ldots + a_1 r^{n-2} + a_1 r^{n-1} + a_1 r^n + \ldots$$

Subtract:

Some factoring:

Solve: $\qquad S_n =$

# Student Activity
## Infinitely Finite

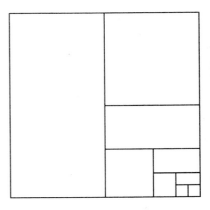

OK, a little to the right...
Just a little more in that direction ...
A little more ...

It can be very difficult to grasp the idea that an infinite series can be bounded by a finite number.  Here are two examples that might help you to clarify that, at the very least, it is possible.

After scratching for what seemed like an eternity, he suddenly understood "infinitely finite."

**Example 1:** If each large square has an area of **1 square unit**, what is the size of each of the **smaller** squares or rectangles below?

**a.** Area: _1_     Area: ___     Area: ___     Area: ___     Area: ___

Now we illustrate the series: $\dfrac{1}{2}+\dfrac{1}{4}+\dfrac{1}{8}+...$ by shading in the appropriate pieces of the square as we add each term of the series.

**b.** Based on this shading, what is the sum of the infinite series $\dfrac{1}{2}+\dfrac{1}{4}+\dfrac{1}{8}+...$ ? ____

**c.** Now show how to find this sum using $S=\dfrac{a}{1-r}$.

**Example 2:** Now we look at the series $\dfrac{3}{10}+\dfrac{3}{100}+\dfrac{3}{1,000}+\dfrac{3}{10,000}+...$

**a.** $a_1 = $ _____ and $r = $ _____ .

**b.** Find the sum of the infinite series using $S = \dfrac{a_1}{1-r}$.

**c.** Now write the same series using decimals instead of fractions:

_____ + _____ + _____ + _____ + _____ + ....

**d.** You should be able to clearly see the sum of the series now without any sort of calculation.  Write the sum of the series as a decimal: _____

**Example 2:** Now we look at the series $\dfrac{7}{10}+\dfrac{7}{100}+\dfrac{7}{1,000}+\dfrac{7}{10,000}+...$

**a.** $a_1 = $ _____ and $r = $ _____ .

**b.** Find the sum of the infinite series using $S = \dfrac{a_1}{1-r}$.

**c.** Now write the same series using decimals instead of fractions:

_____ + _____ + _____ + _____ + _____ + ....

**d.** You should be able to clearly see the sum of the series now without any sort   of calculation.  Write the sum of the series as a decimal: _____

# Student Activity
## Value Remaining

**Directions:** Fill in the missing information in the table below for the application problems involving geometric sequences.

| | What % is lost or gained every year. | What is the value at the end of each year, as a percent of the previous year? | $r =$ | $a_1 =$ |
|---|---|---|---|---|
| **1.** A bank account with a starting balance of $1,000 earns 5% interest every year. | 5% gained | 105% | 1.05 | 1,000 |
| **2.** An endowment of $1,000,000 plans to give away 10% of what's left of the original fund balance every year. | | | | |
| **3.** The population of a town is 50,000 and is increasing by 6% a year. | | | | |
| **4.** The population of a town is 20,000 and is decreasing by 2% a year. | | | | |
| **5.** The value of a $120,000 house increases 3% every year. | | | | |
| **6.** The value of an airplane that cost $1.2 million depreciates 12% every year. | | | | |
| **7.** From studying a population of wolves, biologists determine that from year to year, only 80% of the remaining original population survives. | | | | |
| **8.** Every year Mary spends 20% of the remaining balance of the $20,000 she inherited from her aunt. | | | | |

# Assess Your Understanding
## Sequences and Series

For each of the following, describe the strategies or key steps that will help you **start** the problem. You do **not** have to complete the problems.

| | | What will help you to start this problem? |
|---|---|---|
| **1.** | Evaluate: $\dfrac{9!}{3!6!}$ | |
| **2.** | Find the sum of the first 10 terms of the arithmetic series with general term $a_n = -18 + 3n$. | |
| **3.** | Write the expansion of $(x+y)^5$ using Pascal's triangle. | |
| **4.** | Let a sequence be defined by $a_n = 6 + 5n$. Find $a_{10}$. | |
| **5.** | Find the 7$^{\text{th}}$ term of the expansion of $(2x-1)^{12}$. | |
| **6.** | Find the first four partial sums of the series: $1 - \dfrac{1}{3} + \dfrac{1}{9} - \dfrac{1}{27} + \ldots$ | |
| **7.** | Insert two arithmetic means between 6 and 30. | |
| **8.** | An endowment containing $200,000 plans to pay out 5% of the amount remaining in the original fund every year. What will the payment be in the 5$^{\text{th}}$ year? | |
| **9.** | Write the general term for the sequence: $32, 8, 2, \dfrac{1}{2}, \ldots$ | |
| **10.** | Find $S$ for the geometric series: $\dfrac{2}{5} + \dfrac{4}{25} + \dfrac{8}{125} + \ldots$ | |

# Metacognitive Skills Assessment
## Sequences and Series

Metacognitive skills refer to the ability to judge how well you have learned something and to effectively direct your own learning and studying. This is a self-evaluation tool designed to help you focus your studying and to improve your metacognitive skills with regards to this math class.

Fill the 1st column out **before** you begin studying. Fill the 2nd column out after you study for your test.

Go back to this assessment after your test and circle any of the ratings that you would change – this identifies the "disconnects" between what you **thought** you knew well and what you **actually** knew well.

**Use the scale below to assign a number to each topic.**
5  *I am confident I can do any problems in this category correctly.*
4  *I am confident I can do most of the problems in this category correctly.*
3  *I understand how to do the problems in this category, but I still make a lot of mistakes.*
2  *I feel unsure about how to do these problems.*
1  *I know I don't understand how to do these problems.*

| Topic or Skill | Before Studying | After Studying |
|---|---|---|
| Finding a binomial expansion using multiplication. | | |
| Replicating Pascal's Triangle. | | |
| Using Pascal's Triangle to predict the coefficients on terms in a binomial expansion. | | |
| Evaluating expressions involving factorial notation. | | |
| Using the Binomial Theorem to write a binomial expansion. | | |
| Predicting the next few terms in a sequence of numbers. | | |
| Identifying the first term and common difference of an arithmetic sequence; writing the general term $a_n$. | | |
| Identifying the first term and the common ratio of a geometric sequence; writing the general term $a_n$. | | |
| Finding any term of an arithmetic or geometric sequence using the formula for the general term. | | |
| Knowing the formulas to find the sum of a finite arithmetic series or a finite geometric series. | | |
| Calculating the sum of a finite arithmetic or geometric series. | | |
| Knowing the formula to find the sum of an infinite geometric series and when the formula cannot be used. | | |
| Calculating the sum of an infinite geometric series (when it exists). | | |
| Finding the arithmetic means or geometric means to insert between two terms. | | |
| Using partial sums of a geometric series to make an educated guess about whether the series has a finite sum. | | |
| Converting a repeating decimal into a common fraction. | | |
| Solving application problems involving geometric series. | | |